SCIENCE, EVOLUTION AND CREATIONISM

NATURAL SELECTION

BIOLOGICAL PROCESSES, THEORY AND ROLE IN EVOLUTION

SCIENCE, EVOLUTION AND CREATIONISM

Additional books in this series can be found on Nova's website
under the Series tab.

Additional e-books in this series can be found on Nova's website
under the e-book tab.

SCIENCE, EVOLUTION AND CREATIONISM

NATURAL SELECTION

BIOLOGICAL PROCESSES, THEORY AND ROLE IN EVOLUTION

JAY R. LYNCH
AND
DEREK T. WILLIAMSON
EDITORS

NOVA BIOMEDICAL
New York

For permission to use material from this book please contact us:
Telephone 631-231-7269; Fax 631-231-8175
Web Site: http://www.novapublishers.com

Library of Congress Cataloging-in-Publication Data

Natural selection : biological processes, theory, and role in evolution / [edited by] Jay R. Lynch and Derek T. Williamson.
pages cm
Includes bibliographical references and index.
ISBN: 978-1-62257-467-4 (hardcover)
1. Natural selection. 2. Evolution (Biology) I. Lynch, Jay R., editor of compilation. II. Williamson, Derek T., editor of compilation.
QH375.N364 2012
576.8'2--dc23

2012025351

Published by Nova Science Publishers, Inc. † New York

CONTENTS

PREFACE

In evolution, natural selection is the gradual, non-random, process by which biological traits become either more or less common in a population as a function of differential reproduction of their bearers. In this book, the authors present current research in the biological processes, theory and role in evolution of natural selection. Topics include evolutionary transitions in mathematical modeling complexity using evolutionary systemic modeling; natural selection applications for algorithmic computation; allozymes, and DNA and natural selection in the mollusk population.

Chapter 1 - Everything evolves but the patterns are displayed at the *evolutionary timescale*, which are not intuitively evident or visible; whereas visible changes are at the *dynamic timescale* but they are evidently unique, idiosyncratic and not suggestive of evolutionary processes. This scenario was even sharper before the emergence of theory of evolution in biology as an *ontological essence* was attached strongly to Man's existence. However, the Darwinian natural selection is pivotal in successfully unearthing the common descent among all species; proves the growth of simplicity into complexity; can explain the trait of altruism; all of which are through the natural selection criterion of being callous, purposeless, unintelligent, and without foresight. Owing to evolutionary thinking in biology, it has become feasible to recognize the evolution of everything but it remains a challenge to devise a *generalized theory of evolution* transparent across all disciplines. This chapter has the aim of meeting the challenge by recognizing the existence of the above timescales and using a modified concept of feedback loops of systems science/cybernetics. The outcome is *evolutionary systemics*, which is a *generalized theory of evolution* applicable to every human endeavor. The applications include philosophy and science as a whole, all driven by evolutionary transitions.

Chapter 2 - Artificial Intelligence (AI) is in symbiotic relationships with computer systems and includes neural networks, expert systems and agents. Each area of AI is complexity in the sense of having hierarchy and interconnectivity but these are overlooked giving a monolithic or unique sense. Therefore, knowledge gained from one area is not transferrable to new areas. Also, some AI applications involve solving complex over-redundant equations using optimisation techniques, implying architectural dissimilarity with natural intelligence. These problems expose major flaws in current thinking stemming largely from overlooking hierarchy and interconnectivity as a part of possible *missing links* in science and their AI paradigms. This chapter invokes possible solutions as a vision suggesting: (i) integrating *evolutionary thinking* with *systemic thinking* expediting the introduction of hierarchy and interconnectivity into AI complexity; (ii) *a new calculus of number sequences,* allowing a mirror image of complexity in the world of integer numbers to possibly removing over-redundancy of the equations and making usage of *bondage* between objects; and (iii) *morality* to discourage the way reductive science, oblivious to morality, facilitated uncontrolled exploitation of natural resources. The vision would contribute to target investments on AI-products towards their potential returns by reusing existing capabilities.

Chapter 3 - The increase in oxygen concentration in the Earth's atmosphere represented an important selective pressure for living organisms and contributed to set up the pace of evolutionary changes in physiological and metabolic systems. Despite the evolution of molecular mechanisms to use oxygen for efficient energy production represented a relevant evolutionary innovation, the reduction-oxidation (or redox) reactions associated with its use have been responsible for the production of free radicals and, more in general, of reactive species (RS). These products of oxygen metabolism may have damaging effects on organism, in essence, due to oxidation of essential cellular components. Consequently, animals have evolved mechanisms to avoid or minimize the production of RS of oxidative metabolism as well as of antioxidant defense systems needed to limit the pro-oxidant activity of RS. The result was a high diversity in molecular and structural antioxidant defenses as well as in redox signaling pathways perfectly integrated in the cellular metabolic machinery. However, the antioxidant machinery is not 100% efficient in mopping RS up, especially when there is a strong imbalance between production of RS and antioxidant response. The consequence of such an imbalance is the generation of oxidative damage to biomolecules. This kind of biochemical stress of cells has been called oxidative stress, which may be defined as the rate at which oxidative damage is generated. Recent studies on animal species suggest that oxidative stress could also represent a universal constraint of life-history evolution, and a modulator of phenotypic development. Oxidative stress could therefore have worked as a relevant selective pressure, and that is why animal taxa have evolved a myriad of mechanisms to regulate and adjust the redox balance. This high diversity in antioxidant mechanisms also suggests that organisms have a great deal of flexibility in how they deal with RS challenges across time, conditions, and tissue types, as well as that different organisms may have evolved different strategies for dealing with similar oxidative challenges. In this chapter, the authors review the high diversity of molecular and structural defenses against RS evolved by animals. In particular, it is their goal to describe physiological mechanisms underlying the production of RS and the biochemical regulation of defense mechanisms, to extend the traditional concept of antioxidant to other structural and functional factors affecting the 'whole' organism, and to review the state of the art about the link between oxidative stress and longevity. It is proposed that evolution of long-lived animals tends to decrease the rate of generation of endogenous damage and to increase the molecular resistance to the oxidative damage in long-lived species.

Chapter 4 - Mathematical complexity is often regarded as an abstract world and sterile due to its formidable status but this reflects the grip of an ontological mindset. This complexity seems beyond the reach of evolutionary processes, where it sounds counterintuitive to explain it by *evolutionary thinking*. This chapter shows that there is nothing unique about mathematics, as it evolves like any other complexity. The modeling approach used is *evolutionary systemics*, which is the integration of *systemic thinking* (the architecture) with evolutionary thinking (the architect). Evolutionary systemics is driven by a set of *axioms*, in which natural selection drives evolutionary transitions through four types of feedback loops from a lower loop to higher ones. The outcome is a modeling capability, which uses historic accounts as evidence. Thus, evidence gathered from historic accounts identifies evolutionary transitions and the approach copes with the roles of *cognitive mathematics of individuals*, *communal mathematical thinking*, and *institutional mathematics*.

The chapter offers an insight into mathematical practices and holds that evolutionary transitions in mathematics are in tandem with other human endeavors. It can be highlighted that, during evolutionary transitions, the past knowledge may become obsolete without being unselected. As mathematics is pluralist by its nature and increasing in complexity, feedforward loops are indispensible for proactively reshaping mathematical practices with the evolving culture. The following strategies are identified. (i) Philosopher's aspirations are in asking stimulating questions but they produce mutually exclusive doctrines, which are contentious and devoid of significant practical benefits. (ii) Practitioners' aspirations are fruitful in axiomatization to overhaul existing knowledge riddled with conflicts to self-evident information but these are few and far in-between. (iii) This chapter promotes goal-orientation by applying evolutionary systemics through identifying missing links in mathematics, proposing a model to resolve their impacts, outlining the elements of a vision to identify its equivalent cladistics and to manage expectations on mathematics. A new calculus is presented to support the thinking behind goal-orientation in the sequence of natural numbers primed from the building block of "1" by the virtue of new operators.

Chapter 5 - Since the emergence of computational mathematical modeling in the 1950s, its developments have been suffering from fragmentary visions, if any. Still there is no insight into, or foresight of, modeling complexity as a whole. This chapter formulates a vision for modeling complexity using evolutionary systemics, which integrates evolutionary and systemic thinking. It recognizes cognitive model thinking of the individuals assisted by their instincts; communal model thinking due to cohesion of individual mindsets; and institutional modeling practices capable of evolutionary transitions. The following transitions are identified: (i) primitive modeling, prevailed until the 17th century; (ii) prototype modeling until 1950, which started afresh by unselecting its past primitive modeling practices; and (iii) computational modeling since 1950, which largely reinvented past capabilities under the computational environments. Over the years since 1950, modeling complexity has seemingly been growing organically but whilst this signifies healthy growth, it underlines potential problems. The practice-data is used to analyze this complexity by applying evolutionary systemics. This helps referencing a vision of modeling complexity to the full evolutionary contexts and identifying relevant issues. In this way, the paper shows that modeling complexity is infested with increasing complexity and increasing dynamism. The emerging complex systems suffer from a new kind of entropy retarding growth. The chapter also studies the structure of complexity exploring the feasibility of deriving a cladistics for modeling complexity. This is a radical departure from conventional thinking, as new investments on model developments can avoid bottlenecks. The gained insight invokes the need for rethinking and making a research case for the formation of the science of modeling .

Chapter 6 - Natural selection is one of the cornerstones of modern biology and one of the great intellectual revolutions of human history. The term was first introduced by Charles Darwin who put forth a coherent theory of evolution and amassed a great body of evidence in support of this theory. In the last few decades, the continuing advance of modern technology has brought about something novel. Evolution is now producing practical benefits in a very different field, and this time, the creators can claim that their explanation fits the facts perfectly. This field is computer science, and the benefits come from a programming strategy which incorporates the notion of natural selection within algorithmic computation, and builds the so called evolutionary algorithms. The main goal of this survey is to summarize the previous research on two of the most important evolutionary algorithm techniques, which are Genetic Algorithms (GAs) and Differential Evolution (DE). Both algorithms are population-based, which means that a population of candidate solutions is optimized through the inspiration of biological motivations. GAs are efficient search methods based on principles of natural selection and population genetics. DE took a serious attempt to replace the classical crossover and mutation operators of genetic algorithms. The authors focus on these algorithms to guide us through a review of many of the most important innovations that they have provoked to science. In this chapter, different variations of operations used by these algorithms such as fitness function, mutation, selection and crossover are discussed. The authors try to identify some of the problems of these methods that have not been studied systematically yet. They also describe the parallel potential of the evolutionary algorithms by focusing on the higher computational costs that big populations causes for some hard problems. As astonishing and counterintuitive as it may seem to some, evolutionary algorithms have proven to be an enormously powerful and successful problem-solving strategy, dramatically demonstrating the power of evolutionary principles.

Chapter 7 - Mollusks are a group in which some interesting examples have been described over the years concerning effects of natural selection over populations. A classical example is the observed clines of *Lap* alleles and salinity gradients in mussels. With the introduction of DNA techniques in the field of population genetics, it has been proposed that comparing the amount of geographic variation in allele frequencies for protein and DNA polymorphisms is a powerful method for detecting the effects of natural selection. Once more, mollusks are important models in these studies. These issues will be reviewed in this chapter and the importance of the molecular markers (allozymes and DNA) for studies on natural selection discussed in the light of the results obtained from mollusks which includes the Brazilian cockle *Anomalocardia brasiliana*, the brown mussel *Perna perna* and the blue mussel *Mytilus edulis*.

Chapter 8 - Soil microbes are important component of the terrestrial ecosystems that play the critical role in decomposition of organic substances and mineralization of nutrients. Since soil microbes influence the nutrition of plants in the nitrogen-limited areas there has been an increased interest on using soil microbes as an important indicator of soil quality.

It is also widely recognized that changes in soil microbial biomass and community composition can influence decomposition of organic matter and nitrogen cycling in soil. Grasslands constitute about one thirds of the earth´s area and are regarded as one of the most productive ecosystems on earth.

Since management measures are often taken in order to optimize the productivity of this system, enhanced knowledge about the functional differences of the soil microbes under various management activities is relevant for proper management of these systems. This chapter attempts to review studies on the structural and functional differences of soil microbial communities in various grassland ecosystems. This chapter also presented data on the differential responses of soil microbial functions such as decomposition rate and nitrogen mineralization rate to litter inputs in order to reveal the characteristic differences of the semi-natural and improved grasslands with special references to grasslands of Shiriyazaki in the northern Honshu island, Japan.

Chapter 9 - Darwin's well-known idea of evolution by natural selection was complemented during the last 100 years: evolutionary changes depend on adaptive evolution (mutation + natural selection), but chance and historical contingency also contribute to evolution. The effects of chance are usually due to founder events, bottlenecks and genetic drift. Historical contingency can become important if certain genetic changes of adaptive value in the past constrain or promote evolutionary outcomes. The importance of genetic adaptation, chance and history on evolutionary change can be empirically tested by the robust laboratory experiments using haploids and asexual microalgae. The experiment followed the idea of "replaying life's tape" by means of the simultaneous propagation of 10 independent isolates of each strain which were grown under increasing nutrient concentration and temperature (resembling global change) during c. 100 generations. Evolutionary changes in fitness (acclimated maximal growth rate), photosynthetic capacity and morphology (size and form) were analyzed in species from continental freshwater-bodies, coastal and oceanic habitats (using two strains per species). Adaptation by new mutations that resulted in the enhancement of growth rate arose during propagation of derived cultures under the new environmental conditions was the main component of evolution of fitness. Mutations were selected, displacing the wild-type ancestral genotypes. However, chance was the pacemaker of evolution of photosynthetic capacity. In contrast, changes in morphology and size were due to the contribution of history and chance.

In: Natural Selection
Editors: J. R. Lynch and D. T. Williamson

ISBN: 978-1-62257-467-4

Chapter 1

LEARNING FROM NATURAL SELECTION IN BIOLOGY: REINVENTING EXISTING SCIENCE TO GENERALIZE THEORY OF EVOLUTION – EVOLUTIONARY SYSTEMICS

Rahman Khatibi*
[1]Consultant Mathematical Modeller, Swindon, UK

ABSTRACT

Everything evolves but the patterns are displayed at the *evolutionary timescale*, which are not intuitively evident or visible; whereas visible changes are at the *dynamic timescale* but they are evidently unique, idiosyncratic and not suggestive of evolutionary processes. This scenario was even sharper before the emergence of theory of evolution in biology as an *ontological essence* was attached strongly to Man's existence. However, the Darwinian natural selection is pivotal in successfully unearthing the common descent among all species; proves the growth of simplicity into complexity; can explain the trait of altruism; all of which are through the natural selection criterion of being callous, purposeless, unintelligent, and without foresight. Owing to evolutionary thinking in biology, it has become feasible to recognize the evolution of everything but it remains a challenge to devise a *generalized theory of evolution* transparent across all disciplines. This paper has the aim of meeting the challenge by recognizing the existence of the above timescales and using a modified concept of feedback loops of systems science/cybernetics. The outcome is *evolutionary systemics*, which is a *generalized theory of evolution* applicable to every human endeavor. The applications include philosophy and science as a whole, all driven by evolutionary transitions.

Keywords: natural selection, generalized theory of evolution, evolutionary-systemics, entropy, hierarchy, interconnectivity, fixed ontology, self-organization, feedback loops: d-loop, ℓ-loop, e-loop

*E-mail: rahman_khatibi@yahoo.co.uk

1. Introduction

1.1. Statement of the Problem

"Evolution in, fixed ontology out" is the message of this paper, each reflecting a different worldview or mindset in relation to existence. Evolutionary thinking, driven by *natural selection*, explains existence through selections from small and random variations, but this worldview seems counterintuitive. The *fixed ontological mindset* regards organisms and the natural world as fixed creations and imposes permanency, a deceptively self-evident and rather intuitive idea. Fixed ontology was overturned by evolutionary thinking in the second half of the 19th century, when biology was unshackled and evolutionary thinking emerged as the only scientific basis to explain the existence of organisms. Fixed ontology has become irrelevant forever within biology. In spite of claims of evolutionary processes outside biology, evolutionary thinking has failed to reach its full potential. This paper builds on the rigor of evolutionary biology and presents a generalized model of evolution applicable to science, culture and complexity.

Evolution processes are better differentiated on the time dimension, as follows. (i) *Dynamic timescale*: time starts at an origin, processes are triggered on continuous or discrete time intervals of fractions of a second to hours, days and so on, the process ends. (ii) *Long timescale*: this is normally large compared with dynamic timescale, during which adaptation may take place. (iii) *Evolutionary timescale*: changes take place in many generations at the genotypic level. There is also a fictitious ontological timescale, in which time is regarded as eternal and ontological substances attached to entities are regarded as fixed throughout with primacy over their *accidental* properties. Evolutionary timescale ought to be parts-and-parcels of the wider science and its importance is underpinned by resorting to the following strategy: if evolutionary thinking is not explicit in any particular human endeavor, it is in the grip of ontology, even if inadvertent. Fixed ontology unduly retards the wider sciences and is irreparable. Thus, this paper aims to undo its grip.

1.2. Evolutionary Systemics in a Nutshell

Scientific enquiries are focused largely on explaining the complexity of life, where the role of theory of evolution is pivotal. The enquiries through evolution started from the physical world (phenotypic resolution) ever since the 17th century; consolidated in the 1920s by the emergence of neo-Darwinism; and came to its culmination in 1953 when it penetrated the chemical world (genotypic resolution). Thus, by drawing on from many disciplines, theory of evolution inadvertently opened the way to holistic thinking independent of systems science, where the latter emerged since 1950. Holism is now one of the tenets of science.

The existence of three timescales for any *evolvable entity*, as stated in Section 1.1, is the statement of the obvious. Evolutionary scientists deal explicitly with the evolutionary timescale with a focus on both phenotypic and genotypic behaviors and tacitly with other timescales but without any formalized approach.

Scientists observe widely the three loops of positive feedback, negative feedback and feedforward but only at the dynamic timescale and as such, if system scientists have a tacit perception of the long and evolutionary timescales, this is tacit but not explicit.

Table 1. Proposed Timescale Subdivisions of Feedback Loops

Loops	Drivers at		
	Dynamic Timescale	Long Timescale	Evolutionary Timescale
Zero⁺ Feedback See Box 1.2	d-loop: Randomness	ℓ-loop: Obscure	e-loop: Spontaneity
	Zero⁺ Feedback Loop gives rises to simplexes, which then grow.		
Positive Feedback (Boxes 1.3-1.4)	d-loop: Performance	ℓ-loop: Adaptation	e-loop: Natural Selection
	$\boldsymbol{Positive\ Feedback\ Complexes = d\_loop + \ell\_loop + e\_loop}$		
Negative Feedback (Boxes 1.4-1.5)	d-loop: Internally-consistent performance	ℓ-loop: 'Discretion'	e-loop: Natural selection of rules for internal consistency
	$\boldsymbol{Negative\ Feedback\ Complexes = d\_loop + \ell\_loop + e\_loop}$		
Feedforward See Box 1.6	d-loop: Goal-driven for external consistency	ℓ-loop: Mission for external consistency	e-loop: Natural selection of rules for external consistency
	$\boldsymbol{Feedforward\ Complexes = d\_loop + \ell\_loop + e\_loop}$		
Note 1: Positive feedback has three integral components of d-loop (real dynamic state of the system), ℓ-loop (manipulation of the phase), or e-loop (changes to hereditary codes) Note 2: The same is said about negative feedback or feedforward loops. Note 3: In discussing a loop, the focus is on all three types of d-loop, ℓ-loop e-loop. Note 4: These loops operate pluralistically and emerge (or are pump-primed) one after another through evolutionary transitions due to inherent entropy normally in the above order. Note 5: The emergence of the next evolutionary transition is not deterministic but a potential driven by natural selection. Sound entities may be pump-primed but not selected.			

This paper integrates theory of evolution with systems science and therefore it is expedient to institute the roles of these feedback loops at each of these timescales. As shown in Table 1, the following sub-loops are proposed: *d-loop* at the dynamic scale, *ℓ-loop*, at the long timescale, and *e-loop* at the evolutionary timescale. This synthesis goes a long way towards a generalized theory of evolution making every discipline transparent. Although there are generalized Darwinian theories of evolution, they are problem-specific and not transparent to one another. This paper takes a critical view of science and presents an original approach to synthesize existing scientific knowledge to develop a new generalized theory of evolution. It fully embraces neo-Darwinism without getting involved with its ongoing scientific debates. The 'leitmotif' of this paper is: building blocks and rules. It builds on Khatibi (2011). The building blocks are not fixed but vary in each discipline. However, the arrangement of building blocks is not arbitrary but they are 'fashioned' by the concept of feedback loops, where each loop has three manifestations corresponding to each timescale. The feedback loops comprise: zero$^+$ feedback, positive feedback, negative feedback and feedforward loops, where zero$^+$ is proposed by Khatibi (2011). The systems science account of these loops makes up the basis for the generalization of theory of evolution.

Box 1.1 – Axioms on System

The term 'system' immediately and unequivocally means:

(i) A set of inputs; a set of building block modules; a set of outputs;

(ii) Workable connection from inputs through building blocks toward the outputs;

(iii) Flow of material and information from inputs towards the outputs and depending on the evolutionary stage of the system the flow of information from the output or even the environment to or toward the inputs;

(iv) System encompasses: feedback loops, hierarchy, institutionalization, various timescales, pluralism and working together holistically.

Box 1.2 – Axioms on Zero$^+$ Feedback Loop

- **Tailback**: zero$^+$ feedback is the tailback of *evolvable entities* to uncertain origins;
- **Connectivity**: no connection between building blocks, no output for a given input, no dynamic timescale, no evolutionary timescale and no system but a potential;
- **Spontaneity**: motion or motivation exists to hone abundantly existing potential building blocks by shaping and reshaping in the environment; in the natural environment, the honing is spontaneous as they freely float or available but in man's world, there are a host of activities including having a hunch;
- **Emergent property**: At one instance, the potential materialises through self-organisation for the production of an entity by a particular intertwining of building blocks with rules triggering an emergent properties in the natural environment or through man's activities, e.g. design or trial-and-error;
- **Priming prototype (3 if-s)**: (i) if the entity has an emergent property; (ii) if the emergent property creates a selective advantage; and (iii) if the selective advantage primes a selection, the entity should be called a *prototype*;
- The prototype may emerge at one or many locations, at one or many times;
- **Emergence of purpose**: The *prototype* at the instant of emergence; triggers the purpose for survival through uncontrolled reproduction, characteristic of one-way flow of information of connecting inputs to building blocks to outputs;
- **Critical mass**: if the prototype proliferates in the environment, then it can form a critical mass to be transformed into an *evolvable entity*;
- **Triggering selection**: the zero$^+$ feedback stage is completed by reaching a critical mass, from which time the evolvable entity kicks off the next stage;
- **EvSyM** characterises this period by the term the Zero$^+$ Feedback. "Zero" signifies that during this long period, there was no viable connection; "+" signifies that a new feature has emerged from the 'no-existence' of the past; Feedback signifies that the prototype connected to its past;
- Other terms are equally valid: *open loop, spontaneous potential system with zero-flow of information*.

The classical science of feedback loops is confined to d-loops and this is why it is unwittingly ontological; whereas ℓ-loops and e-loops are all that define the horizon of theory of evolution.

This makes the culture rife for doubt on the law of natural selection, e.g. Karl Popper, who was even ambiguous on the law of natural selection. In between systems science and theory of evolution there is a gap, but bridged by Table 1, their integration. This is the integration of building blocks with rules or the architect with the architecture, to be referred as *evolutionary systemics* or *evolutionary systemic modeling* (EvSyM).

Box 1.3 – Axioms on Positive Feedback Loops

Routemap: survival, proliferation, diversification, adaptation, co-evolution, flow of information in one direction and instigation of entropy;

Positive feedback d-loop

- Inputs are connected to building blocks in the direction from the inputs towards the outputs and the whole arrangement produces outputs from given inputs;
- The interactions between the building blocks instigate internal entropy and between evolvable entities instigate external entropy;
- The survival is ensured by uncontrolled reproduction under entropy;
- In EvSyM, the positive Feedback refers to, *open loop*, closed system with 1-way information flow; closed systems becomes rather vague;

Positive feedback ℓ-loop

- Different configurations of the building blocks gives rise to different traits and this is at the core of adaptation, which may be attributed to the behaviours at the long (adaptive) timescale for the individuals of the *evolvable entities*;
- As there is no one way of reproductions, variations are inevitable in the long-run stemming from mutation but in man's world, there can be many reasons for variations contributing to *diversity*;

Positive feedback e-loop

- The e-loop is the connectivity of the building blocks as the architecture or blueprint for reproduction, including the know-how on the building blocks;
- The e-loop is simple but over many generations, natural selection transforms simplicity to complexity by selections dictated by selection pressures.

Box 1.4 – Axioms on Transition to Negative Feedback

- As no system is perfect, the connections between the building blocks are one source for potential inconsistencies and waste of some of the inputs;
- Often, the higher the dynamism, the greater is inconsistencies. This triggers a selective advantage for the emergence of a system to reduce entropy;
- A positive feedback complex is unable to cope with its inherent entropy unless it operates at a slow dynamic rate. An additional subsystem is needed to modulate both inputs and the working of the modules to maintain the output at the required level and maintain the integrity of the system. Entropy triggers the selective advantage for the emergence of such a subsystem;
- If such a subsystem fails to emerge, entropy undermines the system's efficacy;
- Due to entropy, evolvable entities dwindle, reach critical masses and perish;
- Critical masses may show resilience and revive through cyclic fluctuations.

Box 1.5 – Axioms on Negative Feedback Loops

Negative feedback loops emerge to maintain homeostatic conditions.

- The emergence of negative feedback is not automatic but is triggered by entropy, selected by natural selection and is operated by a small subsystems;
- This subsystem comprises: transmitters, a fact engine, actuators;
- The positive feedback complexes are reinvented to flexibly accommodate the supervision by negative feedback subsystems;
- Targets are essential for creating consistencies within the internal systems;
- Two-way flow of Information is essential;
- Traditionally, these subsystems are referred to as regulating subsystems, *closed loop* and open system with two-way flow of information but EvSyM distinguishes three negative feedback sub-loops:

Negative feedback d-loop: This is the reactive fact engine and works ubiquitously

Negative feedback ℓ-loop: This provides adaptive provisions against stagnation to snap out by "*discretionary*" powers overriding set targets or by resetting them;

Negative feedback e-loop: This is the routemap of resetting fact engines;

The role of natural selection: Negative feedback is a product of natural selection in the first place but the set targets are subject to natural selection in the long runs.

Box 1.6 – Axioms on Feedforward Loops

Feedforward loops emerge to maintain external consistency with the environment

- The emergence of feedforward is not automatic but is triggered by external entropy and selected by natural selection;
- The systems are reinvented to flexibly accommodate goal-seeking capabilities;
- Goals are essential for creating consistencies within the external environments
- Free-flow of Information is essential;
- Traditionally, these subsystems are referred to as goal-seeking systems, *closed loop* and open system with free information flow but EvSyM distinguishes three feedforward loops:

Feedforward d-loop: This is the proactive fact engine focused on the environment;

Feedforward ℓ-loop: This adaptive provision periodically articulates the goals;

Feedforward e-loop: This is the routemap of resetting vision/mission;

The role of natural selection: Feedforward is a product of natural selection in the first place but the goals are subject to natural selection in the long runs.

For a lay scientist the stimulus of "feedback/feedforward loops" would bring immediately physical systems to focus, i.e. *d-loops*. By the same token, the focus of same lay scientist on hearing "adaptation" or "hereditary codes" would be theory of evolution or natural selection, i.e. *ℓ-loops and e-loops*. It would hardly cross anyone's mind that science is ignorant on the integrity of these three sub-loops. Evolutionary systemics unearths this integrity – "so near and yet so far."

This paper offers EvSyM as the transformation of the state-of-the-art on science into building blocks; it then reworks and reassembles them. The rules for reassembling are original and reworking of some of the building blocks involve originality but the rest is the state-of-the-art in science.

For instance, the sequence of the four feedback loops is an original proposition but the breaking down of feedback loops into three constituent sub-loops is original.

An understanding of the basic concepts of systems and evolution are the key to EvSyM. A basic amount of information has been axiomatized into a series of Information Boxes and these are presented in Boxes 1.1-1.6. The aim is not to justify EvSyM in this introduction by these Information Boxes but to provide a bird's eye view of EvSyM.

1.3. Scope of the Paper

Attention is focused on the fact that EvSyM is not a review method, it is not a philosophical doctrine and it is not history or historiography; it does not have the historicism of Marxism. EvSyM is a modeling tool. As such, it uses historic/mythological accounts as its input data and provide an insight for the past events and foresight of the future. The paper introduces few new terms, one of which is Evsyming or EvSy-modeling, as shorthand for evolutionary systemic modeling. As some of the terms are smoothened by EvSyM, a list of terms is presented in Glossary and italicized terms in the main text.

Philosophy is the mother of all knowledge but Section 5 exposes the reasons why philosophy defeats its good aims and instead it ends up with contentions. Section 6 carries out EvSy-modeling of science as a whole; Section 7 takes a cursory view of behavior; Section 8 delves to the evolution in theory of evolution. Section 9 merely acknowledges feasibility of EvSy-modelling of social sciences; and Section 10 outlines the need for EvSy-modelling of morality.

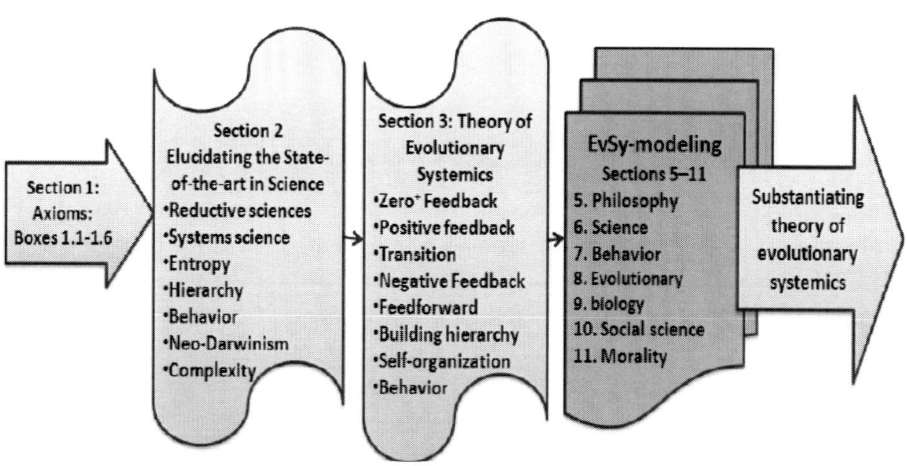

Figure 1. Pictorial Map of this Paper.

2. ELUCIDATING ROOTS OF EVSYM

EvSyM reworks existing science and reformulates them towards a generalized theory of evolution. This section elucidates the state-of-the-art on existing science as used by EvSyM but with a critical thinking.

2.1. Reductive Science

If philosophy was confined to articulating thoughts, this carried on until the 17th century. Since then a new body of knowledge has emerged that is referred to as science with a radical departure from all of the past traditions. The inherent evolutionary transitions and processes are presented in Section 6 but this section suffices to the elucidation of the *new building blocks* associated with triggering the departure from the past traditions.

Science emerged in the 17th century, identified by the working methods of Galileo Galilei (1564–1642), Rene Descartes (1596–1650), Isaac Newton (1642–1727), to name but a few; thrived under the name of natural philosophy; it acquired a formalized methodology by the 18th century; and it acquired the brand name of science by the mid 19th century. According to Cardwell (1957), measuring performance in mechanical systems became possible after the establishment of Galilean mechanics. Also, Kline (1985), states that "Modern science owes its present flourishing state to a (1957) new scientific method which was fashioned by Galileo Galilei (1564-1642)."

As noted by Cardwell (1957) a "distinctive contribution of the 18th century was the technique of systematic improvement." The scientific method is summarized as follows: (i) all components and processes of a system are kept constant except the one, which is varied systematically; (ii) the optimum arrangement of components is selected by noting a particular optimum variation and the procedure is repeated with other parameters or processes.

However, the true character of the scientific method is outlined in Box 2.1 and its explicit building blocks are depicted in Figure 2. The box depicts the methodology in terms of its building blocks.

Notably, in this methodology, outputs become a function of inputs and processes, thus it becomes feasible to select an outcome that meets a certain criteria. This new ability should not be regarded trivial but in fact, it is undoubtedly a negative feedback loop.

The above building blocks of science are notoriously simple and it is almost incredible to think that they were responsible for such an enormous impact. However, Ziman (1978, p88) remarks that "The sources of invention turn out to be extraordinarily subtle and episodic, revealing little more than the diversity of human behavior in unfamiliar circumstances." This is certainly true in reductive science, within which, reduction has a range of attributes:

1. *Assumption*: assumptions are stated all in advance.
2. *Setting explicit performance criteria*: performance criteria are stated in advance. Arguably the performance of a theory is assessed by a set criterion and scrutinized intelligently but the goodness of performance criteria is subject to uncertainty and it is open to natural selection.
3. *Flexibility*: the whole theoretical setup needs flexibility, so that if the performance criteria are not satisfied, appropriate parts are refined.
4. *Openness*: Reductive science nurtured an open culture, where everyone had the right to challenge. This may be inheritance from the past philosophical cultures. However, the new culture differed from the past, as the theories were modifiable due to their flexibility.
5. *Directions of information flow*: Contrary to physical systems with information flowing from inputs towards outputs, in the scientific method, information also flows back from the performance criteria to inputs. Thus, reductive science is a system with

2-way flow of information, which is tantamount to negative feedback loops. Hence, self-righteousness has to be relinquished and the authority has to be passed to agreed performance criteria. This encourages the growth of science from simplicity to complexity, mimicking natural selection.

6. *Integrity of scientists*: Ziman (1991, p42) remarks that "The fundamental principle of scientific observation is that all humans are interchangeable as observers." In the past thinkers were normally mindful of political authorities, who would twist arguments to please authorities.
7. *System thinking* was partially embedded in reductive science by breaking the system down into units. However, due to enormous problems, it was not normally feasible to study all the units together.

Box 2.1 – Key Building Blocks in Reductive Science

Systems thinking can tacitly be found in the scientific method, which lends itself to be reduced to the following building blocks:

1. **Reduction**: This is the ability to break down the system into components, attributed to Descartes (Part V *Discourses*, 1637). In the newly emerging culture of science, problems were reduced into components and each component was treated on its own. When reductive science was established, the analysis was carried out one-at-a-time in isolation. Reduction was a formalization of the parts and their transformation into components. As everything is made of parts, this is not surprising now but in the ontological culture of the past, this was not obvious. Reduction may be referred to as a building block approach or analytical thinking. It is considered as the opposite of "vitalism," a doctrine that claims for the existence of a non-physical inner force called "vital force."

2. **Negative Feedback Loop:** Cyclic reviews in Figure 2 are in fact negative feedback loops. It was not reduction exposing the dismal inadequacy of philosophy but the newly inaugurated negative feedback loops. These loops build on a set of conditions including: assumption, flexibility, setting explicit criteria for performance. This loop is a clear declaration that no assumption is supreme or sacred, as all assumptions must be declared and readily open to challenge. This was an evolutionary transition in the intellectual world and the beginning of the beginning for the demise of the past self-righteous culture. No such transition had taken place in the intellectual history. Notably, the selective advantage of negative feedback loops was well established by natural selection a billion years or so ago.

3. **Theorization**: This was fundamental to the fledgling science as masses of information were contracted by simple statements, often mathematical. Evidence-based qualitative statements are equally credible, e.g. the law of natural selection.

4. **Observation, Data and Measurement**: The scientific methodology uses data/evidence as its currency, which is really a pre-requisite for its functioning. The contribution of data to science must not be underestimated and Barrow (1991, p5) remarks that "Myths do not arise from data or solutions to practical data." The ability to measure data includes inputs, outputs and the state and phases of the system, although some of its phase parameters may be difficult to measure. Measurements are normally costly and may be inconvenient.

There is really no science without any of the above building blocks.

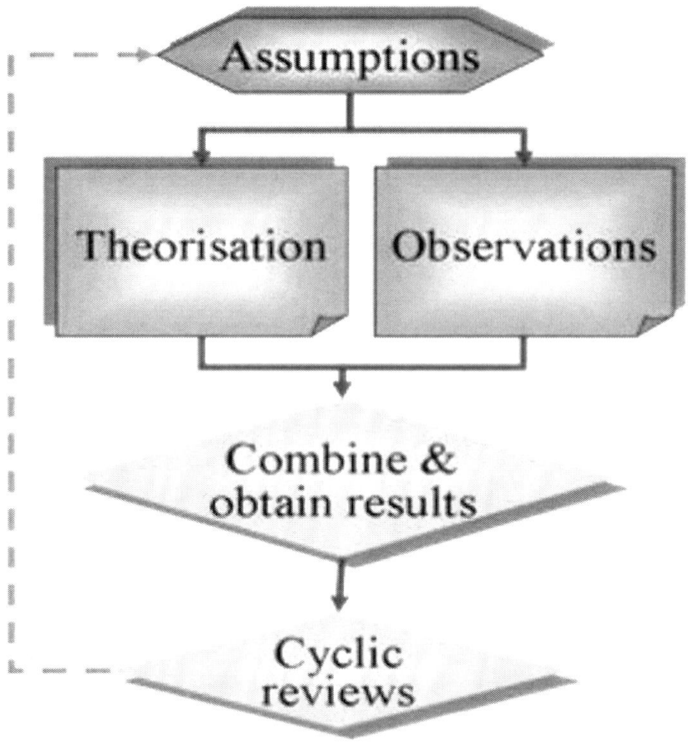

Figure 2. Explicit Aspects of the Scientific Method.

2.2. Systems Science

Systems science emerged in the 1950s and its role on EvSyM is presented in Sections 6.3.5. This section outlines its scope, definition, aims and key features. Its building blocks transferred to EvSyM is presented in Box 2.2. Systems science thrives on the dynamic timescale but its state-of-the-art on the evolutionary timescale is static. Thus, systems science is inadvertently ontological.

Scope: Today science is largely systems science and its scope encompasses formal sciences (e.g. complex systems, cybernetics, dynamical systems theory, and systems theory); it is applied to (the natural and social sciences), engineering (control theory, operations research), social systems theory, systems biology, systems dynamics, systems ecology, systems engineering and systems psychology.

Undoubtedly, systems make up the fabric of science and Kenyon and de Greene (1970, p85) argue that "Every educated person recognizes that a "system" imposes an order or consistency on similar interrelated constituents (for example solar system, nervous system, tax evasion system); yet there is no integrated body of system knowledge acceptable to the educated public as a whole."

For more details references may be made to Mesarović and Takahara, (1975), Checklands (1981), Vickers (1980), Lilienfeld (1978), Kast and Rosenzweig (1970, de Greene (1970), von Bertalanffy (1962).

Definition: A system is a selection of modular units interconnected with one another towards a purpose. Checkland (OSG, 1981, p26) argues that systems "embody a particular purpose." Kenyon and de Greene (OSG, 1981, p.87) state that if the system "constituents are assembled at random, the situation is chaos and a system cannot be said to exist."

The term purpose is important and it needs to be dealt with at this onset. Before putting together a set of elementary building blocks, there is no system and no purpose but a system and its purpose only emerge after putting them together. So purpose is just a connotation of emergent property and evolves with the feedback loop, in which case it is neither *a priori* nor an ontological/ transcendental quality. This is discussed further in Sections 3 and 11.1.

Aims: General Systems Theory (GST), by von Bertalanffy (1940, and 1969), runs as a common thread throughout interdisciplinary endeavors. He founded *the Society for General Systems Research* with others and published *General Systems Yearbook* until 1980 with the goal of: "theoretical systems which are applicable to more than one of the traditional departments of knowledge." Their aims were:

1. There is a general tendency towards integration in the various sciences, natural and social.
2. Such integration seems to be centered in a general theory of systems.
3. Such theory may be an important means of aiming at exact theory in the nonphysical fields of science.
4. Developing unifying principles running 'vertically' through the universe of the individual sciences, which brings us nearer to the goal of the unity of science.
5. This can lead to a much-needed integration in scientific education.

Key Features: GTS runs through all disciplines of systems science by the virtue of a number of key concepts. In the first, it divides the systems to be closed and open systems, where a closed system, in its strict thermodynamic sense, is the one that does not exchange matter or energy with its environment (isolated) and inherently tends towards a static equilibrium or maximum entropy; whereas in open systems, such exchanges are allowed and they are interacting with one another and with the environment. GST is associated with hierarchy, feedback loops, emergent properties and a host of other concepts.

These are nuts-and-bolts of this paper and are discussed where appropriate. The thermodynamic concept of entropy is well recognized in systems science and often treated as negative information but this is rather controversial.

This paper revisits these terms in Section 3 and Box 2.2 highlights its main features. The exchange in physical systems is through matter or energy but in many social-intellectual systems, this can be through information.

The concept of closed and open systems is important in resolving an anomaly associated with the second law of thermodynamics. According to the law of natural selection, species increase from simplicity to complexity over time, whereas the second law of thermodynamics predicts that a system tends towards increased entropy and the eventual death. The concepts of closed and open systems resolve this anomaly.

Adversaries of Systems: Checkland (1981, p38) credits Bertalanffy as the founder of GTS and points out that "The general theory envisaged by the founders has certainly not emerged".

Box 2.2 – Building Blocks of Systems Science

The state in closed systems is determined by an understanding of its inherent entropy and given this, the final state depends on its initial condition; whereas in open systems energy/information is supplied at a rate that enable a final state to be reached from many different initial conditions and thus homeostasis becomes possible. These concepts fail to discern the dynamic timescale from the long and evolutionary timescales. Equally, traditional concepts can be clearer on the direction of information flow and feedback loops. This paper revisits these concepts and argues that the feedback complexes defined in Table 1 can reshape science for better than using the closed and open system concepts. Khatibi (2003) combines the scientific method (Figure 2) and systems concepts, as follows:

- Complexity is organized hierarchically.
- Lower hierarchies are 'closed systems' in the sense that they contain inherently increasing entropy for a further discussion on entropy). Higher hierarchies are open in the sense that there is a feedback loop to regulate the inherently increasing entropy associated with both closed-system lower hierarchies and open-system higher hierarchies.
- Feedback loops maintain the flow of information from lower to higher hierarchies and within a higher hierarchy, enabling control and regulation.
- Each hierarchy is a synthesis of a finite and often a few simple components, which are bound together by certain rules. These components may be referred to as 'simplicities'.
- The components of a higher hierarchy are a composition of lower hierarchies components but there are emergent properties at a higher level which are absent in the lower ones.
- The complexity of a system is a composition of its hierarchies. In this sense, complexity is a system without referring to its goals or purposes.

Although Checkland is a leading figure on systems science and has introduced the term systems thinking, he is somewhat justifiably critical of GST but there are criticisms too e.g. by Buck (1965, pp223-38) from the vantage of positivism, other criticisms are presented by Lilienfeld (1978). Whilst these are directed against GST, some philosophers of science are uneasy towards the concept of systems from the vantage of existentialism. The followings are examples:

- The term system is used pragmatically by Abbott (1995, p63), stating: "the whole notion of 'system' is a local and contemporary one, peculiar to the 'self-centred' and so anthropocentric mode of modern thinking". He regards system as synonymous with 'mathematisable'.
- Dostoevsky (1956, p69) argues: "But man has such a predilection for systems and abstract deductions that he is ready to distort the truth intentionally, he is ready to deny the evidence of his senses only to justify his logic."
- Jaspers (1938, p164) argue that for German idealists system is "a detour from reality and is, therefore, lies and deceptive. Kierkegaard granted that empirical existence could be a system for God, but never for an existing spirit; system corresponds with what is closed and settled, but existence is precisely the contrary. The philosophers

of systems are, as a man, like someone who builds a castle, but lives next door in a shanty." T Nietzsche, "The will-to-system is for him a lack of honesty."

- Heidegger (1977, cited by Abbott 1995) states: "In the Middle Ages a system is impossible, for there a ranked order of correspondences is alone essential, and indeed as an ordering of whatever is in the sense of what has been created by God and is watched over by His creatures. The system is still more foreign to the Greeks, even if in modern times we speak, though quite wrongly, of the Platonic and Aristotelian 'Systems'."

Evidently, Abbott or Heidegger does not pity the Middle Ages deprived of the conception of systems but modernists being apparently menaced by it.

2.3. Entropy

The concept of entropy emerged in the 19th century during the study of thermodynamic systems and the term was coined by Rudolf Clausius (1822–88) in 1865 but Ludwig Boltzmann (1844–1906) revisited the subject in the 1870s by introducing his statistical treatment connecting microscopic and macroscopic dimensions of entropy. It is the key for the Second Law of Thermodynamics, stating that in a "closed" thermodynamic system, entropy will increase but cannot decrease. Thus, the state of the system tends towards rest or dissolution, signifying the absence of available energy to do work. EvSyM relies on entropy as a force instigating the selective advantage for evolutionary transitions. The concepts behind entropy are therefore elucidated below and the key features transferred to EvSyM are highlighted in Box 2.3

The concept of entropy in thermodynamics is explained by Max Planck (1991, p163-169) that "the principles of conservation of energy do not suffice for a unique determination of natural processes." Conservation of mass and momentum are fundamental laws of nature underpinning a deterministic framework for scientific approaches. However, if these are insufficient, does it mean that nature is indeterminate? According to Max Planck (1991, p163-169):

1. The conservation laws do not express the direction of energy flows.
2. The conservation laws of mass and energy are insufficient.

The second law of thermodynamics was developed to describe the processes in between the initial and final states by quantifying a proportion of the total energy that is dissipated by gas particles through unrecoverable random movements. This internal energy is referred to as entropy and is expressed by the second law of thermodynamics. It proves that entropy in any system will increase over time unless energy is added to the system. Further energy is only made available at the expense of further fuel as energy is degraded by heat transfer down a temperature gradient, (Joel, 1996).

In searching for a new fact/information by moving from the first to the second law of thermodynamics, Max Planck argues that the only fact is that there is no new fact. This authoritative statement within the remits of reductive science is explored further and extrapolated by the in Box 2.3.

Box 2.3 – Key Notions of Entropy

The role of entropy in EvSyM is very important. It is not only a feature of the concept of closed systems, as defined by Bertalanffy (1962) in GST, but is also a feature of any evolvable entity. The essence of entropy is that whenever the process plays a role in the transformation of the initial state into the final state and *vice versa*, entropy plays a role so that (i) the conservation laws of mass and energy are insufficient to determine the final state of the system; (ii) the process becomes irreversible; and (iii) without understanding the role of natural selection, the system is relativistic or indeterminate – the latter is completely missed by philosophers.

Khatibi (2011) argues that although thermodynamics may be the birthplace of the concept of entropy, it is observed widely and he provides some evidence from flood risk management. The followings are observed:

- Entropy exists in all systems: inanimate/animate, technical, social and socio-technical; material/symbolic. Its rigorous concept roughly means disorder;
- Entropy is the conversion of some of useful input energy to unrecoverable randomized energy and the concept is generalized by replacing work with information and entropy with uncertainty;
- In closed systems, entropy tends to increase
- In open systems, entropy can be maintained at the same level by the negative feedback loops at the expense of inputting further useful energy.
- This paper regards entropy as a force for evolutionary transitions and recognizes internal entropy instigating the selective advantage for the emergence of negative feedback complexes and external entropy for feedforward complexes.
- The first law is a declaration on (i) existence of universal information to determine order or rules; but (ii) the absence of universal laws on system-specific information, local arrangements or building blocks.
- Within the remits of systems science, the second law is a declaration to drive for knowledge on system-specific information.
- In thermodynamic, aerodynamic or hydrodynamic systems, empiricism is the driver for reducing entropy, turbulence or friction, e.g. empirical streamlining of aerodynamic and hydrodynamic forms.
- In Nature, natural selection is an evolutionary driver for system-specific information, biased toward better workable arrangements.
- Natural selection in higher animals has given rise to fuzzy logic for a more systematic way of reducing entropy, which is a reflection of the first law of thermodynamics; likewise, it has given rise to negative feedback and feedforward loops, as the machinery to reduce entropy.
- Natural selection and entropy are scientific facts; philosophy speculated them, e.g. Marxian dialectics or nihilism in existentialism; many religions; fictionalized them; and mythologies fantasized them.

2.4. Hierarchy

Although the legacy of Greek philosophers has been largely unselected at the dawn of reductive science in the 17th-18th centuries, one of its relics is categorization, which goes back to Plato (424/423 BC – 348/347 BC) but Aristotle developed it further. His categories are no more, but as a methodology, it is still thriving. Normally, categorization is an organized approach to make sense of unorganized objects and one method available for categorizing objects is the vantage theory, see Taylor, and MacLaury (1995) among others. Categorization can easily be arbitrary and adds value if reveals inherent patterns. Successful examples include:

- The Periodical Table in chemistry,
- The cladistic in evolutionary biology, and
- The Comte (1868) hierarchical organization of science (see Lenzer, 1997) is outlined in Box 2.4.

Hierarchy is a complex subject and different disciplines or theories produce their own hierarchical structure. However, the intention of the paper is not to cover the subject, other than its relevance to EvSyM. The above three models are relevance to EvSyM, as they reflect: (i) interconnectivity among mutually inclusive building blocks at a given hierarchy; (ii) interconnectivity among mutually exclusive entities and (iii) hierarchical structures.

Box 2.4 –Science as a Model of Hierarchy

The visionary classification of science by Comte (1868) captures the essence of systems science, although his classification predates systems science, emerged in the 1950s. Checkland (1981, p27) presents a modification of his classification and this paper refines it further, as described below and captured in Figure 2:

Physics is most basic science, as the science of external properties of the elements

Chemistry is the science of compound elements where compounds often have emergent properties absent within the elements

Biology is the science of interacting chemical compounds – emergent property: reproduction

Psychology is the science of biologically-reproduced interacting individuals - emergent property: gaining consciousness of the environment and traits

Sociology: is the science of interacting individuals - emergent property: values;

Anthropology: is the science of interacting societies – emergent property: culture

2.5. Behavior Models: Phase, State, Selection

Behavior is the time-dependent processes of action, reaction, interactions or functioning of a system, under external stimulus. Behavior studies break implicitly or explicitly the subject into (i) the phases (domain, continuum, the environment, cell, body, individual, habitat/biome, context, community, or institution) of a system, (ii) its states (various resolution of time); and (iii) the capacity to act autonomously (decision-making). To the best knowledge of the author, the state-of-the-art on the subject does not explicitly integrate these dimensions together and hence each discipline seems to follow its own tails.

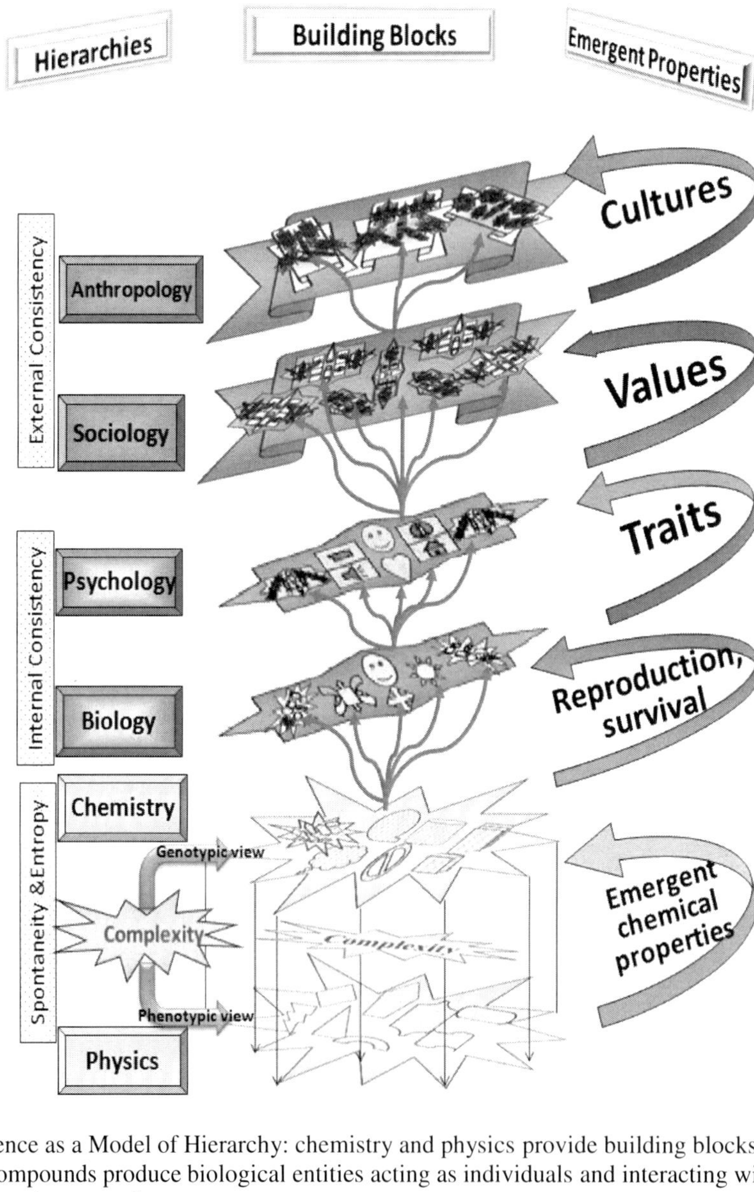

Figure 3. Science as a Model of Hierarchy: chemistry and physics provide building blocks to produce compound; compounds produce biological entities acting as individuals and interacting with the environment; and so on and on.

In pragmatic scientific treatments of behavior, phase and state are intertwined. The concept of closed and open systems (Section 2.2) serves as the phase and some of the varieties of the problems experienced by the author are outlined below. The concept of *control volume* (see Shames, 1982 among others) is used in hydraulics and fluid mechanics in deriving flow equations such as Navier-Stoke equations. In these derivations, normally the whole system is treated as the closed system, where the catalogue of the various forces acting on the fluid are identified. An infinitesimal (or a miniscule) unit of this closed system is selected as a typical unit (referred to as the control volume) and the forces are worked out acting on the control volume. The emerging equations are then generalized over the (closed) system. The derived partial-differential/integral equations are not explicit on the phase but have all the required details to work correctly both in the control volume and over the entire closed system.

The phase has other forms in science. For instance, in fluids mechanics, the phase (continuum) is continuous; whereas it is granular (particulate) in soil mechanics, either the particles are exerting some forces between each other (e.g. cohesive soil, known as clay) or there are no forces between the particulate matters (e.g. sand). The mechanics of such a continuum is very interesting and in soil mechanics, it is dealt with by the Coulomb theory and the Rankine theory, where the medium itself acts an *active force* on its surrounding due to its weight and other forces. Likewise, *passive reactive forces* are set up by the particulate phase when it is acted upon by external forces.

Box 2.5 – Building Blocks of Behavior Models

EvSyM reduces behavior to the dimensions of (i) phase, (ii) state and (iii) decision-making (agency or selection), which is hierarchical. An overview of behavior is outlined below but detailed in Section 7, with the salient points as:

- There are four phases (continuous, particulate, members, and plasma).
- A diversity of time resolutions may be defined: (i) static timescale, when the phase is subject to no movement; (ii) dynamic timescale; (iii) long timescales; and (iv) evolutionary timescales.
- Behavior overarches phase and state and the capacity to act freely but EvSyM treats them in evolutionary/hierarchical context. The capacity to act freely is modeled through evolutionary transitions, as follows:
  - Zero Order (inactive, static) World, i.e. no significant physical motions such as those concerning the objects on Earth or the moon.
  - $Zero^+$ (spontaneous) Feedback: fully random world (e.g. the quantum world) but giving rise to the causality world, i.e. chemical elements with chemical properties. This feedback pluralistic and coexist with loops.
  - Positive (active) feedback: this is the chemical and/or causality world and acts as a cement to connect the particulate world to one another. This coexists with other feedbacks.
  - Negative (reactive) feedback: This is the world of agents, which act autonomously, responsively and adaptively. This is pluralistic and coexists with other feedback loops.
  - Feedforward (proactive): this is the world of decision-makers, acting autonomously, responsively and adaptively and proactively. This feedback is pluralistic and coexists with other feedback loops.
- A further dimension is data/evidence, as discussed in Section 7.

The capacity to act freely is therefore evolutionary.

Solid mechanics and stress analysis (structural analysis) provide another category of phase, where the analysis is applied to individual members of the system, normally ignoring the space within which the members are placed in. Note that the term analysis in the above three cases needs to be clarified. These terms go back to the 18th century, when the term science or behavior was not in common use. The thinking in those days was organized by analytical thinking (reductive approach) and hence the terms. The analysis in the first two types of phases (continuous and particulate) normally assumes homogeneity even though heterogeneity is often the case in particulate phases. Accounting for the behavior of the individual members is of the paramount importance in structural analysis. Heterogeneity is the norm in ecosystems and in communal/social domains, all of which use explicitly the concept of open system.

Pragmatic science is efficient in devising a diversity of theories and modeling techniques to cope with phases, states and decision-making (selection) but they lack a common basis. EvSyM is capable of overarching these dimensions as a common vantage for all behavior models. A glimpse of this is given in Box 2.5.

2.6. Neo-Darwinism

This section elucidates the neo-Darwinian basis for this paper; Section 3.2 reflects a critical thinking on natural selection; Section 8 details the evolution of the theory evolution in biology; Box 2.6 presents some preliminary accounts. The classic view of neo-Darwinism is that: (i) the slowly changing environment exerts a selection pressure on the genes of the individuals within a population; (ii) this is translated to a selection of certain genes during reproduction times that give rise to more suitable traits for the individuals to survive; (iii) the individuals with the traits matching the behavioral requirements of the environment are said to have adapted to the new environment and are more likely to survive and have offspring and hence pass on their favored traits; (iv) when a long-term view is taken through generations, evolution has taken place, which is the routemap of the changes.

Neo-Darwinism integrates Darwinian natural selection (focused on changing frequencies in the gene pool of the population) with Mendelian genetics (focused on changes in particulate genes). Both of these concepts are vantages at the phenotype resolution. The reciprocal of the phenotype resolution is the genotype resolution through the RNA and DNA machinery. Evolution is the collective routemap of the above both as genotype and phenotype resolutions (routemap is not identical to roadmap). Natural selection does not select individual reproducing entities but their traits, which are the outcomes of their genes. It is true that under natural selection, fitness has a selective advantage but fitness does not mean well trained muscles or being brainy. It simply means traits suiting better to a particular environment, although strong muscles and brain help but are not the determining factors. Evolutionary biologists underline the fact as:

$$\textbf{\textit{Natural Selection}} = \textit{Heredity} + \quad \textit{Varirati}\textit{on} + \begin{matrix} \textit{Differential} \\ \textit{Reproduction} \end{matrix} \qquad (2.1)$$

Each of the above may be defined as follows:

1. *Heredity* or genetic codes are encapsulated in terms of the DNA+RNA machinery. This is like a moody sticky material that does not accept new adhesive units all the times but under certain conditions and therefore the replicator can reproduce it intact.

2. *Variation* in the above formula stems from the fact that the gene machinery admits new genes as "recommended by natural selection" thanks to the process of reshuffling. This makes it possible to introduce new genes or change some of the rules from time to time. The outcome is the creation of "diversity pool" or gene pool in the population.

3. *Differential reproduction* means that those evolving entities best adapted to a given environment are most likely to survive, to reproduce and to create the norm for further reproductions.

4. The gene machinery grows from simplicity to complexity over time.

The hereditary codes in living things are capable of (i) producing protein for homeostasis functioning of the species or for the maintenance of the body; and (ii) self-replicate or reproduction (sexual reproduction) involving gene reshuffling. Natural selection reaches from the outer world to the inner world, where, there are inputs, building block compounds and outputs, as well as the rules for synthesis, which are amenable to mutation. These entities have a range of traits when performing in the outer world, by the virtue of which the individuals can be "hooked" to their environment for functioning towards their survival. Survival is the only purpose at this stage. Here, the individual can be a person, any single member of a species, a car, a motor, a scientific theory, a mathematical theorem, a particular painting, etc. If there is no hereditary system, there is no evolution, i.e. heredity connects evolution with genetics.

It is counterintuitive that the hereditary system is the outcome of natural selection, where Nature itself behaves callously, purposelessly and unintelligently and these are also selection criteria. It is even more counterintuitive that these selection criteria are pivotal in natural selection for not working randomly. Natural selection drives slowly simplicity to complexity. For a modern scientist and engineer it is difficult to recondition their minds to the past speculations as for them water is simply regarded as H_2, O and their emergent properties and nothing more. They are often unaware that ontological philosophers still live in past traditions insisting that water exists in the first place, H_2 and O are attached onto it to make its properties.

This way of thinking loomed over any intellectual endeavor including initial evolutionary thinking but the emergence of theory of evolution through the Darwinian natural selection unselected the past ontological mindsets. Now, ontology is irrelevant in biology but it still lurks around where evolutionary thinking has not been fully established, e.g. the wider sciences. Arguably, if evolutionary thinking is not driving the discourse in any area of science, that area of science is exposed to relics of ontology and this is widespread.

Notably, the evolutionary positive feedback stage of a culture is not pleasant most of the times as at this stage there is no co-operation, no passion, no empathy. These attributes are selected by negative feedback and feedforward loops and we are lucky to inherit cultures rich in these attributes.

Box 2.6 – Critical Insight into Neo-Darwinism

The author initially conceived the generality of natural selection by associating it with positive feedback (as a characteristic of closed systems) and associating natural selection with the gene machinery as the theme of the Dawkins selfish gene. It seemed obvious that (i) positive feedback was the signature of closed systems; (ii) natural selection was the signature of the outer world (phenotype) on the inner world (genotype); (iii) positive feedback and natural selection together reflect two faces of the same coin; and (iv) entropy is the instigator of a randomized state and death. However, EvSyM is much more advanced version of these initial thoughts.

Neo-Darwinism applied to the species can simply be generalized by using the feedback e-loops as the architecture for the inherent heredity systems driven by natural selection. Thus, the Neo-Darwinist model of evolution is very much relevant to any evolvable entity of any complexity within the social, cultural and intellectual life. To believe that only living things are capable of evolution is to act as a chauvinist for living things. So evolution of cars or a mathematical theorem and many other evolvable entities are subject to identical architecture to that of living things. The paper offers a hierarchical approach to evolution, where even evolution without natural selection can also be taken on board, e.g. the evolution of mountains, continents and oceans, referred to as physical or *pristine* evolution. Above the physicalistic level, i.e. from chemicalistic hierarchies upward, a hereditary system is likely to emerge and evolve, where natural selection plays the role of free agent. In this context, Nature also has the following roles:

- **Nature is the phase** and therefore in different context it may be taken to mean the cell, domain, continuum, the environment, cell, body, individual, habitat/biome, context, community, or institution. The paper prefers phase as a generic term to overcome the linguistic twists.
- **Nature is the state,** as in evolution in various timescales.
- **Nature is the arbiter (free agent, decision-maker, selector)** i.e. the e-loop

2.7. Complexity

The world of species growing organically is not unique and indeed, it would have been an anomaly if it was unique. Consider the enterprise of science, mathematics, risk management, planning, design, communication, management science, art, sport and so on in man's social, cultural and intellectual worlds. All these too are growing organically. The author calls each of them as enterprises of complexity. There are countless enterprises, each composed of many different areas and sub-areas and may be sub-sub-areas. For instance the enterprise of risk alone prevails in countless areas of human endeavors including floods, food industry, finance/banking, pharmaceutical industry, waste disposal industry and health and safety, to name but a few. The various areas of complexity are jumbled up seemingly growing organically. In reality, they all can have a "fractal" structure, in which a hidden simple structure is replicated by feedback loops. If so, EvSyM stands the chance of producing tectonic shifts in science.

3. Theory of Evolutionary Systemics

EvSyM is a new theory presented in this section, substantiated in Sections 5-11, and this is extended by Khatibi (2012a and 2012b) and Khatibi et al (2012). It describes the routemap taken by any *evolvable entity* through transitions from one stage to the next; it also describes the process of building of hierarchy. As a model of evolution, EvSyM generalizes neo-Darwinism but differs from any other generalized Darwinian evolutionary theory due to using systems science as the architecture for evolution. This paper reworks existing science.

3.1. Stage of Zero$^+$ Feedback Simplexes

Consider the very first living species, language, water treatment plants, electric motors or steam engine cars. Each is an *evolvable entity* from a *prototype* (the first generation product). The focus in this section is on the routemap for the emergence of a prototype and its subsequent *priming* at the point of the very first selection until the emergence of a critical mass as outlined in Table 2. A routemap is a process comprising: (i) a rich environment/culture, (ii) abundant potential building blocks, (iii) the emergence of conditions and/or rules for synthesizing the building blocks together through self-organization when the conditions or specific rules emerge; (iv) the *priming* of a prototype to display its emergent properties; (v) the first and subsequent selection of the prototype by becoming an evolvable entity; and (vi) the proliferation of the prototype until the emergence of its critical mass and the subsequent encapsulation of the process of reproduction. The paper regards these processes collectively as the zero$^+$ feedback loops.

Emergent property and selective advantage: The possibilities for combining the building blocks are countless but the priming of a prototype is not that common. For instance, words can be put together in countless ways but they will often lack a collective meaning. The meaning emerges (note that meanings are emergent property) only when they follow certain rules. The emergence of many different languages also signifies that the rules are not unique but prerequisite (this also invokes the potential for mutation and diversification). If there is no emergent property at the point of priming, there is no prototype but a heap of building blocks. An emergent property serves as a selective advantage and both of these terms express different aspects of the same entity stemming from different disciplines of science.

Systems science is the home for the term emergent property and evolutionary biology is that for selective advantage. These terms bridge the two disciplines together. For instance, emergent property is spontaneous and it emerges as soon as the building blocks are operated according to the rule or when the required conditions are imposed. However, a selective advantage is just a potential and any primed concept or reproducible entity may not be selected at all. The classical example is the heliocentric model of planetary motions, according to which the Earth and planets revolve around the Sun, a notion that was proposed as early as the 3rd century BC by Aristarchus of Samos, see for example Dreyer (1953). This is nearly as old as the geocentric model but was not selected at the time for its intuitive appeal. The main point is that an entity with a selective advantage is not predestined for selection.

Table 2. Zero$^+$ Feedback of a Host of Evolvable Entities

	Function	Possible genotypic processes before priming	Possible Phenotypic processes before the priming	The priming process at the acts of selection	The Process as they it is understood today
• Species	• Reproduction towards survival	• Primordial soup: an environment with abundance building blocks (amino acids) and the conditions for self-organization • The ability to process "food" (from molecules floating in its outside) to create energy • The ability to maintain and expand the cell wall (grow) • The ability to split itself to reproduce • Internalizing self-replication	• Primordial soup: an environment with abundance building blocks (amino acids) and the conditions for self-organization • Plenty of food • Favorable external conditions for the survival	• Self-organization giving rise to a very primitive positive feedback machinery • The stability of the environment for such a length of time that the act of reproduction takes place for enough time to create an appropriate gene pool • When the environment changes, the act of selection itself triggers a ceaseless process	• DNA acting as a genetic blueprint • RNA, an enzyme, copying info of DNA for new proteins/enzymes • An enzyme to manufacture new enzymes and their building blocks • An enzyme to copy the genetic material preparing for reproduction • Enzymes to split one cell into two for reproduction and other operations • Enzymes to produce energy molecules to power other enzymes • A cell wall to contain the cell
• Language	• Better communi-cation for survival	• Physiological developments in the muscles of the face and the mouth • Model thinking neural infra-structure for creating meanings • Appropriate neural infrastructure to remember verbal building blocks	• The culture of association to represent something by another thing sensible to, understandable by, others • The culture of triggering action by the emerging verbal constructs • Clarity in the sound artifacts and hence consonants • The ease to reproduce verbal artifacts	• Self-organization for producing distinct words composed of vowels and consonants • A sufficient number of words to create a selective advantage by improving the communal survival • Changing environment triggers selecting words	• Languages are cultural hereditary system mimicking living things and science. • Languages are evolvable entities • Languages are interoperable but not exactly identical • There are d-loops, ℓ-loops, e-loops • Languages have physicalistic-anthropologistic hierarchies
• Water Treatmen t Plant	• Improving on water quality against visible diseases from water supply	• Basic forms of the followings: • Grit removal, • Primary and secondary filter beds • A host of ancillary works such as pipes and valves	• Water-borne diseases were widespread in industrialized countries • The first such a plant emerged in the 18th and were replicated in many countries	• Water quality was connected to water-borne pathogens in the 18th century, where turbidity was found to be un-aesthetic and bearing fecal matter, could harbor pathogens.	• Automating for reliable operations • Pre-ozonation: inactivate algae, improve micro-flocculation; • Post-ozonation and granular activated carbon beds to remove pesticides; • Dissolved air floatation counter current: remove algae/improve taste and odor

	Function	Possible genotypic processes before priming	Possible Phenotypic processes before the priming	The priming process at the acts of selection	The Process as they it is understood today
• Electric Motor	• Automatic motion by converting electrical to mechanical energy	• basic forms of the following: • Armature or rotor • Commutator • Brushes • Axle • Field magnet • DC power supply of some sort	• An understanding of magnetism and electricity • Conversion of one form of energy to another	• The first motor was primed by Faraday in 1821 or similar ones by other, • Different versions were tried over time for different usages	• Motors are widespread: machine tools, industrial fans, household appliances, disk drives, blowers, pumps, power tools, etc. • Replace fossil-fuel based motors with electric motors • efficient, environmentally friendly and sustainable motors
• Steam Engine Car	• Machines capable of propelling themselves	• Body framework (chassis), wheels • engines (supply of steam, a cylinder, a piston, exhaust ports) • A break system	• Attempts since the 17th until the early 19th century, but a cars emerged by the end of the 19th century	• See Bellis: Automobiles were not invented in a single day by a single inventor • Automobile evolved • Estimated number of patents over 100,000 to reach modern automobiles	• Cleaner, efficient, environmentally friendly and sustainable cars • Energy sources/energy saving • Advanced controls, safety measures • Material

Routemap of zero⁺ Feedback: For natural processes, the act of synthesis is under natural forces; else normally man also synthesizes, invoked by accident, trial-and-error, surreptitious thinking or deliberate innovations and design. The synthesis may lead to a product and this may emerge as a prototype (i) if the product has an emergent property; (ii) if the emergent property creates a selective advantage; and (iii) if the selective advantage primes a selection. Emergent properties are spontaneous and emerge whenever the building blocks are put together by the rule under the right conditions. The emergence of a prototype needs a rich environment/culture teeming with the building blocks. The availability of building blocks alone is not sufficient for the emergence of a prototype, as the act of specific rules/conditions for the synthesis is indispensible. The rules are not "meta" as meta-rules give the sense of someone pulling the strings; whereas the rules are very much intrinsic. If a prototype proliferates in the environment by other compounds or evolvable entities, a critical mass may emerge. This marks the transformation of a prototype into evolvable entity. Zero$^+$ feedback is a potential as depicted in Figure 3. Prior to the priming of a product, it covers an uncertain time when some sort of building blocks are lurching around until the conditions at the environment is about right for the synthesis of building blocks with appropriate rules to prime a product. Until then, relating inputs and outputs (or cause and effects) is not yet obvious but a perception of such a relationship may prevail in the social/intellectual world. Such perceptions are often like a mirage and it is tentatively suggested that only a few of them progress towards a selective advantage. This carries on until the process of the priming of a product is kicked in.

The building blocks used for self-replicating (reproducing) prototypes would be fragile entities and quite variable. It is not impossible to imagine that, the changing environment would trigger a process of selection with the outcome of honing the prototypes towards more stable prototypes. During the period from the priming of a product to its critical mass, the process is immensely time consuming and often fragile, as everything can go wrong. Moreover, the prototype during this period bears little resemblance to its forms in the subsequent generations. In this period, the prototype has to reproduce through an active positive feedback d-loop and it has to be selected until the emergence of a critical mass.

The formation of the critical mass signifies that the prototype is the parts-and-parcels of the environment and has a good chance to survive. It also marks the arrival of the new stage of positive feedback. The phenotypic features are then referred to as a system, theory, theorem or a model but its genotype ones strive to be transformed into a hereditary blueprint. There is a selective advantage to encapsulate the self-replication machinery, as it stabilizes the process from the priming of the prototype until the emergence of a critical mass.

The term priming a product is used here as a generic description. It covers the natural environment, where self-organization is operating; as well as social, cultural and intellectual environments, where the act of priming requires a painstaking effort and a great deal of dedication.

The classic case for building blocks and rules is the DNA and RNA in living things. Before triggering the machinery of natural selection, there was no living species, no DNA, no RNA and no act of selection. Scientific research prior to genetic theories were pioneered by Stanley Miller (1930-2007) and Harold Urey (1893–1981) in the 1950s and now referred to as the Miller and Urey experiment, showing that under particular conditions below the ultraviolet radiations and with the abundance of water, N_2, CO_2, O_2, H_2,, CH_4, and NH_3

organic compounds such amino acids can be created (for electronic experiments see http://www.millerureyexperiment.com/).

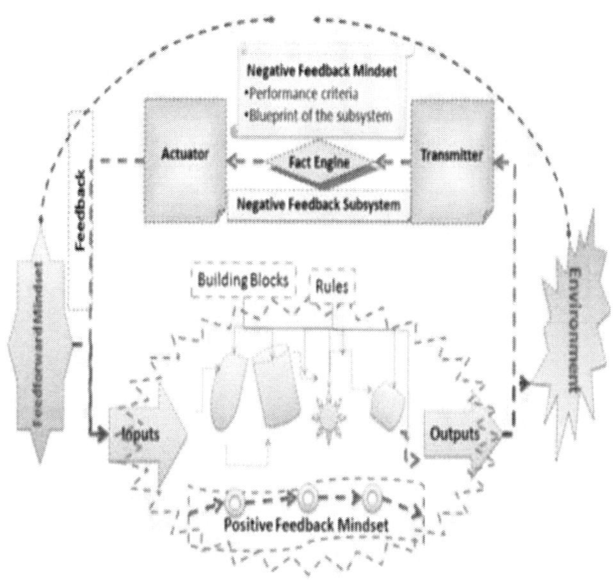

Figure 3. Open Loop Systems Underlying Zero^{+} Feedback Loop Architecture.

An understanding of self-replication of living things is an active area of research (see Smith 1993, pp109-121). It is now known that the reproduction of viruses is based on RNA and this gives some clue to possible primeval evolution processes.

Each of the various inventions or conceptualizations has a unique story. For instance, until the Einstein's photon theory of light in 1905, there were many different hypotheses and Kuhn (1962) recounts various hypotheses since antiquity until the 18th century including: light being regarded as particles emanating from material bodies, a modification of the medium intervening between the body and the eye, or their combinations. Kuhn holds them to be overthrown first by the material corpuscles account of light in the 18th century and then by a transverse wave account of light since the 18th century until the emergence of the photon paradigm. In reality, they represented the thriving environment creating a culture poised for the emergence of an explanatory theory of light, now called the photon paradigm. The past speculations were all 'unselected' when the photon paradigm was selected owing to the selective advantage of its inherent capability to explain the problems related to light. The unselected accounts were not scientific, as they were unselected without influencing the emerging theory.

The term zero^{+} feedback is proposed by Khatibi (2011), signifying no relationship between inputs and outputs. The concept of open loops also describes this process at the d-loop timescale. The entire process of this stage is illustrated in Figure 3. At this stage, there is no feedback loop (positive or negative as to be discussed next in this section), or no architecture and no purpose. The issue of purpose is elaborated in Section 11.1 (case 6) but it needs further explanation. The natural world is devoid of purpose but a scientist doing research to formulate a new concept or an entrepreneur inventing a new prototype has personal purposes and aspirations, where these may be referred to as purposes in the dynamic time dimension; whereas that at the evolutionary timescale is survival and unrelated to the dynamic timescale. Sometimes, these innovations do a great deal of harm to nature, to the species including humanity or to the entrepreneur, as the routemap of the future of the prototype is unknown.

It seems that there is a general ignorance of the evolutionary timescale associated with any *evolvable entity* other than in the field evolutionary biology, who pioneered the understanding of this scale. This paper uses the term *routemap* to describe the evolutionary timescale and *roadmap* to cover both the "mythological" time of emergence and the historic time of after the emergence of the critical mass at the dynamic timescale. EvSyM is only concerned with the routemap for an evolvable entity.

Justifying the New Terms: The term "priming" is chosen to signify the near-end of the zero$^+$ feedback loop, which is analogous to the priming of a pump. The term signifies the input of some effort to create the right conditions for self-organization. One reason to call this process as zero$^+$ feedback is that from time to time the ongoing activities give rise to prototypes by capturing the rules for reproducing prototypes, without which the primed prototype will be haphazardly and will fail to be reproducible. Another reason for the term zero$^+$ feedback is that it may take ages until the emergence of the prototype.

Further Examples: The period of the zero$^+$ feedback loop has various manifestations as follows: (i) quantum scale of matter; (ii) primordial soup is often used to refer to the period during which the first living cells emerged; (iii) the process of inception of each living cells, organism or species; (iv) mythology refers to a similar period in human cultures; (v) the process of self-organization in chaos theory; and (vi) fluid mechanics has also brought to surface the process of chaos at very small spatiotemporal resolution. The mindset of the zero$^+$ feedback loop may be described in terms of the following signatures: "anything goes," the chicken and egg problem, the Catch 22 situation, arrow of time, or the problem of origins. The salient features of this period are captured in Box 2.2, as a set of axioms in general terms applicable to any complexity.

3.2. Stage of Positive Feedback Complexes

Positive feedback and natural selection work hand-in-hand, where (i) a *positive feedback d-loop* describes inputs connected through building blocks to outputs operating at the dynamic timescale; (ii) a *positive feedback ℓ-loop* describes adaptive processes within each generation; and (iii) a *positive feedback e-loop* describes the connectivity (rules or hereditary codes) for the building blocks as registered in the evolutionary timescales, where the selection pressures from the environment are translated into the modification of the hereditary codes by adding/removing new building blocks from time to time.

Positive feedback is a complex of three sub-loops, as depicted in Figure 4. Natural selection is not specific to biological species as it equally applies to any other evolvable entity lending itself to such an order. In a nutshell, if evolution of species by natural selection means descent with modification, so does other evolvable entities displaying their descent with modification arguably directed by natural selection. This section justifies the term positive feedback used to describe encapsulated systems with workable arrangements of building blocks producing outputs from inputs, as stipulated by evolutionary systemic modeling.

The basic concept: The concept of positive feedback is well established in systems science. In its classical sense, it describes the process instigating the escalation of differences between the initial and final states of a system. This outcome stems from (i) the internal arrangement of the system with a potential to create entropy; and (ii) a tendency to proliferate or multiply towards uncontrolled number of reproducible entities. As Milsum (1981) expresses, it is a process "fundamental to all growth processes, whether in living or non-living systems."

Purpose: Under positive feedback, survival is the only purpose and mission. The world under this mindset alone is not a nice place to live in for the absence of values, which emerge at the social level of hierarchy. The values are selected by negative feedback loops, as discussed in Section 10.

The role of the phases in the behaviors of positive feedback systems: The positive feedback d-loop is intertwined with the positive feedback e-loop. D-loops represent individual evolvable entities, each of which is unique at this stage. For instance, each car or any output of the manufacturing industry was unique until they were designed to comply with standards, where standards are outcomes of negative feedback loops.

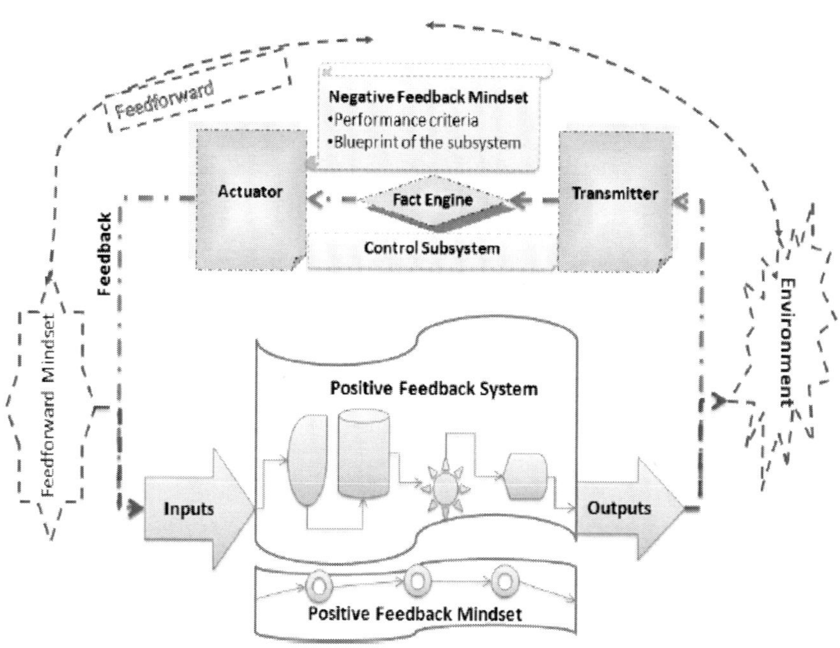

Figure 4. Architecture for Positive Feedback Loops.

The positive feedback d-loops are identified as follows:

- Inputs are connected to outputs through a sequence of building blocks but there is neither any connection of the outputs to inputs nor to any of the building blocks.

Thus, there is little flexibility to adjust the functioning of inputs and internal members to achieve the "required output."

- The required output is rather a fuzzy concept because (i) the connections are not perfect and some loss is inevitable; (ii) outputs are not influencing inputs; and (iii) the quantity/quality of outputs depend on inputs/building blocks and on the efficiency of their connections.
- The order in d-loop means that information flow is one-way: from inputs through the connected building blocks to outputs. This is so even in intellectual endeavors as each of the following is a one-way flow of information: association, analogy, reason, logic, observation, persuasion, consensual agreement (Set 1), but one-way flow of information can also take place in the cultural life by decree, coercion, self-claimed divine revelations, or a lie (Set 2).
- The outer world (or phenotype in biology) encompasses the natural environment, the social, cultural and intellectual worlds, each of which is dynamic and ever changing and unlikely to trigger a corresponding dynamic impact at the evolutionary timescale.

The positive feedback e-loops are identified as follows:

- Heredity: The positive feedback e-loop is the internalized hereditary machinery for the reproduction of the evolvable entities: e.g. genetic codes of species; the scientific method; in mathematics, the e-loops comprises the procedure of definition, operations and arriving at condensed general conclusions; risk analysis procedure in practices managing risks; design documents for scientific/technological products; similar setups in morality, planning, motor cars, sports etc.
- Order (rules): Orders of the building blocks are important and describe a set of know-how rules for synthesizing the building blocks towards their emergent properties. Consider the very primitive water treatment plant comprising grit removal, primary and secondary filter beds and a host of ancillary works such as pipes and valves. The order of the above three main components are essential and this is an intrinsic rule that may be taken for granted and often not articulated. However, if their orders are changed, the effectiveness of the plant will be undermined in most of the cases. Such rules are captured by the positive feedback e-loops.
- Encapsulation/Internalization: There is a selective advantage to encapsulate /internalize the reproduction process through a set of rules describing the know-how on connecting the building blocks. The advantage is the stability of the reproduction and it gives rise to a hereditary system. Note that the alternative is a haphazard approach.
- Open/closed mindsets: The hereditary system is open. If it is a closed system, there will not be descent with modification and this may be referred to as the fixed ontological mindset. Open hereditary systems embrace descent with modification. The hereditary system is a positive feedback complex as it tends to grow at the very slow e-loop tempo, governed by natural selection.
- Traits: Sets of building blocks at the genotype level are responsible for traits at phenotypic level and the selective advantage of these traits create the potential for the individuals to survive within its population.

- Selective advantage: Each trait, order/mechanism (including Set 1 and Set 2 above in a cultural complexity), has a selective advantage. E-loops oversee the addition of new building blocks to existing ones by natural selection from time to time. In the natural world, the variations are triggered by mutation but in the social, cultural or intellectual worlds, mutation and innovation are just few of the mechanism. Just as each one mechanism may be selected; it can also be 'unselected' when challenged. Arguably, those building blocks instigating ignorance or instability are fragile and normally unselected when challenged; whereas, those giving rise to knowledge or stability tend to coalesce together and transform into evolvable entities. Philosophy is an endeavor with one-way flow of information comprising numerous discrete doctrines, and when a doctrine is challenged, the outcome is the emergence of new doctrines without building on the previous one; whereas different ideas in science coalesce together and add to its hereditary machinery.
- Replication is the activation of hereditary codes at a special timescale. Theory of evolution explains replications without ontological essence.

The positive feedback ℓ-loops are identified with Diversity.

- Slightly different rules or slightly different building blocks can give rise to similar or slightly different emergent properties. For instance, a levo or L-sugar is a left-handed version of a right-handed Dextro or D-sugar.
- Dozier (1992, p96) remarks that "The same gene in different environments can produce many variations in the trait, and influence the expression of other genes."
- Another way is through mutation/innovation, which different traits prevail in the population. Positive feedback trigger diversity, which is generated at the genotypic scale but the traits are manifested at the phenotypic scale and accounts for variations in the population referred to as the gene pool.

The role of states in behaviors of positive feedback complexes: Positive feedback in its traditional sense refers to a system in which the differences between its initial and final states are escalated. However, this is very restrictive and only reflects the dynamic timescale. This paper expands the sense of this positive feedback to cover the dynamic and evolutionary timescales, as introduced in Section 1. The idea is that a pure positive feedback system (with inputs connected through building blocks to outputs) works well at the slow tempo (e.g. laminar flows in hydraulics, or all subsistence communal economic systems). If such a system is operated at a faster tempo, it will intrinsically tend to instigate entropy and this will trigger the selective advantage for a two possible regulatory system as to be discussed in Subsection 3.4 and 3.5. In any case, natural selection is intertwined with both of these timescales even at a slower timescale. The state of any system has to be minded with the evolutionary timescale too. Observations at the dynamic timescale identify the following patterns:

- Natural behavior (rhythm or tempo): Each positive-feedback system produces an output for a given input within its natural tempo. The time taken for obtaining the required output is an intrinsic property of the system and this cannot be altered.

- Entropic behaviors: When individuals of evolvable entities operate at a high tempo, increasingly more energy/information will be lost and such a loss may reach to a detrimental state.
- Adaptive behavior: Each positive-feedback system can adapt by reconfiguring the same modules to work in a more appropriate way without the intervention of the positive feedback e-loops. For each input, there may be an adaptive configuration. Adaptive behaviors are slow but less costly than dynamic behaviors.

Entropic behavior instigating oscillatory cycles: The key to positive feedback systems is that their survival cannot be assured and therefore there is a selective advantage for their uncontrolled reproduction, as there is no conception of the environmental capacity for the reproduction at this stage. Similarly, at this stage, the individuals of any evolvable entity are mutually exclusive entities. This is what happens at the level of lower species and plants, where each individual strives for its own survival. This also takes place among the different species, where they by and large act in mutual exclusion with respect to one another.

If the reproduction rate overtakes the rate of natural wastage, the situation will reach an entropic level, at which time the environment will put a halt on the process of proliferation. This can be a collapse of the reproducible entity, as many do not survive at their peak entropic behavior. The absence of co-operation and internal consistency is not peculiar to species as this is observed in any evolvable entity and prolongs the entropic state. This often sets oscillatory cycles, during which the gene pool of the population plays an active role in evolutionary processes.

Entropy also triggers a selective advantage for the emergence negative feedback loops, which can improve the efficiency of the performance of their positive feedback system by modulating specific units. For instance, there are interactions and co-operations among the members of many animal species and between domesticated animals and man, but this requires an evolutionary transition, as discussed in Section 3.3.

Evolutionary (selectional) behavior: Each positive feedback system may work better by adopting (note that adoption is not adaptation) a new module(s) from time to time or by removing an existing module. The timescale for this behavior can take generations. This is an evolutionary behavior through a very slow process. This is the process of building complexity from simplicity by natural selection and because of its slow tempo, the stability of the emerging complexity is well tested before its selection. One of its important products is the mechanism of negative feedback responsible to the world as we know it now.

Capacity of positive feedback complexes to act or behave freely: Natural selection does not select individuals but provides a self-organizational environment, out of which tested traits have more potentials to survive better. When the environment changes over time, the building blocks with matching traits survive but those with adverse traits decline and may disappear over time. This self-organized environment is called natural selection with the emergent property of making complexity from simplicity. To an observer, Nature as a selector acts with the mindset of being callous, purposeless, unintelligent and without foresight and without any past memory[1]. In reality, both natural selection and the attributes are an illusion but a fitting metaphor to describe the self-organization for making complexity

[1]Some of these attributes were described by Dawkins (God's Utility Function)

from simplicity. It acts as the bridge between the inner world and the outer world. In the pristine motionless nature, such as the moon, there is no natural selection.

Teleology of Positive Feedback Loops: This paper discards the philosophical discourse on teleology and all that goes with design, purpose, or finality in the world. Teleology may be regarded as an evolutionary outcome of feedback complexes. In a pure positive feedback environment, the evolvable entities as a whole interact with the environment through one-way flow of information. Using a reductive approach, consider the teleology of a brand of evolvable entities within its environment. The environment dictates the terms for the survival but it is highly unlikely that the evolvable entities can impact radically on the environment (although this has happened in the past e.g. the fixation of oxygen in the atmosphere, and is occurring now, e.g. climate change). The survival of the evolvable entities is facilitated by their diversity as Nature selects blindly those with the matching traits (in reality, the individuals without the right survival traits will perish and as such the environment neither acts as an executioner nor as a terminator). This survival mechanism was first understood by Darwin in studying biological species. Survival is the common goal of all biological species and all evolvable entities. The following cases are possibilities:

1. *Decline*: If the rate of reproduction falls below that of natural wastage, the evolvable entity can decline to the point of extinction. For instance, if water treatment plants do not proliferated in different cities and countries, this good idea would not survive.
2. *Equilibrium*: It is unlikely that the rate of natural wastages of a system to be spontaneously in equilibrium with its rate of reproduction. This desirable teleological outcome realizes only with negative feedback.
3. *Growth*: If the rate of reproduction is slightly higher than that of natural wastage, growth is likely. Due to "invisible hand" of natural selection and limited resources, output growth does not get out of hand but not always.
4. *Overgrowth towards explosive outcomes*: The familiar manifestation of the positive feedback loop has found a profound metaphor that *there is safety in numbers*. Even intelligent humanity, as a whole, remains somewhat under the spell of uncontrolled reproduction, as one billion is added to its population every dozen years in these very modern times.
5. *Fluctuations*: In the long-run, fluctuation is normally the case but patterns may also emerge, which is described above as oscillatory cycle.
6. *Selection*: The above cases reflect phenotypic behaviors, which stem from genotypic traits, as each individual of an evolving entity has an array of traits to cope with the environment. The genotypic traits are directly related to selections stemming from mutation in the natural world and by mutation, innovation and other factors in the social, cultural and intellectual world at the evolutionary scale.

There are also teleological impacts of the evolvable entities on the environment and the examples include the change in the composition of the atmosphere of the Earth, when the living species emerged. More recently, man is impacting adversely on the environment using the industrial muscle powers. The teleological interactions between the environment and the evolvable entities are also important, e.g. the Gaia theory.

The Routemap of Positive Feedback Machineries: The routemap of positive complexes kicks off upon the emergence of a critical mass. The only goal then is proliferation by keeping up the momentum and being successfully selected generation after generation. Proliferation without diversification is a linear process and unlikely to secure the survival in the long-run. The long-term survival much depends on diversification and thereby adaptation and co-evolution.

On this routemap, the forces of change are (i) the urge to reproduce; (ii) mutation and similar mechanisms in the man's world; and (iii) entropy, to instigate death or trigger evolutionary transitions by the emergence of negative feedback complexes.

The attributes of this routemap comprise: internalized/encapsulated mindset of based on building blocks and rules, gene pool, emergent properties and selective advantage, rich environment in terms of the abundance of building blocks, one-way flow of information, thriving cultures, permissive governance.

3.3. Transition from Positive to Negative Feedback

The concept of evolutionary transition presented here has precedence neither in the theory of evolution in biology nor in systems science. Thus, there is a missing link in science on (i) evolutionary transitions and (ii) the force triggering the transition. However, Khatibi (2011) proposes that evolution itself evolves and regards entropy as a force triggering the selective advantage for evolutionary transitions on the routemap from positive to negative feedback complexes. Entropy associated with positive feedback triggers the need for evolutionary transitions by priming a negative feedback subsystem, as discussed in the next section. Entropy not only arises in all closed systems of thermodynamics, its occurrence is observed widely. Section 3.4 elaborates on negative feedback and Section 5-10 substantiates the existence of evolutionary transitions and Khatibi (2012a, 2012b) provide further evidence. The scientific basis of the idea is based on the classic entropy in thermodynamics, as outlined in Section 2.3.

The basic concept: Entropy tends to increase and acts as a force to reduce the efficacy of the system. Additional energy may be supplied to the system to reduce entropy and to maintain it at a harmless level but this requires a control subsystem. If a control subsystem is feasible, an evolutionary transition has to be primed. If the subsystem is selected, it will supervise the operation of the positive feedback system. This in turn requires some reverse engineering in the positive feedback system, making some of the building blocks flexible by becoming modulatable. The negative feedback subsystem lends itself to the following:

1. Whilst the selection pressure due to the environmental changes triggers the act of selection in the positive feedback e-loop, entropy instigated by imperfect workings of the building blocks instigates the selective advantage for the emergence of a negative feedback d-loop.
2. The negative feedback d-loops are primed by natural selection in natural systems but by design in manmade systems; whereas negative feedback e-loops are primed by natural selection in both environments.
3. As there are many possible negative feedback d-loops for the same function, negative feedback e-loops evolve by natural selection.

Purpose: Evolvable positive feedback systems of natural, social, cultural and intellectual environments lend to control subsystems capable of the reduction of their inherent entropy/internal inconsistencies, without which entropy tends to a randomized state and render the positive feedback system dead.

Routemap of Transition from Positive to Negative Feedback: Entropy is intrinsic and spontaneous to positive feedback systems. It only reaches to explosive states when the system is operated at its extreme state values. The system with endemic entropy triggers the selective advantage for the emergence of a control subsystem through some modifications to the positive feedback system. If both emerge together, a negative feedback subsystem has been primed. If this subsystem is selected, the evolutionary transition is complete.

As a results of such evolutionary transitions, natural selection may be seen to display two "miracles:" (i) *doing things better* by getting most out of existing workable systems and possibly by reconfiguring them; and (ii) *doing better thing* comprising the selection of new building blocks and improved synthesizing rules. Entropy, another dumb feature of Nature, is responsible for the priming of the negative feedback subsystem and the emergence of intelligence and this is quite counterintuitive.

Examples of Entropy: Examples of positive feedback d-loops taken over by entropy include: the culture of uncontrolled reproduction among the species; cancer cells; tendency to increasingly shout by different groups of people having conversations at a closed environment; unconstrained carbon emissions as one of the instigators of climate change; or unregulated land development for residential and industrial purposes in the 18th-19th century in the developed countries.

3.4. Stage of Negative Feedback Complexes

Upon the priming and selection of a negative feedback loop, it is transformed into a powerful control subsystem as depicted in Figure 5. They are ubiquitously active towards set targets. A clear understanding of negative feedback loops was originated in systems science/cybernetics to underpin the internal consistency of the system. The salient features of the figures include:

(i) Negative feedback d-loops are subsystems, known as a *Control subsystem* made up of *Transmitters*, *Fact Engines* and *Actuators*;

(ii) The *positive feedback* d-loops have to be overhauled by acquiring a state-aware capability and having *interfaces* between the positive feedback system and the negative feedback subsystem.

The basic concept: The operation of negative feedback subsystems defies the natural rhythm of the positive feedback d-loops operated at the dynamic timescale. As systems based on purely positive feedback complexes are imperfect arrangements when operated at the dynamic timescale, they quickly get to entropic states. Permanent damage is likely when the operational conditions exceed their allowable limits. If the system is supervised by a control subsystem, its performance within the target band of safe operations (homeostasis) is feasible but by the intervention of a transient solutions. The sweating mechanism is a classical example, as it allows the human body to reach high levels of energy production and maintain the body temperature within the target band. There are countless examples from natural, scientific, cultural and intellectual endeavors. Figure 5 depicts the negative feedback complexes and the minimum requirements are outlined in Table 3.

Purpose: Under the negative feedback subsystems, the purpose of, or the purpose for, any individual of any evolvable entity is to maintain its performance within a prescribed set target. Negative feedback subsystems therefore maintain the internal integrity of the whole system. The set target is not fixed but can vary depending on the encapsulated fact engine.

The role of the phase in behaviors of negative feedback subsystem: Negative feedback subsystem is very simple in its structure but its power is surprising in being able to control often extensive positive feedback systems and one unsophisticated analogy is the role of a shepherd's dog in controlling a flock of sheep. The brain is just a very sophisticated control system.

Behaviors of e-loops: Negative feedback e-loops are simple but there is no single way of control algorithm or performance criteria. It is likely that any initially selected fact engine will go a long way but over a long time the facts have to be revised. The brain is a learning and adaptive negative feedback machine driven by fuzzy logic.

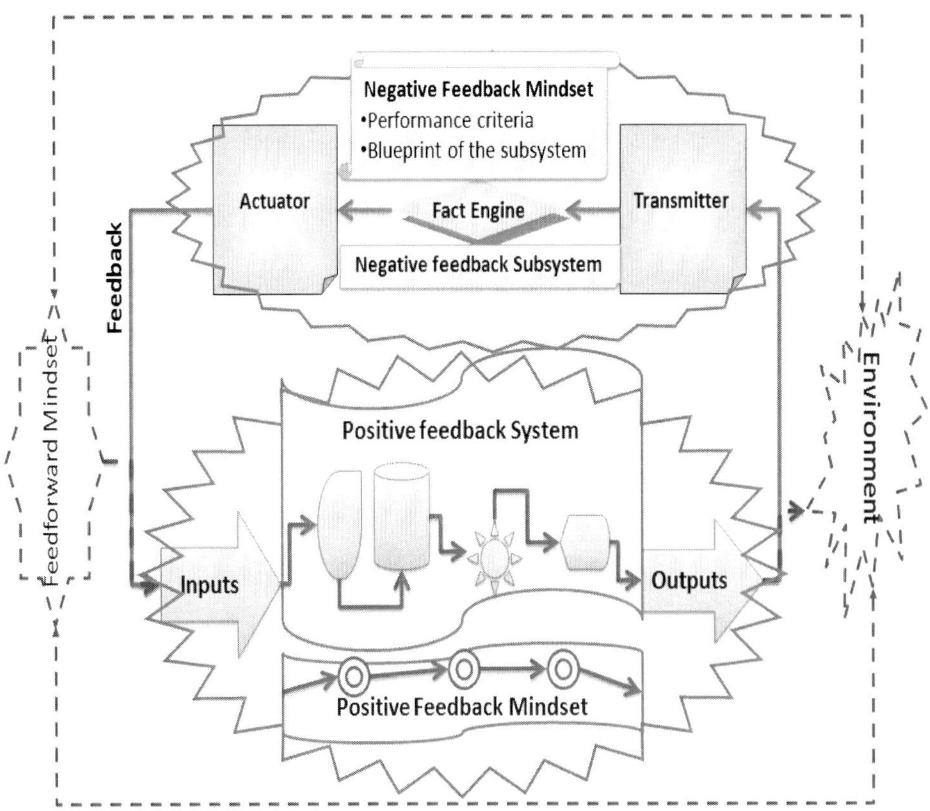

Figure 5. Negative Feedback Complexes Architecture.

Table 3. Basic Makeup of Negative feedback Loops

Attributes	Systems Science
Background	• Classic industrial systems in 19th century • Well-defined inputs, processes, output and interfaces • (i) performing subsystem, (ii) control subsystem, (iii) interface
Interface	• Interface is automated
Data	• Early control machines worked on tacit information • Measurement became essential from 20th century onwards
Transmitter	• Actual unit in control subsystem • Measures output values and passes them to the Fact Engine
Fact Engine	• Target values and computations of state variable for their maintenance • Diversity of techniques, e.g. fuzzy logic or the PID algorithm
Actuator	• Actual unit in Control Subsystem • Physically resets something in the operational system
Emergent Property	• Control over loss of efficacy • Understanding efficiency, reactive operations
Examples	• Homeostatic systems , ecological systems, industrial units: flywheel
Note: Transmitter, actuator and fact engine may have different terminologies, e.g. von Bertalanffy (1981, p64) refers to these as receptor, effector and control apparatus.	

Behaviors of d-loop operations: Under normal boundary conditions, positive feedback systems are not expected to have any problem and they often do not need negative feedback systems. However, the environment is often dynamic and as the boundary values keep changing dynamically, the negative feedback subsystem reacts ubiquitously to maintain the positive feedback systems within the set target. Hunting is observed when the negative feedback subsystem sets the positive feedback system at a certain dynamic state which is not sustainable and therefore oscillatory behaviors may emerge from time to time.

Behaviors of ℓ-loop operations: If the facts engine is not appropriately tuned, the system response can be too quick to too slow. Sometimes there may be trends or unexpected patterns in the boundary conditions and therefore the fact engine needs to be adaptive.

Teleology of Negative Feedback Loops: Owing to negative feedback loops, the evolvable entities interact with one another as individuals through two-way flow of information. Such subsystems are responsible for the concepts of purpose and teleology and they are instruments to overcome the harsh realities prevailing in the environment. The teleology is no longer survival alone. The survival of the evolvable entities has now more tools in disposal and it is possible to accelerate the pace of the processes in Nature. The following cases are possibilities:

- Equifinal behaviors: Individuals of evolvable entities may achieve identical states over time from the different initial states.
- Rapid reactions – d-loops: These are normal identifiers of negative feedback loops, e.g. sweating, the flywheel, hormones, etc.
- Slow reactions – ℓ-loop: These are long processes, e.g. cognition by individuals, or education, training, or calibration of the models, etc.

- Evolutionary changes – e-loops: This triggers genotypic changes in the negative feedback subsystem, transferable to future generations. It is not often visible but the difference can be detected over time.
- Stagnation: This is a very prominent feature of negative feedback loops, as there is no way to get out of the gripping powers of negative feedback d-loops without negative feedback ℓ-loops. Dogmatic beliefs are a classical example, according to the following definition:

$$\begin{aligned} \mathcal{Faith} &= \text{idiosyncratic negative feedback} \\ \mathcal{Dogma} &= \text{Absense of adaptatation} \end{aligned} \tag{3.1}$$

This was in full swing in the Soviet Union, rigorous in the Judaic religions and epidemic among the experts, who recognize no world other than their trained mindsets.

The Routemap of Negative Feedback Subsystems: Negative feedback loops have closed-loop architecture. They give rise to reactive intelligence, which account for the insight into the connections of the components, as well as the experience gained from past operations and the requirements for the present and the future. The ability to maintain steady performance is an emergent property of combining negative feedback subsystems.

3.5. Stage of Feedforward Complexes

Feedforward loops maintain external consistency of the system with the environmental requirements, as depicted in Figure 6. A clear understanding of these loops was originated in systems science/cybernetics to explain anticipation but the author extends this concept to cover the requirements for the external consistency of the system.

The basic concept: The consistency of the operational systems with the external environment is important for survival. Both spontaneous operations of the positive feedback complexes and the reactive operational mode of negative feedback complexes would be necessary for the integrity of the system. However, the environment is not static in the long run, and patterns of change in the environment are significant. For the species alone, as they get older, their required performances keep changing and hence they keep learning new ways of behavior (retuning their fact engines). Similar concepts tacitly exist for any evolvable entity. For instance, in the car industry, the fashion and preferences of the consumers continually change and therefore a car popular in one year may not be so in the near future. There are other countless examples from natural, scientific, cultural and intellectual endeavors. One classical example is horizon-scanning to establish opportunities and threats in the near future. The establishment of feedforward loops ensures appropriate responses for patterns in environmental changes over significantly longer periods.

Evolutionary transitions are normally in the order of zero^{+}, positive, negative feedback and feedforward loops, where these loops are pluralistic in a way that they are not mutually exclusive.

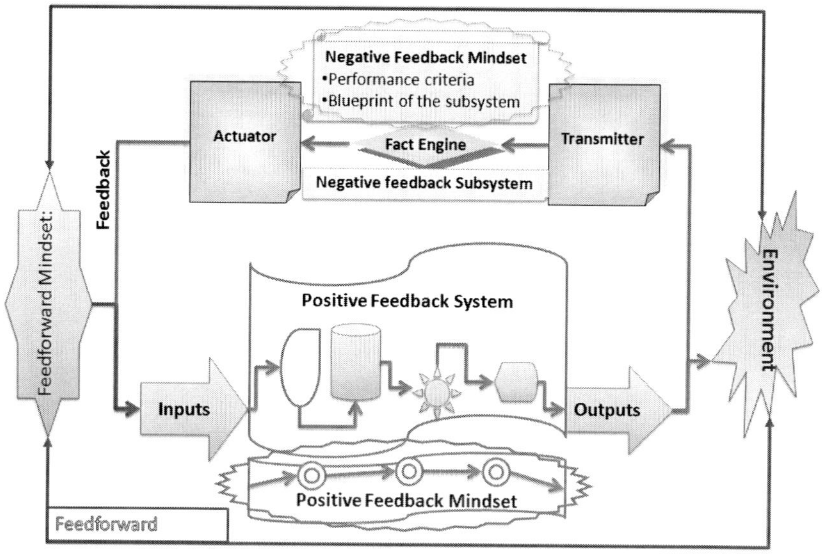

Figure 6. Feedforward Loop Complexes.

However, this paper shows one instance for the emergence of a feedforward complex before the negative feedback complex and that is the unearthing of the law of natural selection by Darwin. Thus, the priming of the negative feedback complexes is more likely to take place before than that for feedforward complexes, although it is not impossible for the feedforward complex to be primed before the negative feedback complex.

Purpose: Under the feedforward loops, the purpose of, or the purpose for, any individual of any evolvable entity is to maintain performance by reducing inconsistencies with the environmental requirements. The emergent property of these complexes is proactive capabilities with many other similar attributes. Feedforward loops therefore maintain the external integrity of the whole system through possibly intermittent modulations of the system. This requires a free-flow of information between the environment and the system.

Behaviors of feedforward complexes: Behavior codes can be tacitly evident behind behavioral acts or explicitly articulated as goals. If these encapsulated facts are insufficient for the conduct of the behavior of the positive feedback systems, problems are likely. The following behaviors have been observed by the author:

- *Normal operations at feedforward d-loops*: Free-flow of information from the environment to the system is necessary but subject to (i) feedforward loops are triggered only for maintaining the direction and this is not instigated too often; (ii) the inherent positive feedback complexes must be flexible to make necessary internal modulations to maintain the external consistency with the environment. Memory plays an important role in the fulfillment of the goals underlying feedforward loops but in the social, cultural and intellectual world, there are a host of other facilities to aid anticipation or feedforward and these include mathematical modeling, vision, etc.
- *Goal-fixation at feedforward e-loops*: The environmental conditions keep changing and therefore the fact engine for feedforward has to evolve with these changes. However, if the fact engines remain unchanged, the performance will gradually look irrelevant or ineffective. This is referred to as goal-fixation and it makes the

experience and learning rather irrelevant. Klein (2009) remarks: "In complex settings, most of the goals are vague and ambiguous rather than clear-cut. Goal-fixation – that is, continuing to pursue goals that have become irrelevant – affects too many managers." "Yet many people take the opposite route. When confronting with complex situations, they try to increase their control over events. They try to increase the details in their instructions instead of accepting the unpredictability of events and getting to adapt. They stumble into goal fixation," Klein (2009).

- *Goal-displacement at feedforward ℓ-loops*: This is defined by Merton (1968) as fixations and lack of adaptability, where the means become ends and more important than the actual goals." This normally prevails in ideological pursuits, e.g. the fate of the Soviet Union, where everything was manipulated to verify the ideological tenets. In philosophy too, the doctrines are defended as if the truth is the doctrine itself and not the subject of enquiry. Another version of this problem is observed in performance management, where the rules are "manipulated" to show that the goals are achieved. Measuring academic excellence in terms of published papers in top journals is the best example, as this impoverishes the scientific culture.

Teleology of Feedforward Loops: The controversial issues of purpose, vision, mission, objective, foresight, find their right context under the feedforward complexes, as these attributes all seek external consistency between a system and its environment. These terms are so familiar that they are not discussed anymore.

The Routemap of Feedforward Loops: If positive and negative feedback complexes operate well, the overall system would perform well, provided that there are no mutual interactions among the individuals of the same evolvable entities but this is unlikely. Interactions among the individuals would instigate external entropy similar to the working of the building blocks of a positive feedback complex. However, appropriate behaviors go a long way towards reducing entropy but they can be a source of less chronic problems. This shows that the loopholes can never be fully closed.

3.6. Building Hierarchy

Evolutionary transitions are intertwined with building hierarchy and this section uses the hierarchies of science to theorize the emergence of hierarchy as evolutionary transitions are triggered.

Quantumistic environments: At the stage of the zero$^+$ feedback loop of any evolvable entity, there is no prototype and no hierarchy. A potential entity is then under the quantumistic (or plasmatic) level of hierarchy, where the conditions are right for self-organizations for the priming of chemicalistic elements. At the instant of the priming, the followings are stipulated:

1. Emergent properties are associated with any element popping out of the quantumistic hierarchy both at the chemicalistic/physicalistic hierarchies.
2. Chemicalistic/physicalistic hierarchies are intertwined, similar to genotypic and phenotypic properties.

3. The physicalistic hierarchy is pristine with no selection but in the chemicalistic hierarchy, complexity can grow from simplicity.
4. As long as the primed chemicalistic elements are stable, they will exist.

Emergence of chemicalistic/physicalistic Hierarchies: Chemicalistic elements primed at the quantumistic hierarchy are either reactant, repellent or inert, triggering the possibilities for compounds, where followings are stipulated:

1. The richer the environment, the greater the likelihood for more compounds with their emergent properties at both chemicalistic and physicalistic environments.
2. Some of the compounds emerge spontaneously but others depend on the prevalence of the right condition. These are collectively equivalent to building blocks and rules.
3. The emergence of compounds is a self-organized way for priming of the chemicalistic and physicalistic hierarchies.
4. More rules make more compounds, which can serve as building blocks for more rules and so on. This is the root of the emergence of positive feedback machinery and the instigation of entropy; it also underlines the loss of efficacy for the emerging positive feedback complexes.
5. The natural environment possibly under entropic conditions has given rise to new conditions of priming self-replicating prototypes only in one period, subsequent to which life emerged and diversified owing to the internalization of the self-replication process. In the social, cultural and intellectual environments, man's efforts (artificial priming) replace spontaneous self-replication (or priming) processes but biologistic entities emerge spontaneously (spontaneous priming). Conversely, artificial priming processes may trigger spontaneous priming under natural conditions, after the lapse of very long times.

Emergence of the Biologistic Hierarchy: The emergence of the first generations of self-replicating prototypes is unlikely to trigger the process of natural selection yet. The first generation is likely to multiply and owing to its fragile self-replication process, the emergence of a population with gene pools is likely. The first change in the environmental conditions would prime the process of natural selection. The complexity in this process may be reduced to the following components:

1. The emergence of biological self-replicating life-bearing prototype organisms ought to have been intertwined with the priming of natural selection; and these ought to have been intertwined with the priming of the biological hierarchy.
2. The biological hierarchy coexists pluralistically with the complexity at the chemicalistic/physicalistic hierarchies and with zero$^+$ feedback.
3. Within the biological hierarchy, there is an evolutionary transition from (i) zero$^+$ feedback entities between chemical compounds and living organisms, e.g. the prokaryotes that do not have a cell nucleus (membrane-bound organelles) or mitochondria; (ii) positive feedback organisms capable of response to stimuli, reproduction, growth and development; such as eukaryotes with a clearly defined nucleus, a nuclear membrane, chromosomes, organelles, including mitochondria; (iii) negative feedback entities capable of homeostasis, such as plant and animal

kingdoms including a diversity of negative feedback loops, where animals exert their homeostatic control by the brain; (iv) feedforward to trigger changes anticipating for dynamic changes in the environment, e.g. humans but this is psychology as discussed next.

The above stretches the author's knowledge to the very limits and therefore errors are likely. However, similar processes are observed on any area of complexity in the social, cultural and intellectual life of man, except that man sets up institutions to carry out the equivalent operations of reproduction.

Emergence of the Psychologistic to Anthropologistic Hierarchies: The author is not sufficiently learned about the psychology of plants and the animal kingdom, as well as their sociology and anthropology. However, with a cursory overview, it seems that these scientific disciplines are collections of unconsolidated empirical constructs. It is tentatively suggested that these four feedback loops are capable of unraveling the hierarchical structure in any complexity.

3.7. Modeling Behavior

Section 2.5 reduces behavior to at least three dimensions of (i) phase, (ii) state, and (iii) the capacity to act autonomously. These concepts are used in generic terms and their usage throughout the paper shows the adequacy of these generic terms. Box 2.5 summarizes the evolutionary systemic basis for behavior. Section 7 adds one more dimension to behavior, namely the currency of the behavior in terms of evidence/data. Other dimensions are likely and their addition is no problem, as evolutionary systemics is open-ended.

3.8. Self-organization

In natural systems, when building blocks come together, there are situations when new patterns or behaviors emerge spontaneously, which seems to be in an organized form and even under no rule. In reality, there are still rules but under the condition, they seem insignificant, trivial or intrinsic. This is similar to natural selection. After self-organization, if the emergent system is stable and if it can be reproduced in a robust manner, the conditions may be captured in the reproduction procedure and trigger a hereditary system. Self-organization has been observed in physical, chemical, biological, economy and many other disciplines. The subject has its roots in the past but the scientific interests are recent. Now it is a thriving research topic but evolutionary systemics is making its limited use. Therefore, a detail elucidation of the subject is not necessary.

4. Evolutionary Systemic Mathematics

A new calculus is outlined that is suggestive of capturing the main features of the theory of evolutionary systemics. This paper has no intention of being mathematical and therefore the new mathematics is presented only in an outline form and in the support of EvSyM. The rest of the paper does not depend on this section. At least, three types of mathematics are envisaged. (i) describing the *state* of the system: this is well developed by the virtue of a sophisticated calculus by reducing the state of physical complexity into its constituent building blocks, e.g. mechanics of fluids reduces the flow of fluids into the forces of gravity, inertia, friction, hydrostatic; and a host of tensor quantities to account for turbulence and mixing, as well as mass balances. A sophisticated example is the Navier-Stokes equations. (ii) Mathematics to aid *decision-making*: this involves the performance, risk and uncertainty issues often using optimization techniques. (iii) Mathematics to account for the *phase* of the system: this paper is focused on the particulate phase of the system and evidently the state-of-the-art on this branch of mathematics is very poor.

The capability of conventional mathematics is well established on dealing with sequences of natural numbers but this is a physicalistic capability and as such, it does not allow an insight into the underlying complexity. The author has developed a new mathematics, providing a chemicalistic/biologistic calculus to the sequences of natural numbers and as such it creates a parallel world with the worlds of the particulate matter and relevant to evolutionary systemics. Nature at the resolution of natural selection displays the following features: (i) natural selection takes place at a particulate environment; (ii) the timescale of natural selection is through generations and not the time measured by ordinary watches or clocks; (iii) at the spatiotemporal resolution of natural selection, genotypic characters are tuned by chemicalistic forces and phenotypic characters by physicalistic forces; (iv) the entities appears to be interacting, interconnected or cumulative. These characteristics are also observed in the new mathematics.

The new mathematics needs a name and the author is inclined to call it the *dumb mathematics*, not so much as for emulating nature, which is dumb i.e. not intelligent, but because the editors of mathematical journals seem to regard this new mathematics as dumb, not novel and so on despite the author's repeated attempts to publish his ideas in mathematical journals since 2000. In reality, the new mathematics should be called *evolutionary systemic algebra*. This paper therefore presents a glimpse of evolutionary systemic algebra but one has to climb up for a fair amount of terminology for the return of an unexpected insight, easily graspable by students at the school level.

4.1. Basics of Dumb Mathematics

A glimpse of evolutionary systemic algebra is outlined here. It is formulated in terms of building blocks and rules for a given hierarchy, similar to EvSyM based on building blocks and some sort of rules. Consider that n-number of building blocks, b, and m-number of rules, r, which are denoted as follows:

$$\{b_1 \quad b_2 \quad \cdots \quad b_{n-1} \quad b_n\} \qquad (4.1.a)$$

$$\{r_1 \quad r_2 \quad \cdots \quad r_{m-1} \quad r_m\} \qquad (4.1.b)$$

If both building blocks and rules are arranged in a certain order, they form sequences; otherwise, they may be regarded as sets of numbers or objects. The author has been engaged with research on the subject since 1995 and offers an alternative analytical calculus to traditional sequences of natural numbers. The mathematical operators below apply equally for both sets and sequences:

Convolution Products

$$\{b_1 \quad b_2 \quad \dots \quad b_{n-1} \quad b_n\}\widehat{\oplus}\{r_1 \quad r_2 \quad \cdots \quad r_{m-1} \quad r_m\}= \{b_1r_1 \quad b_1r_2+b_2r_1 \quad b_1r_3+b_2r_2+b_3r_1 \quad \dots \quad \dots \quad b_1r_{m-1}+b_2r_{m-2}+\dots+b_{n-2}r_2+b_{n-1}r_1 \quad b_1r_m+b_2r_{m-1}+\dots+b_{n-1}r_2+b_nr_1\} \qquad (4.2.a)$$

Direct (Complexity-Creating) Products

$$\{b_1 \quad b_2 \quad \dots \quad b_{n-1} \quad b_n\}\backslash\{r_1 \quad r_2 \quad \cdots \quad r_{n-1} \quad r_n\} = \{b_1r_1 \quad b_2r_2 \quad b_3r_3 \quad \dots \quad b_{n-1}r_{n-1} \quad b_nr_n\} \qquad (4.2.b)$$

$$\{b_1 \quad b_2 \quad b_3 \quad \dots\} \odot \{r_1 \quad r_2 \quad r_3 \quad \dots\} = \left\{b_1 \quad \begin{matrix} b_1 \\ b_2 \end{matrix} \quad \begin{matrix} b_1 \\ b_2 \\ b_3 \end{matrix} \quad \dots\right\} \odot \left\{r_1 \quad \begin{matrix} r_1 \\ r_2 \end{matrix} \quad \begin{matrix} r_1 \\ r_2 \\ r_3 \end{matrix} \quad \dots\right\} = \{b_1r_1 \quad b_1r_1+b_2r_2 \quad b_1r_1+b_2r_2+b_3r_3 \quad \dots\} \qquad (4.2.c)$$

The above definitions are some of the basis for a building block approach to numbers and well within the spirit of EvSyM. Note that the symbols $\widehat{\oplus}$, $\backslash$, $\odot$ are unique to evolutionary systemic algebra but further details are presented in Electronic Supplementary Material (ESM) and this commonly serves Khatibi (2012a) and Khatibi et al (2012). Appropriate details are presented by Khatibi (unpublished). Some of the very basic concepts are:

Factorization: Any synthetic sequence of natural numbers can be factorized into two sequences: (i) *the sequence of invariants*, (the architecture or building blocks) and (ii) *the sequence of counters* (the architect or the rules).

Kernels and invariants: The *kernel* in the world of the sequences of natural numbers is equivalent to quarks or gluons. This can be any number or a set (as opposed to sequence) of numbers. *The sequences of counters* transform the kernels into a *sequence of invariants* to define families of new sequences. Within a particular family of natural numbers, their *sequences of invariants* are interoperable, i.e. sequences of invariants can be *transformed* into others by appropriate transformers.

Counters: The *sequences of counters* can be defined in countless ways. The easiest *sequence of counters* is the simple binomial rule of obtaining new terms in a new generation by adding the previous last two terms together.

Further terms are explained in ESM and include *regenerating products*, *algebraic products* and *recursive products*.

4.2. Zero$^+$ Feedback in Evolutionary Systemic Algebra

The idea is to provide an explanation for the priming and selection of natural numbers. Of course, this is a retrospective explanation using a mathematical logic. Consider the kernel "1", which is not yet a sequence but follow the procedure as:

Step 1 – Generation 0: The *priming* of the sequence of natural numbers from its *kernel*, $\langle 1 \rangle$ is carried out by multiplying the *kernel* with the *sequence of counter* using one of the multiplication rules, represented by an over-bar $\{\bar{1}\}$:

$$\langle 1 \rangle \backslash \{\bar{1}\} = \langle 1 \rangle \oplus \{\bar{1}\} = \langle 1 \rangle \otimes \{\bar{1}\} = \{1\} = \underbrace{\left\{ \underbrace{N}_{G=0} \right\}^{D=0}}_{\varphi=0} \quad (4.3.a)$$

$\{1\}$ is an element of a sequence (a number) but not a kernel or object and therefore a countable quantity, as the *Kernel*: $\langle 1 \rangle$ is not a sequence and can be anything but it is often a building block; $\{\bar{1}\}$ is a self-organized rule and its over-bar discerns its value from that of the kernel. The product's Specification is:

- Kernel = $\langle 1 \rangle$
- *Generation*=0,
- *Degree*=0
- *Order*=0.

Step 2 – *Generation 1*: *Generation* 1 is *regenerated* from (4.3.a), as follows:

$$\{1\} \oplus \begin{Bmatrix} 1 & 1 & 1 & 1 & \cdots \\ 0 & 0 & 0 & 0 & 0 \end{Bmatrix} = \{1 \quad 1 \quad 1 \quad 1 \quad \ldots\} = \underbrace{\left\{ \underbrace{N}_{G=1} \right\}^{D=0}}_{\varphi=0} \quad (4.3.b)$$

(4.3.b) has the following Specification:

- Kernel = $\langle 1 \rangle$
- *Generation*=1,
- *Degree*=0
- *Order*=0.

The Order emerges when the act of summing is carried out, as in below.

Step 3 – *Generation 2*: The sequence of natural numbers is *regenerated*, as:

$$\left\{1 \quad 1 \quad 1 \quad \ldots \quad \binom{n-1}{n-1}\right\} \oplus \left\{1 \quad 1 \quad 1 \quad \ldots \quad \binom{n-1}{n-1}\right\}$$
$$= \left\{ \overbrace{1 \times 1}^{1} \; \overbrace{1 \times 1 + 1 \times 1}^{2} \; \overbrace{1 \times 1 + 1 \times 1 + 1 \times 1}^{3} \; \overbrace{\ldots}^{\ldots} \; \overbrace{1 \times 1 + 1 \times 1 + 1 \times 1 \ldots + 1 \times 1}^{n} \right\} = \quad (4.3.c)$$

$$\left\{1 \quad 2 \quad 3 \quad \ldots \quad \binom{n}{n-1}\right\} = \underbrace{\left\{ \underbrace{N}_{G=2} \right\}^{D=0}}_{\varphi=1} = \underbrace{\left\{ \underbrace{N}_{G=1} \right\}^{D=1}}_{\varphi=1} \quad (4.3.d)$$

By rule, 1×1=1 and purely by convention, there is a selective advantage in the cultural life of man to see 1+1 (and so on) as a whole unit and give it a name. The Specification for (4.3.d) is:

- Kernel = $\langle 1 \rangle$ or
- *Generation*=2, 1
- *Degree*=0 1
- *Order*=1. 1

Notably, the above shows that the kernel $\langle 1 \rangle$ is generated without any assumptions based on an explicitly articulated rule comprising elements all with "1" leading to the generation of the sequence, of "1", "1+1", "1+1+1" and so on. We happen to give them the name of one, two, three, ... and transform any other sequences into names using one, two, ... We could have easily selected the kernel to have the values of $\langle a \rangle$ $\langle a \quad b \rangle$, or $\langle a \quad 1 \quad b \rangle$, where *a* and *b* are any object.

Note that it is now a legend that Bertrand Russell and Alfred North Whitehead spent some 40 pages in their *Principia Mathematica* in the early 20th century to state that "Therefore, 1 + 1 = 2." However, evolutionary systemic algebra here shows that 1+1 is really 1 and 1, to which each language gives a name and that in English is "two".

4.3. Prototypes and their Selections

The idea is to show the priming of prototypes and their selection. This is illustrated through an example.

Example: Consider the example, in which both building blocks and rules are: 1, 2, 3,… both operated by (4.2.a) and (4.2.b):

$$\{1 \quad 2 \quad 3 \quad \ldots\} \oplus \{1 \quad 2 \quad 3 \quad \ldots\} = \{1 \quad 1\times 2+2\times 1 \quad 1\times 3+2\times 2+3\times 1 \quad \ldots\} = \{1 \quad 4 \quad 10 \quad \ldots\} \qquad (4.4)$$

The sequence: $\{1 \quad 4 \quad 10 \quad \ldots\}$) is the tetrahedron sequence and specified as:

- $\mathrm{K} = \langle 1 \rangle$,
- D=0,
- G=4,
- $\varphi = 3$.

Contrast this to the following:

$$\{1 \quad 2 \quad 3 \quad \ldots\} \odot \{1 \quad 2 \quad 3 \quad \ldots\} = \{1^2 \quad 2^2 \quad 3^2 \quad \ldots\} \qquad (4.5)$$

The outcome (i.e. (4.5)) is the square power sequence. Its specifications are:

- $\mathrm{K} = \langle 1 \quad 1 \rangle$,
- D=2,

- G=1 and
- $\varphi = 2$.

Both (4.4) and (4.5) have formed new sequences: (i) the tetrahedron sequence and (ii) the square power sequence of natural numbers, both with obvious emergent properties. EvSyM stipulates that:

(i) If a product has an emergent property; (ii) if the emergent property creates a selective advantage; and (iii) if the selective advantage primes a selection, then the product becomes a prototype. Thus, (4.4) and (4.5) are prototypes.
(ii) If these prototypes proliferate in the environment by other compounds or evolvable entities, then they can form a critical mass to be transformed into evolvable products. Indeed (4.4) and (4.5) have been selected, as they can be found in many other sequences, e.g.

Consider their usage in forming new sequences, e.g.:

$$\{1 \quad 2 \quad 3 \quad ...\} \oplus \{1 \quad 2 \quad 3 \quad ...\} \oplus \{1 \quad 1 \quad 1 \quad ...\} = \{1 \quad 5 \quad 15 \quad ...\} \qquad (4.6.a)$$

(4.6.a) is the Pentatope sequence and its specification is:

- $K = \langle 1 \rangle$,
- D=0,
- G=5 and
- $\varphi = 4$

$$\{1 \quad 2 \quad 3 \quad ...\} \wr \{1 \quad 2 \quad 3 \quad ...\} \wr \{1 \quad 2 \quad 3 \quad ...\} = \{1^3 \quad 2^3 \quad 3^3 \quad ...\} \quad (4.6.b)$$

(4.6.b) is the cubic power sequence and its specification is:

- $K = \langle complex, as\ given\ below \rangle$,
- D=3,
- G=2 and
- $\varphi = 1$.

Hence, both (4.4) and (4.5) are evolvable entities as they can be found in the environment of mathematics contributing to other evolvable entities.

4.4. Interconnectivity

The idea is to express one element of a sequence in terms of basic elements or one sequence in terms of lower order sequences.

Example 1: Consider the square and cubic power sequence of (4.6.a) to be expressed in terms of its square power sequence:

$$\{1 \quad 5 \quad 15 \quad ...\} \wr \{1 \quad 5 \quad 15 \quad ...\} =$$
$$\{1 \quad 25 \quad 225 \quad 1225 \quad 4900 \quad 15876 \quad 44100 \quad ...\} \tag{4.7.a}$$

(4.7.a) can be factorized as:

$$\{1 \quad 16 \quad 36 \quad 16 \quad 1\} \odot \tag{4.7.b}$$

$$\left\{\binom{n+7}{n-1} \quad \binom{n+6}{n-2} \quad \binom{n+5}{n-3} \quad \binom{n+4}{n-4} \quad \binom{n+3}{n-5}\right\}_{n=1,n} \tag{4.7.c}$$

The Specification of (4.7) is:

- $\mathrm{K} = \langle 1 \quad 16 \quad 36 \quad 16 \quad 1\rangle$,
- D=5,
- G=4 and
- $\varphi = 8$.

Likewise, the cubic of (4.6) is expressed as:

$$\{1 \quad 5 \quad 15 \quad ...\} \wr \{1 \quad 5 \quad 15 \quad ...\} \wr \{1 \quad 5 \quad 15 \quad ...\} =$$
$$\{1 \quad 125 \quad 42875 \quad 343000 \quad 200076 \quad 9261000 \quad ...\} \tag{4.8.a}$$

(4.8.a) can be factorized as:

$$\{1 \quad 112 \quad 1828 \quad 8464 \quad 13840 \quad 8464 \quad 1828 \quad 112 \quad 1\} \odot \tag{4.8.b}$$

$$\left\{\binom{n+12}{n-1} \quad \binom{n+11}{n-2} \quad \binom{n+10}{n-3} \quad \binom{n+9}{n-4} \quad \binom{n+8}{n-5} \quad \binom{n+7}{n-6} \quad \binom{n+6}{n-7} \quad \binom{n+5}{n-8} \quad \binom{n+4}{n-9}\right\} \tag{4.8.c}$$

The Specification of (4.8) is:

- $\mathrm{K} = \langle 1 \quad 112 \quad 1828 \quad 8464 \quad 13840 \quad 8464 \quad 1828 \quad 112 \quad 1\rangle$,
- D=9,
- G=5 and
- $\varphi = 13$.

However, it looks that the sequence of invariants: (4.7.b) is different than that of (4.8.b) but this is just a false impression, as evolutionary systemic algebra can express (4.8.b) in terms of (4.7.b), as follows:

$$\{1 \quad 112 \quad 1828 \quad 8464 \quad 13840 \quad 8464 \quad 1828 \quad 112 \quad 1\} = \{the\ wrapper\} \odot$$
$$\{a\ function\ of\ (4.7.\mathrm{b})\} \odot \{weight\} \tag{4.9.a}$$

Each of the above components is given below:

(4.9.b)

$$\text{The wrapper}: \left\{\begin{matrix} \binom{8}{0} & \binom{7}{0} & \binom{6}{0} & \binom{5}{0} & \binom{4}{0} & \binom{3}{0} & \binom{2}{0} & \binom{1}{0} & \binom{0}{0} \\ (\;) & \binom{8}{1} & \binom{7}{1} & \binom{6}{1} & \binom{5}{1} & \binom{4}{1} & \binom{3}{1} & \binom{2}{1} & \binom{1}{1} \\ (\;) & (\;) & \binom{8}{2} & \binom{7}{2} & \binom{6}{2} & \binom{5}{2} & \binom{4}{2} & \binom{3}{2} & \binom{2}{2} \\ (\;) & (\;) & (\;) & \binom{8}{3} & \binom{7}{3} & \binom{6}{3} & \binom{5}{3} & \binom{4}{3} & \binom{3}{3} \\ (\;) & (\;) & (\;) & (\;) & \binom{8}{4} & \binom{7}{4} & \binom{6}{4} & \binom{5}{4} & \binom{4}{4} \\ (\;) & (\;) & (\;) & (\;) & (\;) & \binom{8}{5} & \binom{7}{5} & \binom{6}{5} & \binom{5}{5} \\ (\;) & (\;) & (\;) & (\;) & (\;) & (\;) & \binom{8}{6} & \binom{7}{6} & \binom{6}{6} \\ (\;) & (\;) & (\;) & (\;) & (\;) & (\;) & \binom{9}{7} & \binom{8}{7} & \binom{7}{7} \\ (\;) & (\;) & (\;) & (\;) & (\;) & (\;) & (\;) & \binom{9}{8} & \binom{8}{8} \end{matrix}\right\}$$

$$\text{A function of (4.7.b)}: \left\{\begin{matrix} 1 & 16 & 36 & 16 & 1 & 0 & 0 & 0 & 0 \\ & 1 & 16 & 36 & 16 & 1 & 0 & 0 & 0 \\ & & 1 & 16 & 36 & 16 & 1 & 0 & 0 \\ & & & 1 & 16 & 36 & 16 & 1 & 0 \\ & & & & 1 & 16 & 36 & 16 & 1 \\ & & & & & 1 & 16 & 36 & 16 \\ & & & & & & 1 & 16 & 36 \\ & & & & & & & 1 & 16 \\ & & & & & & & 0 & 1 \end{matrix}\right\} \qquad (4.9.c)$$

$$\text{Weighting}: \left\{\begin{matrix} \binom{8}{0} & \binom{9}{1} & \binom{10}{2} & \binom{11}{3} & \binom{12}{4} & \binom{13}{5} & \binom{14}{6} & \binom{15}{7} & \binom{16}{8} \\ \binom{7}{0} & \binom{8}{1} & \binom{9}{2} & \binom{10}{3} & \binom{11}{4} & \binom{12}{5} & \binom{13}{6} & \binom{14}{7} & \binom{15}{8} \\ \binom{6}{0} & \binom{7}{1} & \binom{8}{2} & \binom{9}{3} & \binom{10}{4} & \binom{11}{5} & \binom{12}{6} & \binom{13}{7} & \binom{14}{8} \\ \binom{5}{0} & \binom{6}{1} & \binom{7}{2} & \binom{8}{3} & \binom{9}{4} & \binom{10}{5} & \binom{11}{6} & \binom{12}{7} & \binom{13}{8} \\ \binom{4}{0} & \binom{5}{1} & \binom{6}{2} & \binom{7}{3} & \binom{8}{4} & \binom{9}{5} & \binom{10}{6} & \binom{11}{7} & \binom{12}{8} \\ \binom{3}{0} & \binom{4}{1} & \binom{5}{2} & \binom{6}{3} & \binom{7}{4} & \binom{8}{5} & \binom{9}{6} & \binom{10}{7} & \binom{11}{8} \\ \binom{2}{0} & \binom{3}{1} & \binom{4}{2} & \binom{5}{3} & \binom{6}{4} & \binom{7}{5} & \binom{8}{6} & \binom{9}{7} & \binom{10}{8} \\ \binom{1}{0} & \binom{2}{1} & \binom{3}{2} & \binom{4}{3} & \binom{5}{4} & \binom{6}{5} & \binom{7}{6} & \binom{8}{7} & \binom{9}{8} \\ \binom{0}{0} & \binom{1}{1} & \binom{2}{2} & \binom{3}{3} & \binom{4}{4} & \binom{5}{5} & \binom{6}{6} & \binom{7}{7} & \binom{8}{8} \end{matrix}\right\} \qquad (4.9.d)$$

$$\binom{7}{0}\times 16\times\binom{5}{1}+\binom{8}{1}\times 1\times\binom{4}{1}=112 \qquad (4.9.e)$$

$$\binom{6}{0}\times 36\times\binom{6}{2}+\binom{7}{1}\times 16\times\binom{5}{2}+\binom{8}{2}\times 1\times\binom{4}{2}=1828 \qquad (4.9.f)$$

$$\binom{5}{0}\times 16\times\binom{7}{3}+\binom{6}{1}\times 36\times\binom{6}{3}+\binom{7}{2}\times 16\times\binom{5}{3}+\binom{8}{3}\times 1\times\binom{4}{3}=8464 \qquad (4.9.g)$$

$$\binom{4}{0}\times 1\times\binom{8}{4}+\binom{5}{1}\times 16\times\binom{7}{4}+\binom{6}{2}\times 36\times\binom{6}{4}+\binom{7}{3}\times 16\times\binom{5}{4}+\binom{8}{4}\times 1\times\binom{4}{4}=13840 \qquad (4.9.h)$$

$$\binom{3}{0}\times 0\times\binom{9}{5}+\binom{4}{1}\times 1\times\binom{8}{5}+\binom{5}{2}\times 16\times\binom{7}{5}+\binom{6}{3}\times 36\times\binom{6}{5}+\binom{7}{4}\times 16\times\binom{5}{5}=8464 \qquad (4.9.i)$$

$$\binom{2}{0}\times 0\times\binom{10}{6}+\binom{3}{1}\times 0\times\binom{9}{6}+\binom{4}{2}\times 1\times\binom{8}{6}+\binom{5}{3}\times 16\times\binom{7}{6}+\binom{6}{4}\times 36\times\binom{6}{6}=1828 \qquad (4.9.j)$$

$$\binom{1}{0}\times 0\times\binom{11}{7}+\binom{2}{1}\times 0\times\binom{10}{7}+\binom{3}{2}\times 0\times\binom{9}{7}+\binom{4}{3}\times 1\times\binom{8}{7}+\binom{5}{4}\times 16\times\binom{7}{7}=112 \qquad (4.9.k)$$

He above is a set of remarkably complex operations but evolutionary systemic algebra makes them extremely simple and obvious, where the author has spent ages over several years to decipher them. Thus, (4.9) is evidence for an analytical interconnectivity among the elements of a complex sequence of natural numbers.

Example 2: The above example has a fair degree of complication. Power sequences, though formidable, are not that much complex. Consider the traditional power sequences up to the 4th degree with their specifications:

$$\{\Re|N^1 = \{1^1 \quad 2^1 \quad 3^1 \quad ...\}\} = \{1\} \odot \left\{\binom{n}{n-1}\right\} \qquad (4.10.a)$$

Where $\{\Re|N^1 = \{1^1 \quad 2^1 \quad 3^1 \quad ...\}\}$ mean reducing the sequence of $N^1 = \{1^1 \quad 2^1 \quad 3^1 \quad ...\}$ and this in turn mean the basic formula that can be used to generate each and every term in the sequence. The specification of this sequence is:

- K = $\langle 1 \rangle$ (Note that $\sum 1$ =1=1!),
- D=1,
- G=2 and
- φ = 1.

$$\{\Re|N^2 = \{1^2 \quad 2^2 \quad 3^2 \quad ...\}\} = \{1 \quad 1\} \odot \left\{\binom{n+1}{n-1} \quad \binom{n}{n-2}\right\} \qquad (4.10.b)$$

- K = $\langle 1 \quad 1 \rangle$ (Note that $\sum\langle 1 \quad 1 \rangle$ =2=2!),
- D=2,
- G=1 and
- φ = 2.

$$\{\Re|N^3 = \{1^3 \quad 2^3 \quad 3^3 \quad ...\}\} = \{1 \quad 4 \quad 1\} \odot \left\{\binom{n+2}{n-1} \quad \binom{n+1}{n-2} \quad \binom{n}{n-3}\right\} \qquad (4.10.c)$$

- K = $\langle 1 \quad 4 \quad 1 \rangle$ (Note that $\sum\langle 1 \quad 4 \quad 1 \rangle$ =6=3!),
- D=3,
- G=1 and
- φ = 3.

$$\{\Re|N^4 = \{1^4 \quad 2^4 \quad 3^4 \quad ...\}\} =$$
$$\{1 \quad 11 \quad 11 \quad 1\} \odot \left\{\binom{n+3}{n-1} \quad \binom{n+2}{n-2} \quad \binom{n+1}{n-3} \quad \binom{n}{n-4}\right\} \qquad (4.10.d)$$

- K = $\langle 1 \quad 11 \quad 11 \quad 1 \rangle$ (Note that $\sum\langle 1 \quad 11 \quad 11 \quad 1 \rangle$ =24=4!),
- D=4,
- G=1 and
- φ = 4.

The kernels of the above sequences seem different but in fact they can be transformed into one another with a very simple rule, as follows:

$$2! = \{1 \quad 1\} = \{1\} \widehat{\oplus} : \{1 \quad 1\} = \{1 \quad 1\} \qquad (4.10.e)$$

$$3! = \{1 \quad 4 \quad 1\} = \{1 \quad 1\} \widehat{\oplus} \left\{1 \quad {2 \atop 2} \quad 1\right\} = \{1 \times 1 \quad 1 \times 2 + 1 \times 2 \quad 1 \times 1\} \qquad (4.10.f)$$

$$4! = \{1 \quad 11 \quad 11 \quad 1\} = \{1 \quad 4 \quad 1\} \widehat{\oplus} \left\{1 \quad \begin{matrix}2\\3\end{matrix} \quad \begin{matrix}3\\2\end{matrix} \quad 1\right\} =$$
$$\{1\times 1 \quad 1\times 3 + 4\times 2 \quad 4\times 2 + 1\times 3 \quad 1\times 1\} \tag{4.10.g}$$

where $\{1 \quad 1\}$, $\left\{1 \quad \begin{matrix}2\\2\end{matrix} \quad 1\right\}$ and $\left\{1 \quad \begin{matrix}2\\3\end{matrix} \quad \begin{matrix}3\\2\end{matrix} \quad 1\right\}$ are transformers. There is a tremendous elegance in the above expressions and the author really engineered the above symbolism after a long intellectual battle. Examples 1 and 2 are a demonstration of the tremendous power of evolutionary systemic algebra but these are just examples, as the author has developed their generic expressions.

4.5. Possible Accounting for Adaptation

The idea is that the kernel of a synthetic sequence expressed as a sequence of invariants can acquire different allele format, all displaying the same emergent property. Both building blocks and rule can be expressed in many different ways: some in standard ways and others in arbitrary ways, while they all preserve the emergent property. Each arrangement may be called as an allele. This is presented by the example of the 4th power sequence of natural numbers, as follows:

$$\{N\}^4 = \{1^4 \quad 2^4 \quad 3^4 \quad ...^4 \quad n^4\} \tag{4.11.a}$$

Using evolutionary systemic algebra, the above may be reduced as:

$$\{\mathcal{A}_0|\{N\}^4\} = \{1 \quad 14 \quad 36 \quad 24\} \backslash\!\left\{\binom{n}{n-1} \quad \binom{n}{n-2} \quad \binom{n}{n-3} \quad \binom{n}{n-4}\right\} \tag{4.11.b}$$

$$\{\mathcal{A}_1|\{N\}^4\} = \{1 \quad 13 \quad 23 \quad 1\} \backslash\!\left\{\binom{n+1}{n-1} \quad \binom{n+1}{n-2} \quad \binom{n+1}{n-3} \quad \binom{n}{n-4}\right\} \tag{4.11.c}$$

$$\{\mathcal{A}_2|\{N\}^4\} = \{1 \quad 12 \quad 11 \quad 1\} \backslash\!\left\{\binom{n+2}{n-1} \quad \binom{n+2}{n-2} \quad \binom{n+1}{n-3} \quad \binom{n}{n-4}\right\} \tag{4.11.d}$$

$$\{\mathcal{A}_3|\{N\}^4\} = \{1 \quad 11 \quad 11 \quad 1\} \backslash\!\left\{\binom{n+3}{n-1} \quad \binom{n+2}{n-2} \quad \binom{n+1}{n-3} \quad \binom{n}{n-4}\right\} \tag{4.11.e}$$

The sequence of counters in (4.11.b)-(4.11.e) are all interoperable, so that each one can be transformed into any other, e.g.

$$\{1 \quad 14 \quad 36 \quad 24\} \widehat{\oplus} \left\{1 \quad \begin{matrix}-1\\1\end{matrix} \quad \begin{matrix}1\\-1\\1\end{matrix} \quad \begin{matrix}1\\-1\\1\\-1\end{matrix}\right\} = \{1 \quad -1+14 \quad 1-14+36 \quad -1+14-36+24\} =$$
$$\{1 \quad 13 \quad 23 \quad 1\} \tag{4.12.a}$$

$$\{1 \quad 13 \quad 23 \quad 1\} \widehat{\oplus} \left\{1 \quad \begin{matrix}-1\\1\end{matrix} \quad \begin{matrix}1\\-1\\1\end{matrix} \quad 1\right\} = \{1 \quad -1+13 \quad 1-13+23 \quad 1\} = \{1 \quad 12 \quad 11 \quad 1\}$$
$$\tag{4.12.a}$$

$$\{1 \quad 12 \quad 11 \quad 1\} \widehat{\oplus} \left\{1 \quad \begin{matrix}-1\\1\end{matrix} \quad 1 \quad 1\right\} = \{1 \quad -1+12 \quad 11 \quad 1\} = \{1 \quad 11 \quad 11 \quad 1\} \tag{4.12.a}$$

The above may be expressed in a direct arithmetic way

$$1^4 = 1 \times 1 = 1 \tag{4.13.a}$$

$$2^4 = 2 \times 1 + 1 \times 14 = 16 \tag{4.14.a}$$

$$2^4 = 3 \times 1 + 1 \times 13 = 16 \tag{4.14.b}$$

$$2^4 = 4 \times 1 + 1 \times 12 = 16 \tag{4.14.c}$$

$$2^4 = 5 \times 1 + 1 \times 11 = 16 \tag{4.14.d}$$

$$3^4 = 3 \times 1 + 3 \times 14 + 1 \times 36 = 81 \tag{4.15.a}$$

$$3^4 = 6 \times 1 + 4 \times 13 + 1 \times 23 = 81 \tag{4.15.b}$$

$$3^4 = 10 \times 1 + 5 \times 12 + 1 \times 11 = 81 \tag{4.15.c}$$

$$3^4 = 15 \times 1 + 5 \times 11 + 1 \times 11 = \quad 81 \tag{4.15.d}$$

$$4^4 = 4 \times 1 + 6 \times 14 + 4 \times 36 + 1 \times 24 = 256 \tag{4.16.a}$$

$$4^4 = 10 \times 1 + 10 \times 13 + 5 \times 23 + 1 \times 1 = 256 \tag{4.16.b}$$

$$4^4 = 20 \times 1 + 15 \times 12 + 5 \times 11 + 1 \times 1 = 256 \tag{4.16.c}$$

$$4^4 = 35 \times 1 + 15 \times 11 + 5 \times 11 + 1 \times 1 = 256 \tag{4.16.d}$$

$$5^4 = 5 \times 1 + 10 \times 14 + 10 \times 36 + 5 \times 24 = 625 \tag{4.17.a}$$

$$5^4 = 15 \times 1 + 20 \times 13 + 15 \times 23 + 5 \times 1 = 625 \tag{4.17.b}$$

$$5^4 = 35 \times 1 + 35 \times 12 + 15 \times 11 + 5 \times 1 = 625 \tag{4.17.c}$$

$$5^4 = 70 \times 1 + 35 \times 11 + 15 \times 11 + 5 \times 1 = 625 \tag{4.17.d}$$

$$6^4 = 6 \times 1 + 15 \times 14 + 20 \times 36 + 15 \times 24 = 1296 \tag{4.18.a}$$

$$6^4 = 21 \times 1 + 35 \times 13 + 35 \times 23 + 15 \times 1 = 1296 \tag{4.18.b}$$
$$6^4 = 56 \times 1 + 70 \times 12 + 35 \times 11 + 15 \times 1 = 1296 \tag{4.18.c}$$

$$6^4 = 126 \times 1 + 70 \times 11 + 35 \times 11 + 15 \times 1 = 1296 \tag{4.18.d}$$

The point behind the above mathematics is that the same element in (4.11.a) can be expressed in terms of different invariants and rules. If one considers that each building block and rules as chemicalistic and their products form a new physical entity, then their sum is a biologistic entity with an emergent property. So the same biologistic entity can be created by different arrangements of the building blocks and rules. This is parallel to adaptation.

4.6. General Expression of Positive Feedback

Consider building blocks and rules expressed by (4.1.a) and (4.1.b). Their direct products by (4.2.b) express some sort of bondage strength of each element. Consider the example:

$$\{a_1 \quad a_2 \quad \cdots \quad a_{n-1} \quad a_n\} \backslash \{r_1 \quad r_2 \quad \cdots \quad r_{n-1} \quad r_n\} = \{a_1r_1 \quad a_2r_2 \quad a_3r_3 \quad \cdots \quad a_{n-1}r_{n-1} \quad a_nr_n\} \tag{4.19.a}$$

$$\{b_1 \quad b_2 \quad \cdots \quad b_{n-1} \quad b_n\} \backslash \{r_1 \quad r_2 \quad \cdots \quad r_{n-1} \quad r_n\} = \{b_1r_1 \quad b_2r_2 \quad b_3r_3 \quad \cdots \quad b_{n-1}r_{n-1} \quad b_nr_n\} \tag{4.19.b}$$

$$\{c_1 \quad c_2 \quad \cdots \quad c_{n-1} \quad c_n\} \backslash \{r_1 \quad r_2 \quad \cdots \quad r_{n-1} \quad r_n\} = \{c_1r_1 \quad c_2r_2 \quad c_3r_3 \quad \cdots \quad c_{n-1}r_{n-1} \quad c_nr_n\} \tag{4.19.c}$$

Consider each of the above forms a positive feedback compound with an emergent property and selected in the mathematical environment, making up the following Kernel:

$$\langle A \quad B \quad C \quad \ldots \quad X\rangle \tag{4.20}$$

Evolutionary systemic algebra is able to keep the exact account of each and every compound in the new kernel.

The kernel may be transformed into a prototype if: (i) the kernels is operated by a set of rules priming an emergent property; (ii) the emergent property offers a selective advantage in the environment; (iii) the selective advantage primes a selection.

This single incident is not still sufficient, as it must have a critical mass before becoming an evolvable entity.

4.7. Three Positive Feedback Sub-loops

The idea is that evolutionary systemic algebra mirrors the image of the three sub-loops assigned to positive feedback by evolutionary systemics: d-loop, ℓ-loop and e-loop. Consider the quadratic power sequence, as an example:

$$\{N^4\} = \{1 \quad 11 \quad 11 \quad 1\} \backslash \left\{ \binom{n+3}{n-1} \quad \binom{n+2}{n-2} \quad \binom{n+1}{n-3} \quad \binom{n}{n-4} \right\}_{n=1,n} \tag{4.21}$$

Each of the sub-loops is represented as follows:

- d-loop: represents each and all of the elements in $\{N\}^4 = \{1^4 \quad 2^4 \quad 3^4 \quad ...^4 \quad n^4\}$. This is the dynamic real world of the quadratic number sequences, where their physicalistic appearance disguises their inner structure.
- e-loops: $\{1 \quad 11 \quad 11 \quad 1\}\Big\backslash\left\{\binom{n+3}{n-1} \quad \binom{n+2}{n-2} \quad \binom{n+1}{n-3} \quad \binom{n}{n-4}\right\}$. This is the generator, blueprint or the architecture (which the author calls it as the reducemental), which may be used to generate any of the terms in the sequence expressed by the d-loop.
- ℓ-loop: the alleles expressed by (4.13)-(4.18). A physical but hypothetical example is as follows. Consider, an evolvable entity expressed by (4.18d). At one time, this allele is favored by the environment but as the environment changes, the favored allele becomes (4.18.c). Whilst both alleles are seen to be equivalent, their inner structures are rather different.

To understand how the blueprint is used to generate any of the terms, reference may be made to Khatibi (2012a), which outlines the nesting mechanism of the basic lower elements in the higher elements.

4.8. An Account of Negative Feedback Feedforward

The idea is to mathematize negative feedback and feedforward sub-loops but this is a challenge and the author still thinks about the subject. One version of the subject is presented here but this is by no means comprehensive.

Negative feedback and feedforward complexes may be taken as simple new additions to the kernel. Thus the kernel may be expressed as:

$$\langle A \quad B \quad C \quad ... \quad X \quad \alpha \quad \beta \quad ...\rangle \qquad (4.21)$$

Another approach would be to create hierarchies, treating positive feedback in the lower hierarchy, negative feedback above it and feedforward above all. Yet another approach would be to envisage specific rules between the hierarchies of feedback loops. The problem is complex but it can be deciphered by starting from a simple problem and trying various models.

The author conceived the ideas in 1995 but does not have the opportunity to apply his ideas even though his gut feeling is that evolutionary systemic algebra is probably the language of e-loops, neurons, growth and possibly of nanotechnology. His applications to academic jobs for the last 17 years have been in vain. It is a strange world that he has not even been shortlisted in any significant way. It looks that academia is spooked by out-of-kilter ideas as the establishment feels safer to follow the crowd. This was very counterintuitive but it underpins the vitality of natural selection and diversity; else the ideas will wither away.

5. EVSY-MODELING OF PHILOSOPHY

Losee (1987, p1) introduces philosophy of science as second-order interpretation of a first-order subject matter. This seems to reflect the mindset of at least most of the philosophers, who treat philosophy as the queen of all knowledge. He also discusses history and historiography in this context and identifies that the doctrines tend to be descriptive, normative or prescriptive. No attempt is made here to summarize his work, as the book reflects a very daring effort to an almost impossible task to make sense of the philosophy of science to show its inherent sensibility. A critical enquiry to any philosophical subject leads to stalemate but leaves one none-the-wiser or a wanderer in the philosophical wilderness. Ironically, philosophy has a full claim over wisdom.

The problem statement is this: rather than philosophy or philosophy of any discipline is regarded as a second-order interpretation of a subject matter, it should be regarded as the zero-order study of fundamental enquiries on life or any other subject matter. Zero-order is defined here as no assumption or an attempt towards it. If an assumption is made, the approach must be scientific to challenge the consistency towards the assumptions.

Science offers the only methodology capable of challenging its assumptions. This section models evolutionary transitions in philosophy and shows that EvSyM can be developed towards becoming a zero-order method of enquiry and as such, it can be a scientific philosophy.

This section intends to illustrate the problem in a thought-provoking way. Its contents are tentative without intending to be exhaustive or rigorous. Figure 7 sets the scene but the dates are broad and not exact.

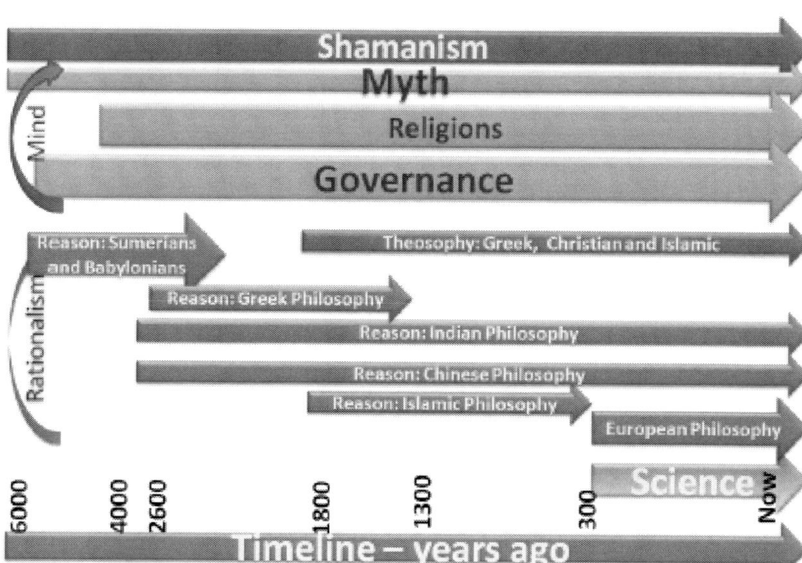

Figure 7. Timeline of Ideas.

5.1. Stage of Zero$^+$ Feedback – Primitive Knowledge

Some 10,000 years ago, Mesolithic cultures emerged at a number of locations in the world. The backdrop for the way of life before this era included: a hunter-gatherer way of life, a verbal language that was probably simple, tool-making, and practicing shamanistic rituals going to some 50,000 years back with possible significance attached to dreams, ancestor-worshipping and ecstasy. A range of modern intellectual constructs can be heritages from such a distant past, including such cultural *simplexes* as association, analogy, reason, logic, observation, persuasion, consensual agreement; decree, coercion, self-claimed divine revelation, dogma, superstition; altruism and love or deception and telling a lie. Some or most of the above cultural constructs were probably selected in the long period of Paleolithic Age. For instance, language is believed to have emerged at the early stages of the Upper Paleolithic Age (some 50,000 years ago). The author is preparing a paper with the premise that languages are nothing other than the association of vowels and consonants with tangible and intangible objects as well as real and abstract concepts by applying evolutionary systemics.

The argument put forward here is that this era was the long period for the emergence of human cultures. Natural selection had already kick-started in the Paleolithic age on a number of aspects, such as technology. This paper is not concerned with them but they still have relevance. It is tentatively suggested that the cultures up to the Mesolithic Age are the backdrop or the long period of zero$^+$ feedback period for priming the simple(x) building blocks of modern cultures. In the era up to the Mesolithic Age, explanatory knowledge was likely to concern causes or effects; input or output but the emergence of cause-and-effects (causative) cultures or inputs-and-outputs were unlikely.

5.2. Positive Feedback Driving Philosophy

5.2.1. Zero$^+$ Feedback Stage

The emergence of the settled way of life during the Mesolithic Age and the subsequent evolutionary transitions unselected most aspects of the past ways of life. Notwithstanding this, semi-nomadic lifestyle survived at the periphery of the settled way of life and in certain parts of the world hunter-gatherer way of life has remained intact until now. Whilst the communal knowledge base with hunter-gatherer ways of life is very healthy and respectful, it has found no breakthroughs to take a bird's eye view of the fundamental questions of life. However, both the settled and semi-nomadic ways of life have contributed to the emergence of the modern knowledge base and the emergence of centers of learning, as well as clusters of civilizations around the world. These have created the right conditions for the priming of a multitude of positive feedback mindsets including communal celebration and ritual occasions, the emergence of private property, organized fighting machinery, artisanship and tool-making, trading activities, taxing systems and keeping their records (triggering number systems, mathematics, arithmetic).

The Shamanistic culture of the Upper Paleolithic era was insufficient to account for new activities, e.g. farming and all that goes with it. It is outside the scope of this paper to address the evolutionary processes on the emergence of farming and the selection of new communal habits and cultures or the retention of shamanistic traditions in certain parts of the world but it is possible to see relationships between the emergence of observation-driven selection of strands of knowledge and new farming cultures. Because of the dream culture rooted in the deep past, the cultures have also been rife for speculation and abstraction. Whilst archaeology reflects evidence for the emergence of new orders in terms of faith and new forms of governance (not addressed in this paper), a number of human cultures were rife for priming philosophy-type knowledge based on observation, speculation, thought, reason, intuition and technology.

The Sumerian culture strived towards reason and was probably on the way to develop philosophy but went extinct before priming fully fledged institutions of philosophy. After the Sumerians, the region remained active on diversity of civilized life but its leadership on the faculties of reason was detracted by the invasion of empires and by the grip of the Judaic religions in the Mesopotamia. The region became a flag-bearer of faith but religions are not studied here. Philosophy emerged in the various regions described below.

5.2.2. Chinese and Indian Philosophies

The author is not qualified to express any opinion on Chinese or Indian philosophies but keen to understand their stability, although this can be explained by their religious connotations. This is less so in Chinese philosophy, which gave rise to systems mindset as opposed to the analytical mindset.

5.2.3. Greek Philosophy Driven by the Positive Feedback Mindset

The aim is not to review the schools of thoughts fermented in Greek wellspring of knowledge but to stress that the thinking then was speculative. The empires of the past were seemingly poor on encouraging freethinking but the Greek culture, like the Sumerian culture, was in a fortunate position of having city states and therefore freethinkers had many safe havens to escape persecutions and hence there were right conduits for ideas in Greek city states, which encouraged the priming and selection of ideas and their proliferation.

It is easy to be immersed into Greek philosophy and drift away from the modern thinking into that of the antiquated of Greek philosophers, as the basis of the modern scientific terminology in science is rooted in their terminology. Self-discipline is needed not to muddle the thinking and causality in Greek philosophy with the modern ones. One should remember that the thinking and causality in Greek philosophy have been unselected by science to the extent that the immense significance of Greek philosophy is now of only a historic value – largely if not wholly. Nonetheless, European philosophy has reinvented Greek philosophy in a modern context by interjections from science. Emerged in the 17th century, science is simply a building block approach (reductive or analytical) but Greek philosophy was largely ontological. The following example highlights speculative nature of Greek philosophy. Consider some of the rules (reproduced in Box 5.1) in Pythagorean order in 6th century B.C.

Pythagorean order in Box 5.1 is purely speculative by modern standards but when one considers the age of the emergence of causality (cause-and-effect thinking) without a building block approach, the statements in the box can have some sense. Also, consider the tetrasomia (the doctrine of the Four Elements: earth, air, fire and water, attributed to Empedocles 5th century B.C. but this has its origins in Mesopotamia and citations in China and India). Science reduced these elements to their building blocks and unselected whatever was said of these Four Elements. Compare this with the Periodic Table by reductive science, which explained effectively the material composition of the world.

Box 5.1 – Pythagorean Order

(1) To abstain from beans.
(2) Not to pick up what was fallen.
(3) Not to touch a white cock
(4) Not to break bread
(5) Not to step over a crossbar
(6) Not to stir the fire with iron
(7) Not to eat from a whole loaf
(8) Not to pluck a garland.
(9) Not to sit on a quart measure.
(10) Not to eat the heart.
(11) Not to walk on highways.
(12) Not to let swallows share one's roof.
(13) When the pot is taken off the fire, not to leave the mark of it in the ashes, but to stir them together.

The nature of Greek philosophy can be explained in terms of the customary challenges expounded by the sages of knowledge against one another, during which they exposed the futility of their speculative knowledge base and this prevented the formation of positive feedback e-loops in the emerging knowledge. The speculative mindset was driven by positive feedback d-loops and there is no way for its growth, unless it is coerced. Thus, it left a rich history on the fundamental questions but it had no hereditary machinery as the problems were approached at much coarser a resolution, too far out from the resolution for their building blocks. Thus, each idea acted as a self-contained unit without coalescing together. Building blocks of earth, air, fire and water or dichotomies of mind and matter are too coarse to start an enquiry and therefore not a correct starting point.

All these notions are often a complex outcome of many simpler elements and starting an enquiry from such high levels was the essence of the problem. It is not fair to criticize Greek philosophy (or any other past worldviews) from the vantage of modern knowledge, as Greek philosophy is one of the treasures of human cultures and the paper does not intend to criticize any past ideas. Equally, it is important to have a critical understanding of the past with the benefit of the present knowledge. Section 6.2 offers a critical assessment of Greek philosophy, arguing that (i) it was not a building block approach and therefore the ideas did not coalesce together but formed self-contained mutually exclusive doctrines; (ii) too many doctrines to the same fundamental enquiry were naturally entropic with a potential to collapse; (iii) it subsequently collapsed, as a whole. However, certain chunks were integrated to the scholastic belief systems in the Middle Ages and as such, they were coerced as the orthodoxy for nearly two millennia, e.g. the Aristotelian ontology and its doctrine of the fixity of the species.

5.2.4. Islamic Philosophy

Islamic philosophy is a generic term referring to the Arabic, Turkic, Persian cultures and to some extent to the Jewish and Assyrian cultures. It produced a long line of philosophical thinkers in the tradition of Greek philosophy. During the reign of Harun Al-Rashid (786–809) of the Abbasid dynasty (762-1258), revolts were becoming a commonplace, e.g. the Zanj Rebellion (Shaban, 1971) in Basra and the Babek Khurrammi Movement (see Lewis, 2002, for background) in Azerbaijan. He was then invoked to revive Greek philosophy and to fuse it with Islam to counter the rebellion movements. This was the pretext but the massive task was the translation of the dusted Greek philosophy and understanding them. This task was undertaken, and Greek philosophy was re-primed in the 9th century to boost the image of Islam. Owing to the works of Al-Kindi (ca. 800–870 A.D.) and Al-Farabi (ca. 872–951 A.D.), rich intellectual activities flourished in the tradition of Greek philosophy with a considerable innovation and a good deal of freethinking encouraged by the existence of many magnets of Turkic powers. Arguably, these were not enough to sustain the thriving freethinking philosophical enquiries and once again, the freethinking element collapsed by the 14th century leaving the room to be fully occupied by proliferating theosophical trends.

The timeline of the main movements are summarized in Figure 8. Precursors of modern evolutionary ideas can reportedly be traced to some of Islamic thinkers. Accordingly, evolutionary ideas have been found in the thinking of Al-Jahiz of Basra (776- 869), the grandson of a black slave, see Zirkle (1941). Similarly, it has been debated that Ibn Rushd, known in the West as Averroés, (1126–1198), has presented precursor ideas on the memes, e.g. see Gatherer (1998). However, the literature on such matters is not often scientific and extracts from the past scholars often does not reveal the past speculative mindsets. A further aspect of Islamic philosophy is discussed in Section 6.2. The aim is not to describe Islamic philosophy but to underline the collapse of rationalism in Islamic philosophy since circa 14th century. The repression by the authorities against freethinkers could not be a reason for this collapse, as the few victims were not from the tradition of rationalism but from Sufi and pantheistic traditions.

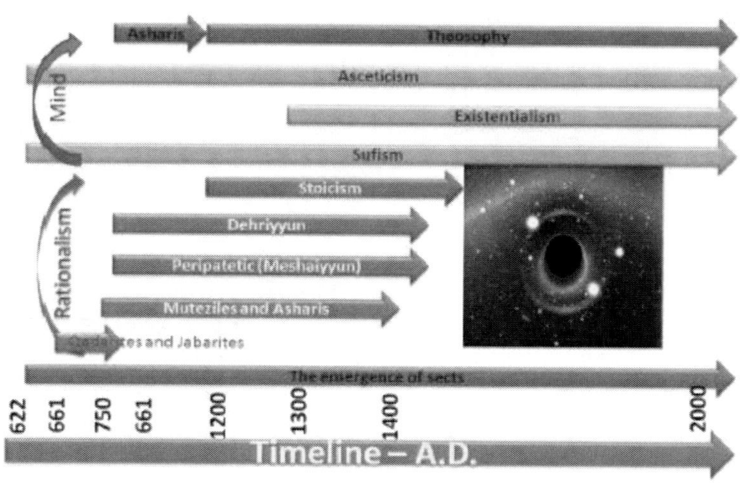

Figure 8. Timeline of Islamic Philosophy with a Focus on Freethinking.

The author argues that Greek philosophy, though a pinnacle in human cultures, was susceptible to entropic collapse and the collapse of philosophical freethinking during the Islamic philosophy was the second time it happened.

5.2.5. European Philosophy

The modern European or Western philosophy emerged from the movement of humanism of the late Medieval Ages and with injections from Islamic philosophy. Similar cultural movements carried on in Europe, as depicted in Figure 9 and this maintained an appropriate level of rich culture for science and philosophy. Whilst the awareness of Greek philosophy was rich, Europe was in the grip of scholastic dogma until the 16th century.

As discussed in Section 6.3.1, the entropic conditions in the 16th century in Italy were destructive and rife for the emergence of a critical thinking. This critical thinking on the cultural front had its ups and downs by going through the periods of humanism, classicism, romanticism, realism, etc. However, an intellectual revolt emerged against Greek philosophy and led to the emergence of science, discussed in Section 6.

Due to the existence of good communication lines for the transfer of ideas, the emergence of science intermittently injected fresh blood to old ideas and this maintained new generations of philosophers in Europe throughout 17th-18th centuries.

The legacy of Greek philosophy was overhauled in successive waves of innovative thinking but had the same fate – stalemate. Schlick (Boyd, 1992, p37) illustrates the problem by giving an account of movements in philosophy as follows: "every philosophical movement is defined by the principles that it regards as fundamental, and to which it constantly recurs in its arguments. But in the course of historical development, the principles are apt not to remain unaltered, ...

At some point the question then arises as to whether we should still speak at all of the development of a single movement, and retain its old name, or whether a new movement has not in fact arisen". Popper (2004, p30) states that "The controversial question whether philosophy exists, or has any right to exist, is almost as old as philosophy itself.

Time and again, an entirely new philosophical movement arises which finally unmasks the old philosophical problems as pseudo-problems, and which confronts the wicked nonsense of philosophy with the good sense of meaningful, positive, empirical, science..."

It should be noted that the author argues that in the Dark Ages and Middle Ages, there was no notable freethinking philosophical movements in Europe until the 16th century. Although there were prominent thinkers but they were largely of theophilosophical denominations that rational freethinkers.

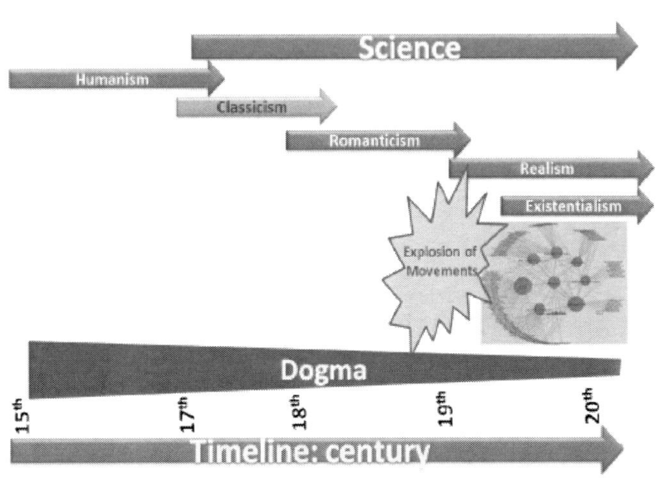

Figure 9. Movements in the Modern Europe: Culture, Philosophy, Science.

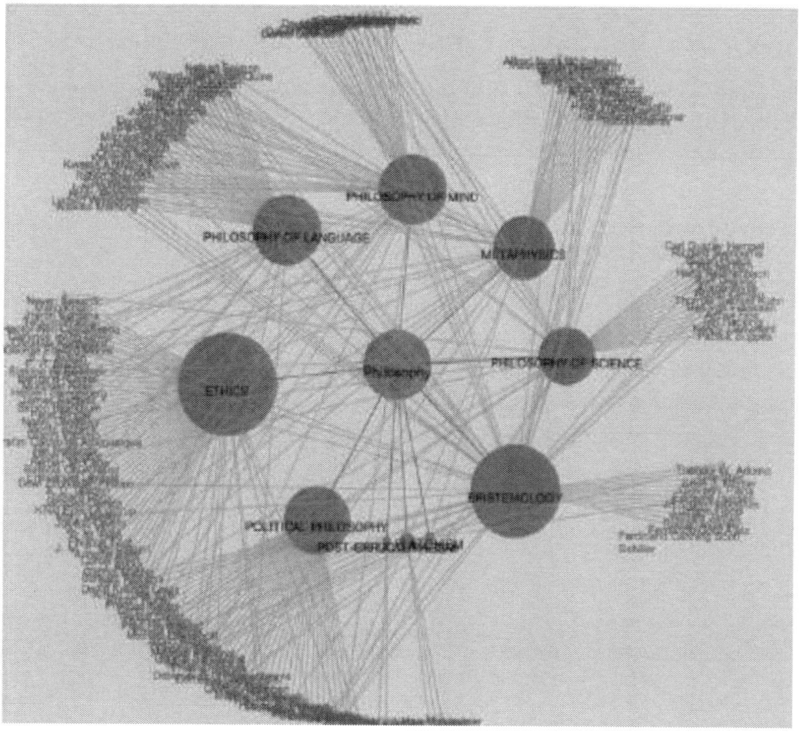

Figure 10. Cladogram of Philosophy (see Acknowledgement).

Figure 10. depicts a cladogram of philosophical doctrines, interpreted as entropy in action: so many doctrines, all interesting, non useful or practical. This may sound unduly strong but Popper (2004, preface) is harsh in stating that “The philosopher ... faces ... something resembling a heap of ruins. ... Indeed it has by now become recurrent question in philosophical circles whether philosophy will ever get so far to pose a genuine problem.” To illustrate the above, consider the journey from the positivist and realist stances of philosophy in the early 1900s through existentialism, relativism, anarchism, constructivism to postmodernism. Consider the problem of demarcation, which Karl Popper (1902–1994) contributed to it. Prior to him the thinking was focused on verificationism and according to Diller (2002) “all philosophical systems in the Western tradition, from the time of Descartes until the twentieth century, have been justificationist ones”. However, Popper (2004) repudiates corroboration or verification and holds that a theory is scientific if survives falsification.

The problem is that everything in science is open to verification and falsification and false hypotheses can be verified (fitting different modeling strategies to the same data) or verified theories can be falsified, e.g. the Newtonian mechanics (which was verified but after the emergence of the quantum mechanics it still retain its status in appropriate problems). Kuhn (1962) had a solution for this through his doctrine of paradigm shifts asserting the importance of the historic integrity of the scientific theories. He defines a paradigm as ‘universally recognized scientific achievements that, for a time, provide model problems and solutions to a community of practitioners,’ Kuhn (1962, preface). The notion of shift highlights the role of the time dimension in scientific theories but then science becomes a landscape dotted with paradigms constructed by the scientific community with known politics and preferences.

Historic integrity of scientific enquiries is a welcomed notion and besides Kuhnian paradigm shift, Hegelian/Marxian dialectics and existentialism have a coherent construct of time. Change in the Marxian dialectics is based on the antagonism of thesis and antithesis driven by the concept of historic inevitability, holding that antitheses increase in quantity and is transformed violently into theses leading to a new cycle of antagonism. Marxian dialectics was intermarried with the Kantian constructivism by Kuhn in 1962 to form the doctrine of paradigm shift, as the Marxian historicism and antagonistic revolutions have no place in science and therefore these were replaced with Kantian constructivism.

The existentialist exposition on time is based on matter (inner world) and mind (outer world) and that mind is determinable through revered individuals but matter is indeterminate though only the revered ones are revealed with an understanding of matter apparently to save mankind from a nihilist prospect. To Heidegger, nihilism is thought in essence and a world-historical movement drawn into the power realm of the modern age. Heidegger (1962, p19) elaborates *Being* and explains its difficulties: "Our provisional aim is the Interpretation of time as the possible horizon for any understanding whatsoever of Being."

Both existentialistic and Marxian forms of explaining changes in history and possibly foreseeing the future are interesting speculations but they are unscientific. The existentialistic doctrine has a mystic mindset but the Marxian doctrine is rather down to earth. The recent political history has been a setback to the Marxian philosophy, not due to the triumph of capitalism (which itself is in turmoil) but arguably for giving rise to a social order without negative feedback loops. The author regards both doctrines as speculations but stimulating and holds that both can be mapped onto EvSyM. In addition, EvSyM subsumes verification (positive feedback complexes) and falsification (negative feedback complexes) concepts and historic inevitability (entropy) historic integrity (evolutionary transitions) advocated by Popper, Kuhn or others.

In the 20th century, philosophy found a new lease of life by trying to explain the main tenets of knowledge in diverse fields of science, e.g. philosophy of science, philosophy of biology, philosophy of mathematics, philosophy of risk etc. It became embarrassing if a modern endeavor was not approached by philosophy, e.g. the philosophy of mathematical modeling is embarrassingly fallen behind in the battles of philosophy. The paper holds that philosophy cannot be overhauled but if there is an alternative, it can be unselected altogether, as suggested below.

5.3. EvSyM – A Vision for Philosophy

Protagoras (490 B.C.–420 B.C.), the protagonist of relativism, has ostensibly said that "man is measure of everything." This is certainly true in philosophy, where each and all of the philosophers regard their presuppositions and predispositions as measure of everything – a tradition that goes back for three thousand years. So, based on Protagoras, it is possible to justify that philosophers are all relativists, although they will certainly deny it, unless they happen to be one. Whilst relativism encourages mutual exclusion and idiosyncrasy, science encourages mutual inclusion. In comparing these two human endeavors with each other by EvSyM, it is obvious that science selected by natural selection has a hereditary core but Greek philosophy does not have such a core and its re-primed lease of life in Islamic philosophy collapsed and European philosophy has become a wilderness. Yet its appeal lives on and this is likely to be due to human curiosity for fundamental questions. Although, it is also likely that this is the case of group selection by like minded individuals who work hard to keep philosophy going by setting the intellectual agenda without nurturing any solution.

Philosophy is an endeavor for fundamental questions and much needed for human cultures. Scientists and engineers are pragmatic and unlikely to lead the way for such questions. Postmodernism is leading the revolt but by normalizing its own entropic image in a much more subtle way than the past relativists. Yet new philosophy has to emerge to discern obsolete doctrines from fresh ideas, to raise new fundamental questions befitting the modern cultures, to challenge the intellect for a critical understanding. The paper puts forward a thesis that the absence of an *evolvable body of philosophical knowledge* can be remedied by EvSyM by making fundamental enquiries on sustainability, climate change, GM products, artificial intelligence and countless other issues by asking questions and formulating visions for the future. This can succeed if there is a movement to unearth the presuppositions and predispositions; and if the visions elucidate adverse impacts of missing a critical thinking on our fragile future; and if a global consciousness emerges demanding zero assumptions.

6. EvSy-Modeling of Science

Man is rightly fascinated with the origin of species and its hereditary machinery and goes a long way to explain it but curiously cannot see that equally fascinating entity is evolving in front of our eyes and yet we cannot see it. This hereditary machinery is science and this section unearths its routemap.

Science encompasses biology and regards its theory of evolution as its pinnacle but is incapable of rolling out this theory throughout its disciplines. This is rather curious despite the fact that evolution seems ubiquitous. EvSyM compensates for this by explaining evolutionary transitions in science, as a whole.

Science is an evolving body of knowledge but not all areas of knowledge are scientific. This then gives rise to a host of questions including: What makes a body of knowledge scientific; when and how did a body of scientific knowledge emerge? Is science evolving? The response to these questions is a subject for books and there are many books approaching these questions through philosophy, history or historiography and some even talk of evolution in science. These are invariably oblivious of the integration of evolutionary thinking with systemic thinking. Thus, this paper presents an old subject in a new framework but only in a cursory manner.

An overview of evolutionary transitions in science is presented first. A rather detailed account of evolutionary transitions are presented by Khatibi (2011) on flood risk management; Khatibi (2012a, 2012b) on mathematical practices and mathematical modeling, and Khatibi et al. (2012) on artificial intelligence. This section identifies three epochal periods: (i) *communal-cognitive knowledge*, prevailed in prehistory and still thriving; (ii) *philosophically-driven knowledge* until 17th century, which was driven by speculation; and (iii) science since 17th century, which started afresh by unselecting its past legacy.

Before presenting evolutionary transitions in science, its definition is essential. However, there is no single way of defining science. Bernal (1954) uses the term science in its broad sense without attempting “to cramp it into a definition.” Indeed, science has so changed its nature over the whole range of human history that no definition could be made to fit.” This paper regards science as a body of evolving knowledge acquired by study, research and training with the following essential attributes: it is reducible; it complies with set criteria through an explicit negative feedback loops, presupposes the availability of data/evidence; and is open to challenge. A body of knowledge without reducibility, without negative feedback and without the openness for challenge is not scientific.

6.1. Stage of Zero$^+$ Feedback – Communal Knowledge

A study of the knowledge gained by man in the prehistory until the Mesolithic Age, (10,000 years ago), is the subject of archaeology and mythology and outside the scope of this paper. In spite of great spatiotemporal variations in the characteristics of this period, its characteristic features include the gradual domestication of animals, farming and emerging settled communities. These evolutionary changes were enriching the culture already capable of making a variety of tools and weapons and communicating verbally on top of having communications through signs and sounds as the other species do. Such attributes marked ostensible differences with other species.

The period up to the Mesolithic Age was the zero$^+$ feedback stage in the evolution of knowledge. In this very slow process, natural selection is evident as the improvements in the fossil artifacts clearly show it. However, the gained knowledge was communal and cognitive. It passed from generation to generation through imitation and without institutionalization. Thus, knowledge gain was haphazard, as due to natural hazards and continual threats from the surroundings, the new gains could have been lost. The potential loss of knowledge due to haphazard elements could have stressed communities and undermined their viability. It is tentatively suggested that these features triggered a selective advantage for the emergence of institutionalization of knowledge. Thus, first quasi-institutions were selected among some of the communal lifestyles by the end of Mesolithic Age. By this time, belief systems had also emerged, possibly due to the selective advantage of creating a routine fitting the rhythm of the emerging way of life. Thus, many of the past habits, knowledge and lifestyles were probably unselected and replaced with new ways and lifestyles in institutional settings. Notably, communal and cognitive knowledge building still thrives.

6.2. Positive Feedback Driving Philosophy/Technology-Based Knowledge

This paper takes a narrow view and only focuses on Greek philosophy by focusing on some of the roots of scientific thinking. These included the emergence of new knowledge, priming philosophy based on observation, speculation, reason and technology. However, faith, occult, magic was other building blocks of intellectual activities. Emerging practices included astronomy, geometry, technology, mathematics, arithmetic, astrology, alchemy.

The environment for Greek philosophy was rich in terms of freely challenging one another and this gave rise to the articulation of many different possibilities on the dimensions of: (i) fact versus value and (ii) inner world versus outer world (or matter versus mind). There were literally 100s of idea-based schools of thoughts, some of which are depicted in Figure 11.

These schools of thoughts did not coalesce together but they were mutually exclusive entities and as such, they instigated their entropy. This is stated in different ways, e.g. Gorgias (485-380 BC): stated that "there is no truth, only argument; the art of rhetoric." Russell (1946, p224) states that: Greek Skeptics argued that "there could never be a rational ground for preferring one course of action over another." Russell (1946, p244) further states "The only logic admitted by the Greeks was deductive, and all deduction had to start, like Euclid, from general principles regarded as self-evident." He states that Pyrrho's disciple, Timon, denied the possibility of finding such principles. "Everything, therefore, will have to be proved by means of something else, and all argument will be either circular or an endless chain hanging from nothing. In either case nothing can be proved." The author finds Timon's to be profoundly deep. Arguably, the above are precursor thoughts on entropy inherent in Greek philosophy.

It is argued that entropy inherent in Greek philosophy caused it to collapse, as follows: (i) at the 2nd century A.D., during which it came to contact with Christianity; (ii) at the 14th century A.D. within the Islamic cultures; (iii) European philosophy emerged since the 16th century has not yet collapsed thanks to interjections from science but it has been undergoing waves of boom-and-bust scenarios.

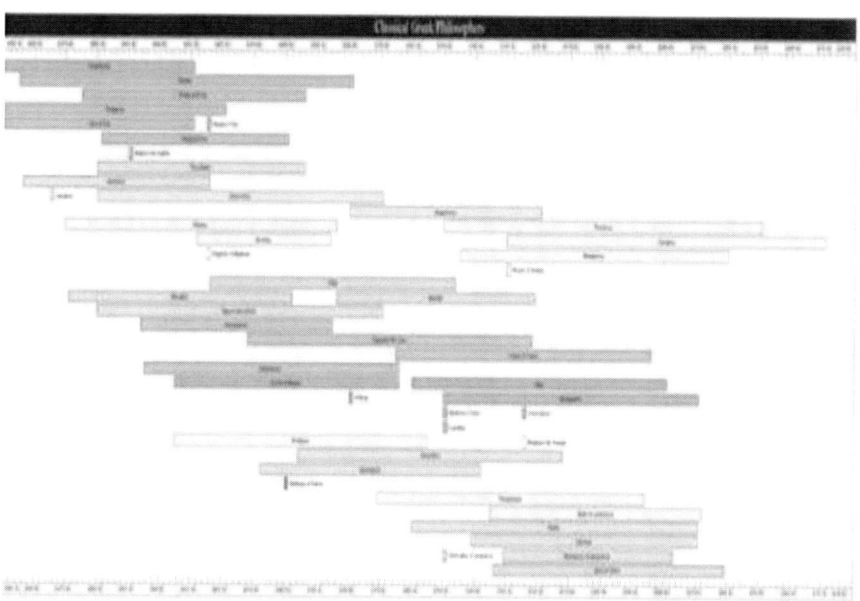

Figure 11. Timeline of Doctrines of Greek Philosophy (see acknowledgement).

The nature of Greek philosophy can be explained by the positive feedback mindset, where it left a rich legacy of fundamental questions but approached the problems at much coarser resolution, too far out from the resolution for their building blocks. Thus, each idea acted as a self-contained unit without coalescing together, e.g. the building blocks of earth, air, fire and water or dichotomies of mind and matter or love and hate are not correct starting point. All these notions are often a complex outcome of many simpler elements and starting an enquiry from such high levels demonstrably led to entropy and the subsequent collapse of philosophical schools of thought then, as a whole. One dimension of Greek philosophy gained longevity by appealing to Christianity, i.e. the Aristotelian ontology and its doctrine of the fixity of the species. This is discussed in Section 8.2.

6.3. The Emergence of Science – 17th Century to 1950

Science is a collective term encompassing reductionism, holism, negative feedback and feedforward loops using the currency of data/evidence, requiring an open culture and with an open invitation to any challenge to any scientific matter as a virtue. It also implicitly refers to the sort of activities that take place in institutions but with deep penetration to communities, as well. The outcomes of science are a win-win culture, the emergence of intelligence and mutual inclusion. This section models evolutionary transitions in science by EvSyM.

The terms reductive science and systems science are in common parlance but the presentation in this paper refers to their evolutionary perspective, as follows:

1. Reductive science emerged in the 17th century, found a better identity in the 19th century by even having the brand name of science (in English) and it remained the norm until 1950. Since its emergence it also capitalized on the credibility of mathematics.
2. Systems science since 1950.
3. In reality, each theory or each area of science is undergoing through these evolutionary transitions.

Before EvSy-modeling evolutionary transitions in sciences, the background is outlined. Bridgman (1929, p277) in articulating uncertainty within the quantum mechanics, says it all: "The conviction that the world was understandable had been science's most important gift to civilization. It had redeemed mankind from centuries of superstition." Medawar and Medawar (1977) regards reduction as 'the most successful explanatory technique that has ever been used in science'. Reductive science is the first human endeavor allowing challenges *to be taken on board*. This is a radical departure from Greek philosophy (and modern philosophy) as their challenges were rhetoric.

Although rhetoric challenges are commendable, they expose the absence of analytical thinking (or a building block approach), which trigger positive feedback d-loops by a spiraling-and-collapsing culture of debate. Reductive science accepts that there can always be differences between observed data and predicted results but the important thing is to reduce the difference. Thus, reductive science produced theories upon theories. Knowing that the difference between observation and prediction is a key notion concerning the integrity of scientific truth, a flexible framework is in place to maintain the integrity by refining the assumptions or even reformulating them, as depicted in Figure 1. This is an explicit instantiation of a negative feedback loop in human endeavors. Its significance should never be underestimated, as this cracked the mind-shell of self-righteousness infesting the scholastic culture in the Middle Ages. The emergence of negative feedback in reductive science marks a paradigm shift or is analogous to the emergence of hereditary codes in biological terms.

Reductive science virtually unselected ruthlessly the past heritages as the arena was ready for a critical thinking to take on board topical issues. It is probably fair to say that Galileo made it a fashion to attack the Aristotelian mindset and he was particularly renowned for such a flare. He also had a flare to make use of mathematics, as it is attributed to him stating that "the laws of nature are written in the language of mathematics," (e.g. Wigner 1960, p532). Whilst science is in symbiosis with mathematics, it is also in a similar relationship with technology. The opportunities spiraled at initial stage when science emerged, as presented below.

6.3.1. Stage Zero$^+$ Feedback Loop

There were four cultures, candidate for a possible priming of science, which were the Chinese culture, Indian culture, European cultures and Islamic cultures. Each of these magnets was advanced in its own rights and had semi-thriving intellectual cultures but stifling freethinking in various ways. They also had well established conduits for the spread of ideas. The Indian and Chinese cultures were rather in equilibrium and unlikely to tip the balance towards finding a more appropriate resolution for the building blocks of fundamental enquiries. It has been said that the Islamic cultures came close to discovering scientific method, e.g. Prof. Jim Al Khalili (http://www.bbc.co.uk/ programmes/b00gq6h7), but the critical thinking did not gather enough momentum to tip the balance towards forming a negative feedback loop in the methods of enquiry. For instance, the epi-circles superimposed on the geocentric model by some of the Islamic philosophers were ingenious but this just prolonged the life of the sinking ship of Greek philosophy and could have not tipped the balance. The European cultures were probably more backward than the others if not equal, but were more stifled with entropy in the intellectual arena than the other magnets.

There is consensus that science emerged in Italy. Although the author is not well versed with the detailed history of Italy, the following account is largely based on the author's interpretation of the accounts given by Russell (1946, Pp479-498). Italy, like Greece, was governed by city states in the 15th century but religious authorities were equally strong and had created a curious situation even giving rise to the expression of Machiavelli. Italy then was stifled with entropy on intellectual grounds, religious superstition, witch-hunting, inquisitions and the conditions were right for the emergence of a negative feedback loop on intellectual activities.[2] In spite of these, the very slow rate of accumulation of knowledge, the accumulation of many millennia of astronomic observations had brought the state-of-the-art to the tipping point. The invention of telescope and thereby its use by Galileo on observing the planetary motions created the grounds for a critical thinking and strengthened the formulated bottom-up heliocentric model to demolish the geocentric model of the planetary movements. The heliocentric model was at last at an appropriate resolution. The names of

[2]Russell (1946, Pp18-19) writes on Europe as follows: "The Middle Ages, though turbulent in practice, were dominated in thought by a passion for legality and a very precise theory of political power. All power is ultimately from God; He has delegated power to the Pope in sacred things and to the Emperor in secular matters. But Pope and Emperor alike lost their importance during the fifteenth century. The ~Pope became merely one of the Italian princes, engaged in the incredibly complicated and unscrupulous game of Italian power politics. The new national monarchies on France ..... The national State, largely owing to gunpowder, acquired an influence over men's thoughts and feelings which it had not had before, and which progressively destroyed what remained of the Roman belief in the unity of civilization.

This political disorder found expression in Machiavelli's *Prince*. In the absence of any guiding principle, politics becomes a naked struggle for power; *The Prince* gives shrewd advice as to how to play this game successfully."

Nicolaus Copernicus (1473–1543), Johannes Kepler (1571-1630), and Galileo Galilei (1564–1642) stand out. However, their different strands of knowledge (building blocks) coalesced together (by a certain implicit rules) and led to the formulation of the heliocentric prototype model (emergent property).

The positive feedback e-loop of science was primed but it also cracked the hard shell of scholastic orthodoxy. There were a sufficient number of intellectual individuals in Europe at the time to form a critical mass and accept this prototype model. A description of the events on the publication of Galileo's ideas is of a historic interest but beyond the scope of this paper and so are the subsequent persecutions or the sources of data, whether they were derived from Greek sources, Islamic cultures or European observations.

This new enterprise even did not have a name for a long time, at least in English. The term science was in currency in the Middle Ages referring to 'the seven (liberal) sciences' (Grammar, Logic, Rhetoric, Arithmetic, Music, Geometry and Astronomy), which were often used synonymously with 'the seven liberal arts'. Notably, its usage as 'to know' with a variety of connotations is even older. The current connotation of the term science is of a recent origin and synonymous with natural and physical science since the 19th century; for instance in the 17th and 18th centuries the notions covered by today's science were commonly covered by natural philosophy.

Virtually, each theory of science started from an uncertain origin and has undergone a period of $zero^+$ feedback loops. Section 3.1 discussed the emergence of the photon theory. Another point should be emphasized is that the emerging culture of science right from the beginning has acquired the character of a clean start by unselecting for good the past speculative approaches e.g. the phlogiston speculation, geocentric model of the planetary movements. They should not even be regarded as a theory or hypothesis. They were knowledge by consent or even by coercion (as the geocentric model was until its demise) and no more.

6.3.2. Stage of Positive Feedback Loop

When a primitive motor was invented by Michael Faraday (1791–1867) in 1821 or by others, automatic motion was an emergent property. It was then a very simple machine and there was no foresight for what was going to follow. The survival of such entities is governed by their proliferation triggered by the natural selection of the products of each of its brands. In scientific terms, this invention was invoked by the emergence of a range of new scientific concepts related to magnetism, electricity and mechanics, without which this invention would have not been contrived. At the same time, each of these scientific concepts was driven by thriving intellectual and entrepreneurial activities. Likewise, a servomechanism was invented to govern constant engine speeds. This was a technological product to regulate steam engines, which was in symbiotic relationship with science. There was no insight on the scientific/ mathematical complexity behind servomechanisms, as the subject of negative feedback emerged in cybernetics in the early 20th centuries. Evidently, the positive feedback e-loop of science was growing in complexity by natural selection.

Almost each and every concept of science is a story on its own but they follow the same routemap of $zero^+$, positive and negative feedback and feedforward loops to differing extents. At the stage of positive feedback, the selected new prototype of scientific concepts strives to be a living entity, owing to its selective advantage by being applied. If it meets all the needs, in time it may be transformed into a scientific theory and carry on serving as the paradigm for the particular subject. Science is a rich ground of examples for this, e.g.:

- Descartes' mechanics was in the old scholastic style but using a modern language. It assumed the universe to be filled with matter which, due to some initial motion, settled down into a system of vortices carrying the sun, the stars, the planets and comets in their paths. Despite the problems with the vortex theory, it was championed in France for nearly one hundred years even after the success of Newtonian mechanics discarding such impossible dynamical systems (O'Connor and Robertson, 1997). This is reminiscent of *group selection*, which but was ultimately unselected; as science is a human enterprise but not a national one.
- It is likely that many approaches will emerge. For instance, Copernican heliocentric model using circular orbits provided a new perspective; Kepler's laws of planetary motion improved them with elliptical orbits and served as a celestial model; the Galilean mechanics was a great advancement compared by the Aristotelian mechanics (including the laws of lever) as it took on board an increasingly large number of subjects, e.g. the inclined plane, the pendulum, the stability of matter, the spring, etc. Whilst these were fit-for-purpose and capable of explaining the available data or supporting evidence, there were an increasing pressure to understand other types of motion. This led to the emergence of mechanics (of solids, liquids and gases), dynamics, discussed in the next section.
- The emergence of the concept of friction is a fascinating story and that in open channel hydraulics is described by Leliavsky (1965), according to which a number of fallacies entered to the state-of-the-art at various times, until they were corrected by the succeeding generations. Whilst friction formula for pipe flows have gained general acceptance, still there may be local preferences in using friction formula for open channels, e.g. Chezy equation in France, Manning equation is often used in English speaking countries, Gaukler-Strickler equation in Germany and Agroskin equation in Russia.
- Calculation of flows through compound channels was once an active area of research and in 2000 a survey by Knight (2001) showed that their outcomes were 13 formulae besides rules of thumb approaches but these new ones were not being used by the hydraulic modelers. There was a massive campaign to consolidate these formula lead to a £0.5m pound project developing a software tool, CES, but the modeling community still prefer to use their old approaches.

The full range of attributes during this stage for each scientific concept includes the proliferation of the areas of scientific studies and concepts (e.g. pipe flows, open channel flows, groundwater flows, gas flows, Newtonian and non-Newtonian fluids, etc.); diversification of alternative methods, adaptation by finding ways of applying emerging methods (e.g. after the derivation of the Navier-Stokes equations, they were adopted to a diversity of practical problems to produce specific solution methods), co-evolution (e.g. the derivation of the Navier-Stoke equation was in a culture where similar equations were derived for open channels (St-Venant equations; transient flows in elastic pipe, sometimes referred to as the Aleivi equations, etc.) and entropy. These are detailed by Khatibi (2012a) and Khatibi

(2012b) for mathematics and mathematical modeling, which are equally applicable to other fields of science. However, the overlooked notion of entropy is very important in each and every area of science and acts as a force to trigger the selective advantage for evolutionary transitions, as discussed in next.

6.3.3. Evolutionary Transitions to Negative Feedback

Throughout the period of positive feedback of an emerging scientific concept, similar scientific methods tends to multiply, each of which are often fit-for-purpose but strong in certain aspects and poor in others. Arguably, this is entropic and continually reduces efficacy of the particular practices and as such, this triggers a selective advantage for seeking an overarching method or the development of a theory. If such a theory emerges and is selected by the scientific community, an evolutionary transition takes place towards a more consistent technique but this may not occur. The followings are examples:

1. Consider sediment transport, which is an area of hydraulics that remains susceptible to a high degree of uncertainty and numerous empirical equations. Simons and Şentürk (1992) tabulate many of them in one table but the uptake of all of them is doubtful. This signifies that there is a selective advantage for the emergence of an overarching theory, to consolidate all of these techniques but until the emergence of a better solution, there is a conceptual crisis on sediment transport.
2. Consider mechanics. The initial contributions by Galileo and Kepler and later contributions by researchers fused together through Newton laws of motions, which provided a backbone to the emergence of classical mechanics. Reference may be made for more details of the various contributions to Dugas (1988).

The point is that when a particular area of research is consolidated by a theory or a law with a critical mass of researchers subscribing to the emerging mindset, the theory is replicated throughout, the problem areas proliferate and new contributions are continually selected. The arrival of a new theory marks an evolutionary transition, after which many of the past concepts or methods will be unselected. For instance, after the emergence of computational numerical models for solving partial differential equations, the past simplified techniques were largely unselected and became obsolete. In hydraulics, these included a diversity of often ingenious tabular and graphic solutions or just simplification of the equations to facilitate quicker manual solutions. Similar processes occur throughout science.

6.3.4. Stage of Negative Feedback – Emergence of Reductive Science

The postulate put forward here is this: for science to transform into hereditary machinery, it had to be at least internally consistent; to be internally consistent, it would not be logical to have many different formulations of the same thing; a single formulation of the same thing is called a theory and must be reductive, have negative feedback loops, amenable to challenge, and supported by evidence; such a framework emerged one after another and often in bundles at different corners of science but they coalesced well together; this became the hereditary machinery of reductive science, which grew from simplicity to complexity, mirroring that of living organisms. According to EvSyM, following the emergence of positive and negative feedback d-ℓ-e-loops, the enterprise is very strong indeed. Below is the detailed background.

The sciences from their inceptions in the 17th century up to 1950 were reductive. For instance, since the Kepler and Galileo's celestial and terrestrial mechanics, today's classical mechanics is the outcome of selections from virtually 100s of dedicated contributions during the 17th-19th centuries. These comprise 100s of definitions, concepts, equations, some of which play a paradigmatic role, e.g. Newton's laws of motions. The emergence of a theory in a particular area of science is equivalent to the emergence of its negative feedback d-ℓ-e-loops as described in Section 3.4.

Negative feedback loops inherent in reductive science give rise to a range of emergent properties, which may be referred to as attributes of reductive science. These include internal consistency; mutually inclusivity among different areas of science, where different areas of science interact or overlap; explanatory and impressive predictive capabilities. EvSyM copes with explaining any area of reductive science with the following overview:

- *The following theories are examples of those selected*: Big Bang Theory (though this is scrutinized now), Atomic Theory, Conservation of Matter and Energy, Chaos Theory, Catastrophe Theory, Continental Drift and Plate Tectonics Theory, Theory of Evolution, Theory of Special Relativity, Photon Theory of Light, Theory of Electromagnetism, Theory of Homeostasis, Cell Theory.
- *Inconsistency*: Inconsistency creates cultural shock waves until they are rectified, e.g. the inadequacy of the Newtonian mechanics to explain the motion at atomic and subatomic resolutions. The intense research on the subject led to the development of quantum mechanics. The scientific methodology allows a flexible mechanism to cope with inconsistencies by a better definition and creating new concepts. Scientists are open-minded and are prepared even to write off past concepts and if necessary take up new ones. Scientific concepts are open to challenge to unravel inadvertent inconsistencies and anomalies. Archaeology and mythology are areas of science susceptible to anomalies and normally when new evidence emerges, past theories are readily revised.
- *Skepticism*: Some theories are under skeptic scrutiny and these include the big bang theory, string theory and the theory of everything.
- *Theories or facts*: Some theories are not regarded theories but fact by an overwhelming majority of the scientists, e.g. theory of evolution.
- *Vested interests*: Certain constructs sometimes are pushed by vested interests as a theory and one example is intelligent design.
- *Unhealthy trends*: Certain scientific activities sometimes suffer from equivalents of kin selection or group selection and these includes transformation of certain models to software tools and promoted by commercial interests against others.
- *Unselected hypothesis*: Examples include psychoanalysis, social Darwinism and eugenics and the latter are unselected two for good. Those commonly referred to as rejected theories were in fact hypotheses in the first place but were unselected, as soon as they were challenged and these include: the phlogiston theory, Newtonian model of corpuscular theory of light, geocentric model of planetary motions, flat Earth, alchemy, astrology, ether.

The strength of reductionism was on simplicity and objectivity conferring the ability to challenge the inherent assumptions. These together with mathematical descriptions of many problems created an outstanding transformation and gave a true meaning to modernity but there were prices to pay, as studies next.

6.3.5. Transition from Reductive to Systems Science

Reductionism overlooks impacts of a system on the environment. Thus, it is opportunist and lacks a strategic approach. It had two main impacts: (i) liberation of many degrees of freedom by harnessing a wide range of energy forms through the industrial muscles which were created by science and technology in the first place; and (ii) the creation and accumulation of adverse impacts, which emerged as conflicts inherited by the present generation, e.g. carbon emission, risk of climate change, ozone hole, mounting wastes and diversity of pollution forms, to name but a few. When an overall view of reductive science is taken, over the years up to 1950, reductive science could be seen as an enterprise which overlooked its impacts and instigated external entropy at an unprecedented scale. The underlying mindset was helped with by the utilitarian philosophy, as discussed in Section 10 and thus unleashed hazards impacting the fragile Earth since the industrial revolution. McNeill (2000) takes an overview of these impacts and states that "in the twentieth century, humankind has begun to play dice with the planet, without knowing all the rules of the game." This was one of the contributing factors to triggering evolutionary transitions, as discussed below. The external entropy triggered for evolutionary transition from reductive science to systems science.

The paper does not discuss the opposition to reductionism but it may be remarked in passing that the emerging systems mindset, though rejected pure reductionism, it did not unselect it. In fact, reductionism was reinvented within the systems thinking by putting more systems together and adding more appropriate feedback loops. Oppositions to reductionism were staged by ontologists claiming that reductionism loses the ontological essence. However, systems science has not found any such essence, other than uncertainty and patterns that were not anticipated in the past. Even harsher attacks on reductionism are staged from dogmatic quarters. Reference may be made to (http://www.drury.edu/ess/philsci /KleeCh5.html) for an academic account of reductionism and antireductionism.

6.3.6. Stage of Feedforward Loops – Emergence of Systems Science

The term "system" is quite old but treating a system as interconnected units towards a goal or purpose emerged in the 1940s. This point is elicited by Chapman (1977) expresses his observations on the launching of the paradigm of models in the 1960s also remarks on systems in a historic perspective that: "At about this juncture an attempt was made to incorporate greater power, greater generality, by adding systems theory to the existing framework."

Systems science did not emerge from the single source of biological sciences but also evolutionary biology and cybernetics. The Bertalanffy's account reveals some of the background: "I presented it first in 1937 ... in Chicago. However, at this time theory was in bad reputation in biology. ... So I left my drafts in the drawer, and it was only after the war that my first publications in this respect appeared. Then, however, something interesting and surprising happened. It turned out that a change in intellectual climate had taken place, making model building and abstract generalization fashionable. Even more, quite a number of scientists had followed similar lines of thought. So General Systems Theory, after all, was not isolated or a personal idiosyncrasy as I have believed, but rather one within a group of parallel developments" (OSG, 1981, p.61). Thus, a selection of a theory depends on the rich culture of the scientific community and even if a theory may have a selective advantage, still it may not be selected.

The experience in cybernetics is described by Wiener (1948, p.viii) by likening the emerging control theory to that of the Copernican system unselecting the "overstrained Ptolemaic system, with a simple and natural heliocentric description of the motions ... so the study of non-linear structures and systems, whether electric or mechanical, whether natural or artificial, has needed a fresh and independent point of commencement."

An appreciation of inter-component synergy emerged in cybernetics and biology, giving rise to the emergence of systems science. This may be regarded as evolution by speciation in biological terms. However, the entropy accounted above associated with the impacts of utilitarianism and reductive science triggered the need for a wider evolutionary transition, as discussed in this section. The systems experience in evolutionary biology is quite different, as the researchers started all with their narrow line of research but their results increasing crossed into the adjacent disciplines, as studies in Section 8. However, it should be noted that although systems science fully embraces theory of evolution, it has no evolutionary basis but EvSyM aims to compensate for it.

Systems science nurtured a new culture for 'holism' since the 1950s and set the pace of scientific developments since then. Also, the developments of classic basis of positive and negative feedback loops are attributed to Norbert Wiener, see Weiner (1948). Hoos (OSG, 1981, p317) argues that "Systems analysis has come to represent the dominant paradigm." Lilienfeld (1978) discusses the rise of the concept of systems in the middle of the 20th century due to contributions from Ludwig von Bertalanffy's GST; Norbert Wiener's cybernetics; W. Ross Ashby's related works leading to feedback and automation; Shannon, Weaver, Cherry developing information and communication theory; and Neumann and Morgenstern's games theory.

The emergence of systems science triggered debate on reductionism versus holism within the scientific community, e.g. Grene (1974), Medawar and Medawar (1977) and the shortfalls of reductionism within the philosophical community often overlooking holism, e.g. Trout (1992), Garfinkel (1992), where it is often impossible to make sense of these philosophical indulgences. The main point is that after the emergence of systems science, there was no unselection of reductive science but the evolutionary transition comprised of the reinvention of reductive science within the framework of systems science. Checkland (OSG, 1981, p31) puts this in a different way "systems thinking and analytical thinking will come to be thought of as the twin components of scientific thinking."

6.4. Hierarchy in Science

Systems science capitalizes on hierarchy and Checkland (1981, pp32-33) argues (citing Simon, 1962) that "the age of planet Earth is such that only hierarchically organized entities have had time to evolve! The structure of the world we inhabit could only be hierarchical." Although the EvSy-modeling of hierarchy in science will reveal how science came to be on a limb, only a cursory overview is presented in Table 4.

7. EVOLUTIONARY SYSTEMIC BEHAVIOR

This section presents evolutionary systemic view of behavior without actually modeling its evolutionary transitions as part of the effort to observe the limit on the size of the paper. At least four dimensions of behavior are identified. *The first dimension* of behavior is phase or media where evolution takes place. As the media for evolution is particulate, the potential building blocks may fall into the following particulate phases:

(i) Zero interactions: comprising inert entities, static or dynamic, engaged in action-and-reaction but no interactions, e.g. physical objects such as stone;
(ii) Mutually exclusive individuals of entities: interacting with each other;
(iii) Communal type interacting individuals: interactions can be theorized in terms of passive and active pressures, similar to granular soil media;
(iv) Institutional settings: undergoing evolutionary transitions.

Table 4. Evolutionary Transitions in Hierarchy

<table>
<tr><th>Hierarchies</th><th>d-loop</th><th>ℓ-loop</th><th>e-loop</th></tr>
<tr><td>Quantum scale</td><td colspan="3">• No prototype, no evolving entity, no feedback, no hierarchy but a potential
• Heisenberg's Principle of Uncertainty
• An understanding for this resolution emerged in the first half of the 20th century</td></tr>
<tr><td>Physics</td><td>• Visible/testable complexity at the d-loop level</td><td>• Variability in physical properties</td><td>• Laws of nature; empirical rules
• Mutation, uncertainty</td></tr>
<tr><td>Chemistry</td><td>• Elements, compounds
• Positive feedback d-loops</td><td>• Alleles (tentatively suggested)</td><td>• Periodic Table + atomic theory
• E-loops: e.g. DNA+RNA</td></tr>
<tr><td>Biology – autonomy</td><td>• Positive feedback d-loops
• Negative feedback d-loops</td><td>• Positive feedback ℓ-loops
• Negative feedback ℓ-loops</td><td>• Natural selection: positive and negative feedback e-loops</td></tr>
<tr><td>Psychology – traits</td><td rowspan="3">• The 4 feedback d-loops are functional pluralistically
• Functions are driven by the hierarchical characteristics
• Hierarchies have emergent properties of traits, values, cultures, respectively</td><td rowspan="3">• Adaptive: +ve/-ve feedback and feedforward ℓ-loops
• Tentatively suggesting: adaptation is the reconfiguration of building blocks for nearly the same selective advantage</td><td rowspan="3">• Mutation
• Natural selection: +ve/-ve feed-back and feedforward e-loops
• Purpose for: survival, internal / external consistency emerging hierarchically</td></tr>
<tr><td>Sociology – values</td></tr>
<tr><td>Anthropology - cultures</td></tr>
<tr><td colspan="4">Note: The science underpinning this table goes through evolutionary transitions driven by feedback loops.</td></tr>
</table>

Arguably, the potential for evolutionary transitions among the individuals (Item ii) is internalized but that of the institutions (Item iv) is externalized. The dynamics of the communal interactions is very interesting and the author is quite surprised that the analogy with soil mechanics has not been taken on.

The second dimension of behavior is the state of evolvable entities. This is subject to at least three timescales of: d-ℓ-e-loops related to positive and negative feedback and feedforward complexes. See: Section 1.2 and Table 1.

The third dimension of behavior is the capacity to act as free agents. The author argues that it corresponds to the concept of truth, which long been a contentious topic in philosophy and very muddled. However, the many of loose concept can simply be linked together in Table 5 above, which reflects the core of evolutionary transitions in behavior, albeit in very succinct terms.

Table 5. Evolutionary Transitions in the Capacity to Act as Free Agents

	d-loops – governance	ℓ-loops – charity	e-loops - morality
Zero$^+$ Feedback	• Stabilizing: e.g., love, just • Destabilizing: e.g. self-righteous	• Instinct • Self-sacrifice, altruism • Selfishness	• Spontaneity • Mutation
positive Feedback	• Traditional discourse on truth • Wise governance – tyranny	• Charity • Self-help groups	• Morality • Utilitarianism, many '.isms'
negative Feedback	• Performance criteria, • Accountability	• Think tanks to horizon scan the future truth	• Ethics • Trait Selection
Feedforwar d	• Democracy • Voting system	• Risk-based decision-making	• Sustainability, • Decision-making by participation
Pluralism	• "Horses for courses"	• Versatility	• Rich culture

The fourth dimension of behavior is sustaining the behavior through the flow of currency or evidence/data comprising material, energy or similar entities, as well as information. The paper does not cover evolutionary transitions in evidence/data. The author's preliminary thoughts were articulated by Khatibi et al. (2004). For now, it suffices to mention that, indeed evidence/data also undergo evolutionary transitions and currently, the entropy created by data under ontological mindsets is very serious. The solutions are technological and this is just to postpone evolutionary systemic solutions.

8. EvSy-Modelling of Theory of Evolution in Biology

This paper shares the mainstream view on the evolution of biological species regarding theory of evolution as a fact supported by the overwhelming evidence gathered over time. The gathering of evidence is an ongoing process and the trend is that as more evidence emerges the understanding of the past evolutionary processes of species becomes clearer. Sometimes there are debates on how to respond to specific questions stemming from the details of the evolution of species and as such, evolution is just like many other theories of science but one of its most successful ones. This body of scientific knowledge is now outlined briefly through EvSyM. Although the past thinkers are regarded with a great admiration and not responsible for the lack of the future knowledge, care is needed not to evade the space of the past thinkers with the modern thinking and EvSyM creates a framework for this important consideration.

Biology is the birthplace of evolutionary thinking and the thinking in this discipline of science is now under the banner of Neo-Darwinism, which is driven by the law of natural selection. The most striking feature of Neo-Darwinism is that it deals with variations. This is not any type of selection but selecting positive features, where the act of selection has no foresight. The body of neo-Darwinian knowledge is applied to species rendering the following: (i) genomes concerned with the entirety of hereditary information of the organism encoded in DNA or, in RNA for many type of viral organisms; (ii) genetics concerned with traits associated with genes embedded in the genome; (iii) natural selection, which is a focus from the phenotype resolution to genotypic study of the genes and their variations over time; (iv) Mendelian inheritance studying inheritance at the phenotypic resolution; and (v) cladistics that is concerned with groups of organisms that share derived characters. The outcome of these activities is that they clearly point towards common descent. As stated by Lipscomb (1998) "That all of life is connected in an immense *phylogenetic tree* is one of the most significant discoveries of the past 150 years."

This section models the theory of evolution in biology by EvSyM. The focus is to show that different strands of knowledge emerged since the 17th century were readily coalesced together to form an evolvable body of knowledge but this was a painstaking work, in which the Darwinian natural selection played the role of a gelling agent.

8.1. Stage of Zero+ Feedback – Primitive Knowledge

Preoccupation with the existence of man, animals and plants goes to prehistory and the mythology of all cultures has almost something to say about the origins. It is often the case that the mythological accounts strike deep notes from time-to-time in the mind of modern man. However, they normally relate to: outcomes of observations, at their best; perceptions related to life experiences; stories stemming from shamanistic ascetic experiences; or speculation, dream or plots to preserve self-interests. Such accounts should not be contaminated with the modern thinking to arouse undue overexcitement. The mythological accounts on existence simply illustrate the point that a particular enquiry started a long time ago serving as a prototype to modern times. Taking species as analogy, the particular mythological accounts are like species that have gone extinct long time ago but their study is important to understand the evolutionary processes. As this subject is vast (and the author is not qualified on the subject), the only point to be made is that, the very mythological accounts clearly show an evolutionary transition from the past archetypes to the present human scales. Those archetypes were figments of imagination (or *a gala concert to an obsolete way of life*).

8.2. Primitive Evolutionary Thinking Driven by Positive Feedback – Mesolithic Age to 17th century

Enquiries through scholarly observations and speculations or hypotheses on primitive notions of evolution or natural selection could be found among Mesopotamian, Greek, Chinese, Indian and Islamic thinkers and many other cultures. It should be remarked that the ancient Egyptian civilization apparently did not explore ways of reason, as Russell (1946, p26) expresses, "The Egyptians were preoccupied with death, and believed that the souls of the dead descend into the underworld." However, studying the past heritage is a slippery slope as historians and philosophers often diffuse unwittingly modern concepts to the past.

The past thinkers had noticed change but even the most basic notions were yet to emerge (e.g. the hydrological cycle was learned in the 17th century) let alone understand very complex mechanisms of evolution. They had not gathered building blocks of evidence to explain life. Even great thoughts of those times were merely based on observations, perceptions, intuition or speculations.

Greek philosophers thought about changes and the literature often credit Thales for taking water as the origin of all things; Anaximenes for taking air as the principle of all things and regarding the subsequent process as a thinning or thickening; Anaximander of Miletus (c.610–546 BC) for deducing living things to be in a gradual development from moisture under the influence of warmth. Many other accounts are often well into dreamlike worlds or myth. Modern interpretations attached to the legacy of these thinkers are not helpful.

Greek philosophy had a lasting impression through Aristotelian fixity of species which was turned into orthodoxy through Christianity. Modern scientists or engineers would not be persuaded by the Aristotelian mindset, as it is devoid of any building blocks approaches, although it is rich in speculative and intuitive thoughts. One of these thoughts is ontology and the fixity of species. It was transformed into orthodoxy throughout the Middle Ages (5th to the 15th) by appealing to Christianity.

In Aristotle's time there was no reductionism and no evolutionary thinking through evidence. Ontology is elaborated by Aristotle in his *Categories* and these come with a baggage of such terms as ontology, metaphysics, axiology, accidents, eminence, universals etc. However, Smith (2001) gives an accessible account, as follows below. He argues that ontology is one of Aristotle's remaining legacies often applied to objects of the various types and categories, whichever the discipline it may be. Aristotle approached these commonsensical objects in relation to everyday reality but formalized them. At the heart of Aristotle's ontology is a hypothesis (or more appropriately speculation) of substances (e.g. organisms) and accidents (qualities, events, processes) with attributes given in Box 8.1 (as Aristotle conceives).

The penultimate article in Box 8.1 is concerning for Smith (2001), who sees it causing problems for standard modern approaches to ontology based on physics. However, this paper regards all of the articles in Box 8.1 as speculative, to say the least. In fact, it is not necessary to discuss accident or the other notions above, as they are purged from the modern terminology and have become irrelevant.

Box 8.1 – Aristotelian Ontology

(i) Substances existing on their own but accidents require a support from substances for their existence. Substances are the *bearers* or *carriers* of accidents.
(ii) Substances remaining numerically one and the same, can admit contrary accidents at different times (to be hungry sometimes; sometimes tanned).
(iii) Substances are 'one by a process of nature.'
(iv) A substance has a complete, determinate boundary.
(v) A substance has no proper parts which are themselves substances.
(vi) A substance is similarly not the proper part of any larger substance.
(vii) A substance takes up space.
(viii) A substance is self-identical from the beginning to the end of its existence.
(ix) The existence of a substance is continuous through time: substances never enjoy intermittent existence.

Ontology remains very strong among deists and theists even in modern times. Whilst they often appreciate the explanatory powers of reductive science, they react differently to evolution. Some completely write it off and some cast doubt on the so called "theory" of evolution. These opponents of evolution are normally comfortable when science is applied to technological things such as airplane or automobile but they are uncomfortable with evolution applied to species; and all hell breaks loose when man is regarded as just another species. By and large, they do not want to break off from their Aristotelian ontological mindsets and claim that reductionism leaves out ontology. This is not well-defined to a modern scientist.

Ontological philosophers claim that it may be that stable wholes of atoms, molecules, bio-chemicals, cells or organism explain a lot of things but these are arbitrary combinations of parts and leave out their ontological essence. The paper is not minded with allaying such concerns as their arbitrariness is evident by comparing Box 5.1 with 8.1. Evidently, both are arbitrary speculations but the latter is more sophisticated than the former. To grasp the core idea, consider the following four cases:

$$Evolvable\ Entity = Positive\ Feedback\ (d\_loop + \ell\_loop + e\_loop) \qquad (8.1)$$

$$Defy\ Selection = Positive\ Feedback\ (d\_loop)\ with\ e\_loop\ being\ inadvertently\ absent \qquad (8.2)$$

$$Group\ Selection = Positive\ Feedback\ (prescribed\_zero\_e\_loop) \qquad (8.3)$$

$$Inert\ to\ Selection = pristine\ natural\ objects\ used\ in\ (d\_loop)\ with\ no\ e\_loop \qquad (8.4)$$

Examples of (8.1) are species, science, mathematics; an example of (8.2) is speculations in philosophy; examples of (8.3) are fixed ontology, postmodern philosophies; and an example of (8.4) is pristine natural objects. The author realizes that the term speculation may be a retrospective expression for the mindsets prior to the emergence of science and possibly unfair. The mindset in pre-science periods is better characterized by paraphrasing Maurice and Duhem (1980, p2-3) who use the expression of penetrating thoughts using language to express it.

It may be noted that the author builds a front against fixed ontology but the author has no problem with pragmatic ontology practiced in modern computer sciences and data management, although they too feel the burden of a fixed mindset. Equally, even fixed ontology is itself a front against divinity.

8.3. Negative Feedback Driving Biological Sciences

The thinking on evolution was invoked by the emerging critical thinking and by modern thinkers in the 17th century. Initial works on evolution displayed a significant departure from the past. The emerging ideas from this early period until the establishment of neo-Darwinism in the 1950s provided the necessary rich culture for the emergence of the science of evolution identifying its fundamental building blocks and pointing towards common descent of species. The emergence of a full understanding was evolutionary with the following routemap:

1. Pulling the string started with the phenotypic studies by movements in the 17th-19th centuries: (i) taxonomy of the species; (ii) explaining similarity in the species; (iii) explaining the hereditary features.
2. Phenotypic studies put the pointers towards the genotypic studies and this evolved through research on cells, genetics and genomics.

8.3.1. Stage of Zero^{+} Feedback – the Rise of Evolutionary Thinking

The 17th century may be the source for the emergence of initial understanding on evolutionary thinking but the coalescing evolutionary thinking was also helped by the ideas on the age of the Earth, continental drifts and many other terms and concepts. The culture in Europe in the 17th-18th centuries was rife for a critical thinking and the phenotypic regularity patterns among the species provided the primitive building blocks to form an evolvable biological knowledge. Still the Aristotelian ontology overshadowed the broader culture and essentialism retarded biology. There were many programs of gathering evidence from all over the world by the European scholars, although some of which were tinted with the strong racist sentiments of the time. From the 17th century until the merge of the Darwinian natural selection with the Mendelian genetics by the 1920s, the growth of evidence went through phases. The emerging ideas displayed a significant departure from the past, as the Aristotelian ontology was unselected. The various movements of the time may be categorized as follows:

1. *Taxonomic activities*: The taxonomy of species was thriving; pioneering works include Carl Linnaeus (1707–1778) and John Ray (1627-1705).
2. *Explaining Similarities among the species*: Hypotheses emerged to explain the phenotypic similarity patterns among the species by: Pierre Louis Maupertuis (1698–1759), Georges Louis Leclerc (1707–1788), Jean-Baptiste Lamarck (1744–1829), Erasmus Darwin (1731–1802), Alfred Wallace (1823–1913), Charles Darwin (1809-1882).

3. *Heredity*: Extensive experiments were carried out to hybridize providing evidence for a wide variety of inheritance phenomena, including hybrid sterility and the high variability of back-crosses. Examples of such experiments include: Carl Linnaeus, Joseph Gottlieb Kölreuter (1733–1806), Carl Friedrich von Gärtner (1772-1850), and Charles Naudin (1815-1899) and Gregor Mendel (1822–1884).
4. Many other activities were going on at the same time that were contributing to the growth of scientific culture from simplicity to complexity and enriching the cultural environment, e.g. paleontology began with the work of Georges Cuvier (1769–1832) and geology by James Hutton (1726 –1797) and Charles Lyell (1797–1875).

The above activities were modern in the sense that they were interpreting a considerable amount of evidence (though the evidence could have been anecdotal too) and were theorizing their subject to seek an expression of a law of nature. However, they resembled philosophical doctrines by not coalescing with one another and failed to form a positive feedback e-loop except that for the taxonomy of species by Carl Linnaeus (1707–1778). His works did not meet the necessary and sufficient conditions for theory of evolution but provided a backbone for the emergence and sustenance of evolutionary thinking. Thus, Linnaean Taxonomy in Theory of Evolution (LTiTE):

$$LTiTE = Priming\ positive\ feedback\ (d\_loop + e\_loop) \qquad (8.5)$$

8.3.2. Stage of Positive Feedback – Evolvable Theories of Evolution

The law of natural selection by Charles Darwin (1809–1882) played a key role in consolidating the evolvable body of theory of evolution at the time. He published his book: *On the Origin of Species* in 1859. Natural selection is very simple but it is a breakthrough to explain the shared ancestry of all the species, known as the common descent. This epochal event led to unselecting Aristotelian fixity of species. The EvSyM description of the law of natural selection is:

$$Law\ of\ Natural\ Selection = Positive\ Feedback\ (d\_loop + \ell\_loop + e\_loop) \qquad (8.6)$$

This naturally coalesced with Linnaean taxonomy but was still without genetics and therefore was not a robust law. As theory of evolution gained a scientific status, its d-loop reached inevitably other disciplines. Although, the evolvable body of the law of natural selection delicately captured the evolution of species, it did not necessarily apply to individuals. The expressions such as "the survival of the fittest" by Herbert Spencer (1820–1903) opened up the way for the abuse and misinterpretation of science, hence social Darwinism. Spencer's conception of evolution advocated progressive developments and his sociology of progression from simplicity to complexity through three stages was a philosophical position and therefore speculative, rather than being an evidence-based scientific work.

Darwinian evolution was also the subject of "speciation" as it drifted to other intellectual areas of thought, such as economics and psychology, in the early part of the 20th century but without displaying any similarity between the various disciplines. Darwinism without genetics was science but very vulnerable one, as demonstrated by social Darwinism. Their synthesis emerged in the 1920s by the rediscovery of Mendel's genetics.

In the 18th century, there were various hypotheses to explain heredity all with problems, as opposed to vitalism which was implicated with soul or some sort of "self-determinism." The culture was rife for a more scientific understanding in the 19th century and Gregory Mendel (1822–1884) studied variation in plants by experimenting on peas during the years 1856-1863 by cultivating and testing some 29,000 pea plant. He studied the traits and predicted their inheritance patterns. He presented his "particulate inheritance theory" in 1865 and published it in 1866. His work was scientific and capable of priming the science of genetics from the phenotypic vantage but it was criticized and not selected at the time. The EvSyM description of the law of the Mendelian genetics is:

$$Mendelian\ Genetics = Priming\ Positive\ Feedback\ \left(e_{loop} for\ genetics\right) \quad (8.7)$$

Mendelian Genetics was rediscovered in 1900 by Hugo de Vries (1848–1935), Carl Erich Correns (1864-1933) and Erich Tschermak von Seysenegg (1871–1962). The emerging research works on cells also played a role on selecting the Mendelian genetics. If the law of natural selection was the necessary condition for theory of evolution, the Mendelian genetics was the sufficient condition but their synthesis emerged in the 1920s-1930s by the contributions of J.B.S. Haldane (1892–1964), Sewall Wright (1889–1988), Ronald Fisher (1890–1962) and others. They developed the population genetics and re-coined the term neo-Darwinism for the synthesis of Darwinian natural selection and Mendelian particulate inheritance, where August Weismann (1834–1914) already coined in 1900 though not selected. The EvSyM description of neo-Darwinism is:

$$Neo - Darwinism =$$
$$Positive\ Feedback\ \left(d_{loop} + \ell_{loop} + e\_loop\right)\ of\ Theory\ of\ Evolution \quad (8.8)$$

Neo-Darwinism primed the science of evolution by synthesis. It also acted as the gelling agent for genetics, cytology, systematics, botany, morphology, ecology and paleontology. The routemap of theory of evolution is from phenotypic to genotypic world. Without the benefit of an insight into the genotypic complexity, the route was fragile. Not surprisingly, it erred and gave rise to social Darwinism. Neo-Darwinism brought robustness to theory of evolution and made it a mature science but possibly, due to bad experience with social Darwinism, other disciplines of science remained tacitly or explicitly ontological. Notably, Popper (1976, p.151) states that "Darwinism is not a testable scientific theory, but a metaphysical research program." This exposes the utility of philosophy and tells more about philosophy than Darwinism.

8.3.3. Stage of Negative Feedback

The routemap of theory of evolution was driven by phenotypic features and pointed to the genotypic side of the theory evolution. Although (8.8) states the priming of the science of evolution but arguably this science still did not have the machinery for internal consistency. Internal consistency of the science of evolution emerged from the theory of cells, which was the culmination of the opportunities created by the invention of microscope. Its detailed understanding is beyond the author's experience but the milestones included (i) Robert Hooke (1635–1703) coined the term cell in 1665; (ii) Antonie van Leeuwenhoek is regarded as the first to witness a live cell in 1674 and others contributed to the development of the theory of cells; and (iii) The theory of cells emerged by 1850 due to the works by Theodor Schwann (1810–1882), Matthias Jakob Schleiden (1804–1881), and Rudolf Virchow (1821 –1902).

An understanding of the working of the cells brought theory of cells nearer to genetics, e.g. the works by August Weismann (1834–1914) and Hugo de Vries (1848–1935) and others. The pointers became stronger by understanding the structure of cells to explain genetics since the 1920s. Without EvSy-modelling of evolutionary transitions on the emergence of molecular genetics and genomics, it is sufficient to mention that the explanation of the genetic codes by James Watson and Francis Crick led to an evolutionary transition from positive feedback to negative feedback in the science of evolution. The EvSyM description of the science of evolution is:

$$Science\ of\ Evolution = Positive\ Feedback\left(d_{loop} + \ell_{loop} + e_{loop}\right) + Negative\ Feedback\left(d_{loop} + \ell_{loop} + e\_loop\right) \quad (8.9)$$

The science of evolution (which still may be referred to as theory of evolution) led to the formation of molecular genetics, as an internally consistent science. One outcome of this science is the Human genome Project (1990-2003), which identified all the approximately 20,000-25,000 genes in human DNA among other things.

8.3.4. Stage of Feedforward

Biological evolutionary knowledge has spread out to other disciplines, e.g. memetics, evolutionary economics, evolutionary psychology, evolutionary anthropology, neural Darwinism, evolutionary linguistics, Quantum Darwinism, cosmological natural selection, genetic algorithms and evolutionary computation. Other ongoing activities include generalized Darwinian algorithm, universal selection theory (universal selectionism), universal Darwinism. These are all good signs but they are not transparent to one another and not attacking the real issue of developing science mindful of its own external inconsistencies and entropies.

James Watson was recently embroiled in racism. His opinion was reported by Milmo (2007) among other, as: “One of the world's most eminent scientists was embroiled in an extraordinary row last night after he claimed that black people were less intelligent than white people and the idea that “equal powers of reason" were shared across racial groups was a delusion.” According to him, James Watson stated: “All our social policies are based on the fact that their intelligence is the same as ours - whereas all the testing says not really.” The gut feeling of the author is that James Watson is wrong but that is not the issue.

This “Watsonian” mindset underpins the existence of chasms in science due to the absence of feedforward loops. Racism, egoism greed (remember the film: greed is good) and many other similar traits are external entropies looming over the world culture as destabilizing factors and encouraging strange repellents in the sense of chaos and catastrophe theories. If science remains mindless, these entropies can bring the world to the point of destruction. There remains the hope that, if stabilizing building blocks are selected, saving the world from destruction may be possible. Racism is a destabilizing factor and science ought to stay clear of it. The generalized theory of evolution has a great deal to contribute for this ideal. The EvSyM description of the generalized theory of evolution is:

$$Science\ of\ Evolution = Positive\ Feedback\ (d_{loop} + \ell_{loop} + e_{loop}) + Negative\ Feedback\ (d_{loop} + \ell_{loop} + e\_loop) + Feedforward\ (d_{loop} + \ell_{loop} + e\_loop) \quad (8.10)$$

Evolutionary systemics is a theory, like any other, according to which knowledge of any subject evolves through evolutionary transitions. The positive feedback mindset forms our survival machinery, but the negative feedback mindset forms our internal consistency and responsibility, and yet the feedforward mindset is the basis to constitute external consistency and values. The latter two may have a small material basis but they steer the survival machinery for equity, co-operation, altruism and many other equalities.

9. EvSy-modelling Social Sciences

The author has developed a critical view of social sciences by some involvements with the literature on management sciences connecting to psychology and sociology but has no specific experience or training on the subject. However, the brief exposures are sufficient to indicate that these sciences are empirical without having a theoretical depth. It is quite possible that some sense can be brought to the subjects by the wisdom of Table 5. According to this, values of a society are selected through negative feedback loops by the society who also unselects some from time to time.

10. EvSy-modelling Morality

Morality and ethics are another battleground of philosophy, where self-righteous religious authorities would like to dictate the terms and arguably, both are late arrivals, as this section tentatively suggests that morality is a product of natural selection ever since man became communal. In this sense, there may even be some legacy from much deeper pasts. To Socrates, morality was “no small matter, but we ought to live,” (Rachels, 2003, p.ix) but this prescriptive mindset does not conform to a naturalist mindset of morality. The author views morality as a dimension of human culture, and it is the mindset that individuals avoid voluntarily and do not commit harmful activities (say, to steal) against others even when can get away. The act of stealing, for instance, has a short-term selective advantage for the individual concerned but in the long-run, this encourages everyone to steal, which is not mutually beneficial. This section takes a cursory view of morality as it is a very important dimension of human cultures and as expressed by Harris (2009, p.xxix) “Obviously, human evolution has made us more social-beings.” The breadth of this subject by EvSyM would be to model morality through the evolutionary transitions of (i) zero$^+$ feedback for the making of morality from prehistory until the emergence of philosophy; (ii) positive/negative feedback making of morality since the emergence of the settled way of life; (iii) positive feedback d-ℓ-e-loops of philosophized morality, ethics and aesthetics; (iv) an EvSyM vision for morality, ethics and aesthetics.

Of the above breadth of the subject, the role of utilitarianism, as a doctrine in European philosophy, is important and therefore it is outlined here. David Hume (1711–1776) was probably one of the first protagonists of utilitarianism but it was developed by Jeremy Bentham (1748–1832) and John Stuart Mill (1806–1873). These were influential and probably helped to set the cultural mindset for the Industrial Revolution (mid-s 17th-20th century). One version of utilitarianism by John Stuart Mill is stated as:

- Actions are judged right or wrong by their consequences, nothing else matters
- The only measure of consequences is happiness or unhappiness
- Each person's happiness counts the same

The above mindset was a consequentialist but from the vantage of anthropocentric stance. The outcome of such a mindset was to overlook the impacts of the emerging powerful industrial revolution, as discussed in Section 6.3.5, which instigated the production of external entropy and the emergence of adverse environmental effects. Further aspects of morality in relation to AI are discussed by Khatibi et al. (2012), arguing that philosophical traditions contain hidden agenda (presuppositions and predispositions), often driven by "universal truth" and fail to set out core principles on a case-specific basis.

On the basis of EvSyM, the author tentatively suggest that, the cohesion of the society has three fundamental evolutionary transitions based on positive, negative feedback and feedforward d-ℓ-e-loops, giving rise to the system of laws and governance at the dynamic timescale and charities at the long timescale and morality at the evolutionary timescale. The e-loops for morality are sets of behavior codes that a community/society adopts or removes through natural selection at e-loop level. When internal or external entropy is excessive, evolutionary transitions may take place, e.g. the 1953 Universal Human Rights Declaration, or 1992 Earth Summit culminating in the principle of sustainability.

In order to do justice for the subject, a full scale paper is needed to open up the vision on morality based on EvSyM covering such subjects as GM products, euthanasia, risk, as well as looming catastrophes such as climate change etc.

11. Discussion and Vision

It is the statement of the obvious that evolution is general and not unique to the living organisms. However, existing generalization approaches are not crosscutting through all the disciplines. This paper substantiates EvSyM by explaining any existing data/evidence related to any social, cultural or intellectual area of complexity. The ability of EvSyM to transparently explain these diverse evolutionary areas of human endeavors is quite unique. The paper presents a significant amount of evidence to the wider applicability of natural selection but this section catalogues anecdotal evidence for generalized evolution, potential harms to science in its absence and articulates the principles of a vision.

11.1. Interpreting some Anecdotal Evidence

This section articulates anecdotal cases in the form of excerpts from the literature to illustrate a number of observations articulated as propositions. The section then presents a collective context to these anecdotal cases and finally offers a resolution to the conflicting issues.

Proposition 1: Without a generalized theory of evolution dogma is looming around, which is driven by the fixed ontological mindset.

Case 1: Abbott (1991, p87) grants a 'dismal' status to science and views its outlook as "disastrous" thrashing out Carnap, Popper and Habermas; but then he is encouraged to see the rapid arrival of 'hermeneutic' phenomenology glorifying Heidegger and even "reaches beyond" to "Christian Church" by "exercising *credo ut intelligam*" until he is "truly on a limb"; fortunately there is an advice for "merely clever" ones who are not prepared to go under the umbrella of "*credo ut intelligam*" and wish to stay away from "dogmatics." His solution is that "The problem really then comes down to this: that one cannot even begin to talk about the informational revolution and all that is entrained with it, without resorting to dogmatic categories, even though this approach is highly uncongenial to the usual ways of ways of thinking of modern science."

Case 2: Dozier (1992, p.86) implicates Aristotle with the discovery of evolution and DNA stating that: "Two general lines of biological thought have been evolving since the time of the Greeks. Aristotle explored both of them in his treatises ... The first line focuses on the proper way to classify the diversity of living things, the science of taxonomy. The second line focuses on the riddle of reproduction. The first gave rise to the modern theory of evolution. The second resulted in the discovery of DNA and the genetic code."

Case 3: Laird and Roux (2008) regards modern mechanics as inheritance from antiquity by stating that "Modern mechanics was forged in the seventeenth century from materials inherited from Antiquity and transformed in the period from the Middle Ages through to the sixteenth century. These materials were transmitted through a number of textual traditions and within several disciplines and practices, including ancient and medieval natural philosophy, statics, the theory and design of machines, and mathematics."

Case 4: Kuhn (1962, p.2) is concerned with development of science by accumulation, where there are out-of-date theories like: Aristotelian dynamics, phlogistic chemistry, or caloric thermodynamics. He describes the historians' dilemma in two scenarios: (i) should these be labeled as myth? then today's science can be regarded as myth in the future; (i) should they be called science? Then science has included bodies of belief quite incompatible with the ones we hold today. He says that the historians' solution must be the latter but he has a better idea that there are paradigm shifts and the arbiters of the paradigms through the decision of the scientific community. Kuhn (1962, p77) states that "The decision to reject one paradigm is always simultaneously the decision to accept another and the judgment leading to that decision involves the comparison of both paradigms with nature *and* with each other."

Case 5: Von Bertalanffy (http://www.panarchy.org/vonbertalanffy/systems. 1968.html) contrasts Kelvin's degradation with Darwin's evolution stating that "According to the second principle of thermodynamics, the general trend of events in physical nature is towards states of maximum disorder and leveling down of differences, with the so-called heat death of the universe as the final outlook, when all energy is degraded into evenly distributed heat of low temperature, and the world process comes to a stop. In contrast, the living world shows, in embryonic development and in evolution, a transition towards higher order, heterogeneity, and organization. But on the basis of the theory of open systems, the apparent contradiction between entropy and evolution disappears."

Case 6: "Purpose" is intuitive and philosophers often treat it as a pregiven or preordained notion. System scientists and evolutionary scientists rightly ignore top-down philosophical doctrines and articulate their own notion of purpose. Likewise, teleology and vitalism are other unselected notions rooted in Greek philosophy. Checklands (1981, p31) argues that "Aristotle's teleological outlook, in which objects in the world fulfilled their inner nature or purpose, seemed tainted by quite unnecessary speculation. Nevertheless the history of modern biology is the history of the reinstatement of purpose as a respectful intellectual concept." Likewise, Vickers (1981, p21) argues that "That dreaded heresy, "vitalism", can at last be acknowledged to be a special case of a general exhibits laws and principles not to be found at the level below, though bounded by limits which lower-level impose."

Context: The above cases are really related to the same core problem. Case 1 reflects an unhealthy culture when science ignores the manifestation of its external entropy; Cases 2 and 3 may stem from obligation to pay homage to the past giants of knowledge but more likely there is ignorance on unselection of past speculations; Case 4 is a failure to understand the actual mechanism of evolutionary transitions; Case 5 signifies that despite the correctness of von Bertalanffy's statement, the concept of open system is reductive for treating open and closed systems in an exclusive way; Case 6, underlines the fact that emergent property and self-organization are yet to find philosophical interpretations.

Of the above cases, the author is particularly alarmed with postmodernism, which promotes dogma stemming tacitly or explicitly from the ontological mindset breeding on dogmatic self-righteousness. Fixed ontology is the running theme through all of these cases. The issue is that Dozier is unlikely to subscribe to fixed ontology as he proposes a thought provoking theory of unified selection to encompass quantum codes, genetic codes and synaptic codes. His attribution of evolution, DNA and genetic codes to Aristotle signifies the strength of the ontological mindset in the prevailing scientific culture.

Natural selection is not an authority or an institution to select. The term simply means that when the environmental conditions change, the evolvable entity with unsuitable traits will perish as they can no longer cherish in the new environment. This is true on species and equally true on any scientific theory. When Lavoisier's chemistry explained fire, no one simply believed phlogistic chemistry but also an insight was gained why no one should believe in phlogistic chemistry. The Kuhn's presentation above is simply not true, as there are hardly any formal decisions made, and when they are made (e.g. in the use of a particular mathematical model, see Khatibi, 2011b), it is a group selection, which will ultimately fail. It is one thing to compare the Galilean mechanics with Aristotelian and another thing to compare the Newtonian mechanics with that of Galilean. Newtonian mechanics and Lavoisier chemistry are scientific and better explanations, which are used widely; Galilean mechanics is scientific but obsolete; the Aristotelian mechanics and phlogistic chemistry are unselected and not even obsolete. These choices are not by individual decisions but through self-organization, a process that is commonly referred to natural selection.

Resolution: The above cases should not be dismissed out of hand as they have underlying issues. For instance, dogmatism may be dreadful but it offers a certain level of stability, which is much needed in a world where old stability-creating values were unselected by the seemingly modern commercialism. The author was brought up in a vineyard and farming community using the traditional qanat irrigation system but in less than one generation everything was swept away by the encroaching modern commercialism. In a cultural world that even languages fall a victim to racial, national and global aggression, there is a great deal of food for thought to be fearful of the future. The above cases may be anecdotes but are also expressions of external entropies instigated by technology and the mindless science of the past. The irony is that science may have instigated its external entropy but it is science that has to find the resolutions. Arguably, no scientific solution is due without a generalized evolutionary theory and this paper shows EvSyM can be the starting point to address real issues.

The idea of resurrecting purpose, teleology and vitalism is very important and needs to be explained by evolutionary systemics. One must be clear that speculative mindsets of purpose, teleology and vitalism or any other pregiven notions (e.g. anything with 'meta' adjective) are dead and buried, i.e. they are unselected for good. Some of these terms are often used in the discourse of natural language, sanitizing philosophical terms by deactivating their speculative nature. Arguably, it is the discourse of natural language which serves as the basis for terminology and hence their usage in this paper. Likewise, the author has no problem with the pragmatic use of ontology in information systems, although he is aware of the underlying problems with data ontology.

Proposition 2: It is a myth to confine natural selection to living things but this may stem from the myth of the "survival of the fittest" instigated by social Darwinism.

Case 7: Dawkins (http://www.guardian.co.uk/science/2008/feb/09/darwin. dawkins1) states that "Although Darwin's theory can be applied to much beyond the evolution of organic life, I want to counsel against a different sense of Universal Darwinism. This is the uncritical dragging of some garbled version of natural selection into every available field of human discourse, whether it is appropriate or not. Maybe the 'fittest' firms survive in the marketplace of commerce, or the fittest theories survive in the scientific marketplace, but we should at very least be cautious before we get carried away. And of course there was Social Darwinism, culminating in the obscenity of Hitlerism."

Case 8: The infamous expression of "the survival of the fittest," first used by Herbert Spencer and later by Darwin, has become associated with the term natural selection. This myth has entered to the "food chain" of artificial intelligence and is also used in other disciplines.

Context: The running theme in Cases 7 and 8 is the hesitation to generalize theory of evolution. The concern is justifiable if it stems from a precautionary measure against improper application of natural selection, e.g. social Darwinism/eugenics.

Resolution: Failure to formulate an all-out pervasive evolutionary thinking throughout every discipline is an encouragement for the resurrection of past ghosts. It cannot be that evolutionary thinking works in biology (one discipline) but not in others. This paper shows that indeed natural selection can be generalized to the same rigor as in biology. Those who think natural selection only applies to organisms, they hold an aristocratic view of evolutionary biology.

Attention is drawn to the difference between an idea and a concept. An idea is the starting point or a wellspring giving rise to concepts, constructs, notions, theories, theorems, hypotheses, conjectures, systems, etc. Ideas are human treasures and never bad. It is the subsequent conceptualization that can go bad. Thus, when the paper opposes a certain concept and regards them obsolete or unselected, this is not to be extended against their original ideas.

For instance, Marxism was constructed on the high ideal of social justice but erred badly on the solutions by engineering the dictatorship of a class against the inherent entropy (antagonism of classes). The Marxian solution was unselected as it instigated its own downfall by its inherent entropy but the high ideal of social justice must remains intact, as an idea and inspires the author.

11.2. General Use of the Term Natural Selection

The author was encouraged by pervasive occurrence of feedback loops. Forrester (OSG, 1981, p268) states that "The concept of feedback system behaviour applies sweepingly from physical systems through social systems." But it may be asked that is the author using these terms too sweepingly? This paper shows that not only Forrester's statement is justified but there are further potentials in the concept of feedback loops to serve as the architecture for a generalized theory of evolution. But then the author was concerned with stretching the term natural selection to social and intellectual constructs. Although this stretching is done by the virtue of feedback loops, as clearly defined in Sections 1-3, the author's personal experience in publishing his mathematical developments was also helpful and this is discussed below.

As mentioned in Section 4, for more than 10 years, the author has been unable to get the new calculus published, where the readers of this paper will be the judge of its originality. In the last submissions, after nine prolonged months of waiting for the review with few reminders, the single reviewer's report was: "The objective of this paper is to study the sequence of natural numbers from an analytical point of view. The author introduced a series of symbols, definitions and operators, and tried to factorize any synthetic sequence of natural numbers into more basic sequences: one is the sequence of counters (playing the generic role), and the other is the sequence of invariants (playing the specific role)." [This was a simple reflection of abstract signifying that the reviewer had not even read the manuscript.] "The manuscript does not seem to be an original research paper, as there are no substantial results. Overall, I do not think this manuscript should be accepted for publication in *this journal*." [*Italicized words* replace the name of the journal.] This was all that the author received after nine months of waiting.

The above level of comment and the editor's acceptance of the comment are curious to say the least. It is acknowledged that there are caring reviewers and hopefully more than 10 years of bad experience by the author is not common. Nonetheless, it is a reflection of politics and self-righteous characters among the scientific/mathematical community. If science, mathematics and other intellectual disciplines are purely dependent on such individual behaviors, the future of science would be very grim indeed. However, being at the receiving end of such self-righteous determinations, the author reached the tipping point to realize the futility of the d-loops and the need for e-loops and also to see the underlying patterns. The benefit of this realization is Table 1, which is an expression of these patterns learnt from an almost arbitrary world. Societies or institutions solely dependent on individuals (d-loops) will be unstable but those will be stable who have set up the machinery for natural selection with ℓ-loops and e-loops for their positive and negative feedback and feedforward complexes. Even at the communal level the selective advantage of interactions, co-operations, altruism and many other traits have been recognized and selected. Humanity is a construct by natural selection of these novel traits, which also define the essence of humanity but these are threatened by self-righteous acts.

If a particular product is propped up by an authority or an agent, this is a group selection and will succeed as long as the support continues but very likely to collapse as soon as the support ceases to exist, after which natural selection will take its own course. Arguably, natural selection should not be regarded as Mother Nature or Mr. or Mrs. Nature or whatever. The term nature in natural selection conveniently describes a particular condition, as a result of which an aspect of complexity pops out of chaos and gives rise to the growth of simplicity to complexity over many generations. Natural selection is the one to decide on the addition or removal of building blocks. The quality of objective and unbiased selection stems from the attributes of being callous, purposeless, unintelligent, dumb and without foresight and without having any past memory has the virtue of unbiased.

Besides natural selection, there are many other types of selections and those in biology include: sexual selection, kin selection, group selection, dreaded racial selection, artificial selection and population selection. Only natural selection has a lasting effect and can wipe out the outcomes of any other selections at a stroke or in few generations. By the same token, if it was not due to natural selection, the values selected throughout human cultures would have not existed. Arguably, man in the sense of Nietzschean egoism can be the most perilous agent for selection.

11.3. Tentative Vision by EvSyM

Science without a transparent evolutionary theory is now deeply troubled. The reason is that anti-science elements are well organized but scientists remain largely pragmatic, inert and do not mind the shortfalls. Pragmatism worked so far when the external entropy of science was not significant or overwhelmed by the past entropic cultures. This is not the case anymore and any persistence on pragmatism is an expression of naivety. Thanks to science rationality is strong but its external entropic manifestations are endemic: one billion individuals are added to the world population every dozen years; a single person with an electric chainsaw can wreak havoc into a mature jungle which has survived for millions of years; risks are increasing; complexity is increasing at an alarming rate and therefore they may display chaotic or catastrophic behaviors; GM products reflect today's troubled science; the world is inflated with bogus politics, where even racism is being swept under the carpet and languages are driven to extinction. Is then the salvation of science with mysticism or with science itself?

The author has a vision to response to all of the above concerns and this is a rejection of mysticism. This vision is not even based on science but on the science of evolutionary systemics, according to which natural selection operated at the e-loop level is the universal force operated only at the local level; it invents simplicity in the form of a positive feedback e-loop, not one simplicity but a countless number of them; transforms them into complexity by adding/removing through generations and a never ending process; it has mechanisms to maintain its internal and external consistency. The time has come to rethink science and build a moral science. Evolutionary systemics offer appropriate tools for both rethinking science and developing the moral science.

If the maxim of "Know thyself" was inscribed in the forecourt of the Temple of Apollo at Delphi, the message of this paper is Know Evolutionary Systemics.

CONCLUSION

Edelman (1992, p33) remarks: "Ignoring the origins of things is always a risky matter." This risk is still looming over science by failing to formulate a generalized theory of evolution in the wider sciences, philosophy and in other social, cultural and intellectual human endeavors. The outcome is an amorphous growth of knowledge, without much transparency among different areas of knowledge. This encourages the prevalence of a postmodernist mindset, where the Feyerabendian mindset of "anything goes" is a real risk. The paper searches for the fundamental causes and attributes the emerging intellectual wilderness to the absence of a *generalized theory of evolution* in wider sciences.

Rephrasing Timon (see Sec 6.2) the paper holds that everything has to be proved by means of something else (transparency between different disciplines) by virtue of their evolutionary transitions, else all argument are either circular or an endless chain in a chaotic wilderness. Fixed ontology is not a solution as it is analogous to the dictatorship of proletariats, which collapsed in the Eastern blocks for fixing the social order. Now that the wheels of human intellectual endeavors have gained an unstoppable momentum, any fixation is likely to bring mankind to an explosive outcome. Perhaps we are at the brink of such a cataclysmic prospect and therefore the generalized theory of evolution has a serious message. *Evolutionary systemics* is just one theory now, which seems capable of explaining the delicate balance of the modern state of knowledge and against the aristocratic mindset of restricting neo-Darwinism to the organisms.

Like von Bertalanffy's General Theory of Systems, evolutionary systemics synthesizes existing scientific knowledge. An integral part of existing scientific knowledge is the foregone conclusion that theory of evolution is fact. Likewise, systems science is pervasive and brings under its umbrella a whole range of disciplines. Both suffer from unwieldy governance by not being integrated with each other and therefore they are not entirely open-ended. Evolutionary systemics removes the problems associated with both of them to become an open-ended enterprise to overarch any scientific discipline. The key for this is an original contribution by (i) adding zero^{+} feedback loops to the list of the three known loops: positive feedback, negative feedback and feedforward loops; and (ii) breaking down each of the latter loops to three more sub-loops, driven by the time dimension, which are:

- *Dynamic timescale*: Building blocks are assembled together to form positive feedback d-loop, negative feedback d-loop and feedforward d-loop. This is the visible system and its multifaceted operations starts at a time origin, processes are triggered on continuous or discrete time intervals of fractions of a second to hours, days and so on, the process ends. These operations can be accommodated concurrently for multiple stimuli. These make up the common knowledge base of feedback loops but the evolvable entities encapsulate their inner existence, which is visible under different timescales, described below.

- *Long timescale*: this is normally large compared with the dynamic timescale, during which adaptation may take place mostly by reconfiguring the existing building blocks available to the d-loop scale.
- *Evolutionary timescale*: The encapsulated inner system is a blueprint (architecture) driven by natural selection to form hereditary machinery through positive feedback e-loop, negative feedback e-loop and feedforward e-loop. This machinery drives the d-loop complexes. The e-loop machinery at this timescale replenishes diversity and gene pools. The lesson learned from the knowledge gained over the last two-three centuries is that diversity is very important, as it is through this mechanism that the resolution to problems may be found in the long run.
- The paper uses historic, archaeological and mythological accounts as evidence for the implementation of evolutionary systemics and refers to it as *evolutionary systemic modeling*, or EvSyM.

Evolutionary systemics makes use of entropy and exposes its two faces: internal entropy, which is classical but the paper associates it as the force triggering the selective advantage for negative feedback complexes; and external entropy, which is proposed by the author, as the force triggering the selective advantage for feedforward complexes. Owing to negative feedback and feedforward complexes, the role of natural selection becomes rather sophisticated. These loops emerge in sequence in pluralistic way.

The paper presents further aspects of evolutionary systemic in terms of evolution in hierarchy building and proposes the hierarchy of science as a model to understand hierarchical compositions of other areas of complexity. Evolutionary systemics is also applied to explain behavior by reducing it to the dimensions of phase, state, and the hierarchical capacities to act (e.g. decision-making) and currency (e.g. data).

The integration of all the above components of science in evolutionary systemic culminates in the emergence of a mosaic image through EvSy-modelling. This mosaic image emerged in EvSy-modeling of philosophy, science, behavior, evolutionary biology and the philosophy of morality. The emerging images of philosophy and the philosophy of morality are that they are physicalistic entities and lack e-loops and thereby they lack hereditary machinery. Thus, philosophy is personalized or regionalized. The emerging mosaic image of EvSy-modeling of science is that owing to the driving role of negative feedback complexes, science is a human endeavor and no individual, nation or race can have ownership claim over it, even though it emerged from Italy and Europe. This explains why there are multitudes of mutually exclusive philosophical doctrines but science has e-loops operated by natural selection and therefore the new ideas coalesce and evolve undergoing evolutionary transitions. However, social sciences are tentatively seen to be under entropic conditions.

The length of the paper is justifiable due to the search of the above mosaic images, requiring elucidation of the state-of-the-art knowledge, theoretical developments for the integration of the various disciplines of science and EvSy-modeling of a number of human endeavors.

The paper lists a host of destabilizing factors but offers a vision for evolutionary stability by rejecting mysticism and embracing evolutionary systemics. In this vision; natural selection operates at the e-loop level as the universal force but operated only at the local level; whilst self-organization or innovations are responsible for triggering simplicity, natural selection transforms them into countless number of complexity areas by adding/ removing through a never ending process of generations. Evolutionary systemics identifies the forces to maintain the internal and external consistencies. The conclusion of the paper is that time has come to rethink science and build a moral science. Evolutionary systemics offer appropriate tools for both rethinking science and developing the moral science. If the maxim of "Know thyself" was inscribed in the forecourt of the Temple of Apollo at Delphi, the message of this paper is *Know evolutionary systemics.*

It is not rare to see that some researchers still seek Aristotelian solutions to modern problems. Knowing that philosophical doctrines all have speculative basis, this desire to seek Aristotelian solutions is very strange. Aristotle and past thinkers are great sources for inspirational ideas but reincarnating past solutions is just regurgitations. The present generation is better off by seeking innovative and futuristic solutions without using past solutions, which has become unselected or obsolete. Aristotle spent a lot of times to refute his teacher, Plato and therefore he really did not look back. So the question is asked: if Aristotle is resurrected, will he defend his doctrines, e.g. ontology or fixity of species or will he learn to become an evolutionary scientist? It is better to leave this question open ended.

ACKNOWLEDGMENT

I am a hydraulic modeler and far removed from the world of zoology, botany and molecular biology. However, my curiosity led to theory of evolution through my career, the development of new mathematics and philosophy. This paper pools together my thinking in the years since 1995, leading toward a generalized theory of evolution, knowing that I am badly exposed outside hydraulics. Undoubtedly, the concepts presented in the paper can be improved but without academic facilities, this is extremely difficult.

No moral right is given to anyone using the new mathematics in anyway other than for open science.

This paper took approximately 80 man-days in 50 calendar days of intensive focus and non-stop working to write it from the scratch. I am indebted to my wife, Solmaz, to put up with me. I also wish to express my gratitude to Mr. John Little, the president of Swindon Philosophy, for reviewing the segments on ontology and to Prof. Alistair Borthwick (Head of Department, Civil & Environmental Engineering, University College Cork, Cork, Ireland) for his comments on the paper.

I downloaded Figures 10 and 11 from the Internet but cannot find their links anymore for acknowledgement. Also I do not retain some of my past readings materials for reference purposes and apology for those authors that I am unable to make explicit reference; likewise extensive uses were made of books in Azerbaijani that are not cited here but the chief ones include: Cavad Heyet (1996), *A handbook of literature*, Published by the Journal of Varliq; Abasov, E. S., (1999), *Eastern philosophy* (IX-XII centuries), Published by the University of Baku Press; and also Sukurov, A., (2005), *Eastern Philosophy and Philosophers*, Published by Ebilov, Zeynalov and Sons, Baku.

GLOSSARY

Definitions below only stress departures of the terms from classical senses.

Adaptation	Reconfiguration of building blocks without any fundamental restructuring; an outcome associated with ℓ-loops at the long timescale
Anthropologistic	A generalized term reflecting the emergence of culture for each area of complexity as in anthropology
Areas of complexity	Complexity is viewed as a cladistic organization but at this stage, the focus is not at its full branches and hence area of complexity refers to each and all of the branches
	Actions and patterns of actions displayed through time, whereas processes refer to the same activities without a focus on patterns
Biologistic	A generalized term reflecting the emergence of reproduction for each area of complexity as in biology
Building blocks	Abandoning ontology altogether and resorting to the scientific method of reduction, where the scientific entity spirals down from anthropology to chemical hierarchical levels, where the area of complexity cannot be divided anymore
Cladistic	An evolutionary classification technique used in the biological sciences
Chemicalistic	A generalized term reflecting the emergence of cause-and-effect for each area of complexity as in chemistry, rich in the emergent properties of bondage and chemical property
Complex	This term is used when the three d-ℓ-e sub-loops in positive feedback, negative feedback and feedforward are at the focus as they are collectively referred to as d-ℓ-e complexes
Complexity – traditional sense of the term	This refers to entities without any apparent structure, jumbled together or complex systems connected together without any apparent structure
Complexity – evolutionary systemic sense	The absence of hierarchy stems from the physicalistic approaches underlying the thinking and this is a mindset stemming from external entropy of science. This paper suggests that the hierarchy of science can serve as a model of hierarchy to any area of complexity.
Data/Evidence	Data/ evidence serves as the currency for models of behavioral systems with inputs-building blocks-outputs
d-loop	The sub-loop at the dynamic timescale representing real-time activities
Dynamic timescale	Visible activities take place as scenarios, events or incidents, each of which starts at a time origin, processes are triggered on continuous or discrete time intervals of fractions of a second to hours, days and so on, the process ends.
Emergent property	Upon priming a new product from a set of building blocks by a set of rule, the synthesis is either a heap or it may be seen under a new light, which did not exist prior to the synthesis. That new thing is called emergent property and prior to science this was probably taken as the ontological essence.
Entropy, internal and external	A measure of disorder in the system. This paper argues that entropy can arise within the system in the working of the building blocks (internal entropy) or among the individuals of the evolvable entities (external entropy). Internal entropy triggers the selective advantage for the emergence of negative feedback complexes but external entropy for the emergence of feedforward complexes.
Evolvable entity	An evolvable entity emerges if a prototype triggers uncontrolled reproduction and if the prototypes reach a critical mass.
e-loops	The sub-loop at the evolutionary timescale.
Evolutionary timescale	*Evolutionary timescale*: changes take place in many generations at the genotypic level
Evolutionary systemics	The theory of evolutionary systemics is a new theory presented in this paper by integrating systems science/ cybernetics with evolutionary biology.
Evolutionary systemic modeling EvSyM –EvSy-modeling	This is a new modeling capability, where Table 1 and Table 5, as well as Boxes 1.1-1.6 serve as the rational for the modeling of evolutionary transitions.
Feedforward	The connotation of anticipation associated with feedforward loops is stretched in this paper to cover the ability to reduced external entropy

Genotypic	Internally coded inheritable information serving as a blueprint in all living things for reproduction or maintenance
Hierarchy	Hierarchy is a sense of different resolutions in the matter under consideration. This sense often seems vertical or horizontal. The levels of hierarchy tend to form sequences, as opposed to networks, which are located in one level of hierarchy.
Interconnectivity	Connections between the building blocks at a particular level of hierarchy
ℓ-loop	The sub-loop at the long (adaptive) timescale
Long-timescale	This is normally large compared with dynamic timescale, during which adaptation may take place
Mutation	Random faults in gene codes during reproduction or in the building blocks when synthesized by a set of rules
Natural selection and unselection	It is the metaphor for the changing environmental conditions that seems to act in capacity of a selector, where the selector seems to be callous, purposeless, unintelligent and without foresight. Selection takes place at the building block levels and not the individuals of evolvable entities. Just as a building block can be selected, it can also be unselected.
Mindset	e-loops can be referred to as the mindsets.
Negative feedback	A subsystem controlling/regulating positive feedback complex by at least three modules of transmitters, facts engines and actuators, with the prerequisite that the positive feedback complex is flexible to accommodate negative feedback complexes. Thus, it becomes feasible to control outputs by regulating inputs.
Neo-Darwinism	Integration of Darwinian natural selection and Mendelian genetics.
Positive feedback	The bulk of a system comprising the connection of inputs to building blocks and to outputs, where outputs do not influence inputs
Phase	Evolutionary processes operate on particulate phases in generations.
Phenotypic	This is the physical manifestation of the evolvable entity, the sum of everything such as building blocks, structures and the flow of material/energy or data/ information
Physicalistic	A generalized term reflecting the processes that lack hereditary codes, e.g. physics
Pluralism	The four feedback loops can coexist anytime but some zero^{+} feedback complexes tend to destabilize and therefore there are strange attractors and strange repellents in a system and therefore the overall performance of the system is not linear, as chaos and catastrophe is inherent.
Priming	The first act of synthesis of building blocks subject to a set of rules), after which there is a product with an emergent property
Prototype	A prototype is triggered if building blocks are put together by a set of rules triggering emergent properties, if the emergent properties act as a selective advantage, and if the selective advantage triggers a selection
Psychologistic	A generalized term reflecting the emergence of the sense of individuality of an evolvable entity for each area of complexity as in psychology
Purpose	Hierarchical emergent properties characterizing an evolvable entity towards more stable performance
Quantumistic	A generalized term reflecting the order of spontaneity, where if there is a cause, the effect is not obvious and vice versa. Disorder and uncertainty seems to be the main order.
Selective advantage	A primed new product with an emergent property may be such that it can be hooked to other evolvable entities in the environment. If they can be hooked, then this ability is its selective advantage. This term nullifies any presumed role for metaphysics.
Self-replication	Minimalistic and seemingly intrinsic rules under which building blocks produce spontaneously patterns with selective advantages
Sociologistic	A generalized term reflecting the emergence of values among the individuals of evolvable entities for each area of complexity as in sociology
State	Time-dependent behavior in the system.
Sub-loops	This paper divides a feedback/feedforward loop into three sub-loops, as reflected in Table 1.
Timescale	Time resolution
Zero^{+} feedback	This is the very long tailback for the emergence of hereditary machinery of an evolvable entity.

REFERENCES

Abbott, M., (1995), *Hydroinformatics*, Avebury Technical.

Barrow, J., (1991), *Theories of everything*, Vintage.

Bellis, M. (http://inventors.about.com/library/weekly/aacarssteama.htm), accessed on 22 June 2012.

Bernal, J.D. (1954), *Science in history*, Vol. 1, Penguin Books.

Bridgman, P.W. (1926), "Uncertainty and Other Worlds", an article in the book "The Faber Book of Science" edited by John Carey and published by Faber and Faber Ltd, 1995.

Buck, R.C. (1965), "On the logic of general behavior systems theory", in H. Feigel and M. Scriven (eds.), *Minnesota Studies in the Philosophy of Science*, Vol. I, Univer. Of Minnesota Press, Minneapolis, pp. 223-38.

Cardwell, D.S.L. (1957), *The organisation of science in England*, Open University Set Book, Heinemann, London.

Chapman, G. P., (1977), *Human and environmental systems*, Academic Press, New York.

Checkland P.B, (1981) "Science and the systems movement" *systems behavior*, Open University set book, edited by Open Systems Group, 3rd edition.

Lenzer, G., 1997. Auguste Comte and Positivism: The Essential Writings History of Ideas Series. New Brunswick, NJ.

Descartes, R. (1637), Part V *Discourses*, (http://www-history.mcs.st-andrews.ac.uk /Biographies/Descartes.html), accessed in May 2012.

Diller, A., (2002), "Constructing a Comprehensively Anti-Justificationist Position," Jarvie, I., Milford, K., and Miller, D. (2002); Karl Popper, A Centenary Assessment, Selected Papers from Karl Popper, Volume II, Metaphysics and Epistemology, Ashgate.

Dostoevsky, F. (1956), *Notes from under underground*, as collected writings by Kaufmann, W. (1956) *Existentialism from Dostoevsky to Sartre.*

Dreyer, J.L.E. (1953), *A History of Astronomy from Thales to Kepler*, New York, NY: Dover Publications, ISBN0-486-60079-3.

Dugas, R. (1988), *A history of Mechanics*, published by Dovers.

Ferreiro, L.D. (2010), "The Aristotelian Heritage in Early Naval Architecture, From the Venice Arsenal to the French Navy, 1500-1700," THEORIA 68 : 227-241.

Garfinkel, A. (1992), "Reductionism"; in *The philosophy of science*, chapter 20, Pp387-392, ed. Boyd, R., Gasper, P., and Trout, J.D., The MIT Press (1992).

Gatherer, D. (1998), Zygon, Vol. 33, No 2 (http://cogprints.org/4378/1/ zygon.pdf).

Harris, P. (2009), *Toward Human Emergence: A Human Resource Philosophy for the Future.*

Heidegger, M. (1962), *Being and Time*, New York: Harper and Row, 1962.

Heidegger, M., (1977), *The question concerning technology, and other* essays, translation Lovitt W., Harper and Row, New York; cited in Abbott, M., (1995), *Hydroinformatics*,Avebury Technical.

Hoos, R. (1976), "Engineers as analysts of social systems – a critical enquiry," Journal of systems social engineering, Vol. 4, No. 2.

Jasper, K. (1938), *Existenzphilosophie*, as collected writings by Kaufmann, W. (1956) Existentialism from Dostoevsky to Sartre.

Joel, J. (1996) "Basic Engineering Thermodynamics," Longman, 5th edition.

Kast and Rozenzweig (197), *Organization and Management: A systems approach*, McGraw Hill, Inc. with the abridged version in ", *systems behavior*, Open University set book, edited by Open Systems Group, 3rd edition, 1981.

Kenyon, B. And De Greene V. (1970) "Systems and psychology", *systems behavior*, Open University set book, edited by Open Systems Group, 3rd edition, 1981.

Khatibi, R., Lincoln, R., Jackson, D., Surendran, S., Whitlow, C., and Schellekens, J., (2004), "Systemic Data Management for Mathematical Modelling of Environmental Problems", Journal of Management of Environmental Quality; *Vol. 15,* No. 3, Pp. 318-330, (http://www.emeraldinsight.com/journals.htm?articleid=1454620).

Khatibi, R., (2011), "Evolutionary Systemic Modeling for Flood Risk Management Practices," Journal of Hydrology Vol. 401 Issue 1-2, Pp 36–52 (http://dx.doi. org/10.1016/j.jhydrol.2011.02.006).

Khatibi, R., (2012a), "A study of mathematical practices by evolutionary systemic modeling," *Natural Selection: Biological Processes, Theory and Role in Evolution*, e-book published by Nova Publishers.

Khatibi, R. (2012b) "Evolutionary Transitions in Mathematical Modeling Complexity by using Evolutionary Systemic Modeling – Formulating a Vision," *Natural Selection: Biological Processes, Theory and Role in Evolution*, e-book published by Nova Publishers.

Khatibi, R., Surendran, S.S., Everard, M., Ware, J.A. and Wilcox, S.J., (2012), "Vision for Intelligent Agent Capabilities Based on Evolutionary Systemic Rational", *Natural Selection: Biological Processes, Theory and Role in Evolution*, e-book published by Nova Publishers.

Khatibi, R., (unpublished), "Introducing Evolutionary systemic algebra–Evsyal: *Reinventing conventional algebra of sequence of natural numbers with new basic rules, definitions and operators*"

Klein, G.A. (2009), *Streetlights and Shadows: Searching for the Keys to Adaptive Decision Making,* Published by Massachusett Institute of Technology.

Kline M, (1985) *Mathematics for the nonmathematician*. Courier Dover Publications. p. 284.ISBN 0-486-24823-2.

Knight, D.W. (2001), "Conveyance in 1-D River Models" A Scoping study on reducing uncertainty in river flood conveyance, issued by HR Wallingford and the Environment Agency

Kuhn, T.S., (1962), *The structure of scientific revolution*, 3rd edition, The University of Chicago Press.

Laird, W.R., Roux, S. (Eds.), (2008), Mechanics and Natural Philosophy before the Scientific Revolution, Vol. 254; VIII.

Leliavsky, S. (1965), *River and canal hydraulics*, Vol IV, Chapman and Hall Ltd.

Lewis, B. (2002) "The Arabs in History", Oxford University Press.

Lilienfeld, r. (1978), *The rise of systems theory: an ideological analysis*, Wiley, New York.

Leliavsky, S. (1965), *River and canal hydraulics*, Vol IV, Chapman and Hall Ltd.

Lipscomb, (1998), Basics of Cladistic Analysis, http://www.gwu.edu/~ clade/faculty /lipscomb/Cladistics.pdf; access in May 2012.

Losee, J., (1987), *Philosophy of science and historical enquiry*, Clarendon Press, Oxford.

Maurice, P. and Duhem, M. (1980) The Evolution of Mechanics, (date, p2-3): http://books.google.co.uk/books?id=mRvTBgQ-hmECandprintsec=front coverandsource=gbs_ge_summary_randcad=0#v=onepageandqandf=false.

McNeill, J., 2000), Something New under the Sun. W.W. Norton Company

Meddawar, P.B. and Medawar, J.S. (1977),*The life of science*, Wildworld House, London.

Merton, R. K., (1968). Social Theory and Social Structure, revised ed. Collier Macmillan.

Mesarović, M.D. and Takahara, Y., (1975) *General Systems Theory: Mathematical Foundations*, Academic Press.

Milmo, C. (2007) "Fury at DNA pioneer's theory: Africans are less intelligent than Westerners" http://www.independent.co.uk/news/science/fury-at-dna-pioneers-theory-africans-are-less-intelligent-than-westerners-394898.html.

O'Connor, J.J. and E F Robertson (1997), "René Descartes", http://www-history.mcs.st-and.ac.uk/Printonly/Descartes.html (accessed in May 2012).

OSG (1981), Open Systems group, *systems behavior*, Open University set book, edited by Open Systems Group, 3rd edition, 1981.

Planck, M. (1991), "The Second Law of Thermodynamics" pp 163-169 printed in "The World Treasury of Physics, Astronomy and mathematics," edited by Timothy Ferris, Little Brown and Company.

Popper, K., (1976), *Unended Quest*, Glasgow: Fontana, Collins.

Popper, K.R. (2004), *The logic of scientific discovery*, Routeledge.

Rachels,J. (2003) *The Elements of Moral Philosophy,* fourth edition (New York: McGraw-Hill).

Russell, B., (1946), *History of Western Philosophy*, George Allen and Unwin Ltd.

Schlick, M. (1992), *Moritz Schlick, Philosophical Papers, Vol. II* (1925-1936) from *Vienna Circle Collection*, (Boyd, 1992, p37), reprinted in *The Philosophy of Science*, Ed. Richard Boyd, Philip Gasper and J.D. Trout, The MIT Press, 1992.

Shaban M. A. (1971)"Islamic History" Published by the Press Syndicate of the University of Cambridge.

Shames, I. H. (1982),Mechanics of Fluids, 2nd ed. McGraw-Hill, Inc.

Simon, H. (1962), "The architecture of complexity," reprinted in *The Science of Artificial* (1969), MIT Press, Cambridge, Mass.; cited in Checkland P.B, (1981) "Science and the systems movement" *systems behavior*, Open University set book, edited by Open Systems Group, 3rd edition.

Simons, D.B. and Sentürk F. (1992) Sediment Transport Technology: Water and Sediment Dynamics; Water Resources Publication.

Smith, B. (2001) "Objects and their Environments: from Aristotle to Ecological Ontology," *Economic Units* (GISDATA 8), London: Taylor and Francis, 79–97.

Taylor, John R. and Robert E. MacLaury (eds.). 1995. Language and the Cognitive Construal of the World. Berlin and New York: Mouton de Gruyter.

Trout, J.D., (1992), "Reductionism and the Unity of Science"; in *The philosophy of science*, chapter 20, Pp387-392, ed. Boyd, R., Gasper, P., and Trout, J.D., The MIT Press (1992).

Vickers, G. (1980), "Some implications of systems thinking", Pp19-25; OSG (1981), Open Systems group, *systems behavior*, Open University set book, edited by Open Systems Group, 3rd edition, 1981.

Von Bertalanffy, L. (1940) "The Organization Considered as a Physical System"; reprinted in "General Systems Theory", Braziller, New York (1968).

Von Bertalanffy, L. (1962).

Von Bertalanffy, L. (1969) General System Theory: Foundations, Development; George Braziller, N.Y.

Wiener, N.(1948), *Cybernetics or Control and Communication in the Animal and the Machine*, Paris, Hermann et Cie - MIT Press, Cambridge, MA.

Wigner, E., (1991) “The unreasonable effectiveness of mathematics in the natural sciences”, *The world treasury of physics, astronomy and mathematics*, edited by Ferris, T., Published by Little Brown and Company.

Ziman, J (1991), Reliable knowledge, Cambridge University Press.

Ziman (1991, p88).

Zirkle V., (1941) Natural Selection before the "Origin of Species", *Proceedings of the American Philosophical Society* 84 (1), p. 71-123.

In: Natural Selection
Editors: J. R. Lynch and D. T. Williamson
ISBN: 978-1-62257-467-4

Chapter 2

VISION FOR INTELLIGENT AGENT CAPABILITIES BASED ON EVOLUTIONARY SYSTEMIC RATIONAL

***Rahman Khatibi*[1,*]*, Suresh Surendran*[2]*, Mark Everard*[3]*, J. Andrew Ware*[4]* and Steve J. Wilcox*[5]**

[1]Consultant Mathematical Modeller, Swindon, UK
[2]Department of Engineering, Faculty of Advanced Technology, University of Glamorgan, Treforest, UK
[3]Faculty of Applied Sciences, University of the West of England, Frenchay Campus, UK
[4]Department of Computing and Mathematical Sciences, University of Glamorgan, UK
[5]Intelligent Systems Engineering, University of Glamorgan, UK

ABSTRACT

Artificial Intelligence (AI) is in symbiotic relationships with computer systems and includes neural networks, expert systems and agents. Each area of AI is complexity in the sense of having hierarchy and interconnectivity but these are overlooked giving a monolithic or unique sense. Therefore, knowledge gained from one area is not transferrable to new areas. Also, some AI applications involve solving complex over-redundant equations using optimisation techniques, implying architectural dissimilarity with natural intelligence. These problems expose major flaws in current thinking stemming largely from overlooking hierarchy and interconnectivity as a part of possible *missing links* in science and their AI paradigms. This paper invokes possible solutions as a vision suggesting: (i) integrating *evolutionary thinking* with *systemic thinking* expediting the introduction of hierarchy and interconnectivity into AI complexity; (ii) *a new calculus of number sequences,* allowing a mirror image of complexity in the world of integer numbers to possibly removing over-redundancy of the equations and making usage of *bondage* between objects; and (iii) *morality* to discourage the way reductive

* Corresponding Author Tel: +44(0)1793 882414, Email: Rahman_khatibi@yahoo.co.uk

science, oblivious to morality, facilitated uncontrolled exploitation of natural resources. The vision would contribute to target investments on AI-products towards their potential returns by reusing existing capabilities.

Keywords: artificial intelligence, hierarchy, interconnectivity, evolutionary thinking, systemic thinking, new calculus, morality

1. INTRODUCTION

Artificial Intelligence (AI) has challenged imagination by crossing the limits of past mindsets and has triggered great debates. However, there have been concerns that its delivery does not match with early expectations e.g. Yudkowsky, 2006, Crevier (1993), as there are gaps between them. This paper reflects the state-of-the-art on the subject and its peripheral cultures and presents a vision on the premise that there are *missing links* in science and thereby in AI, without which solutions are unlikely to deliver the full return for investments on AI.

The missing links of science suggested by this paper include: (i) the absence of a model of hierarchy in science, applicable to the AI complexity to account for interconnectivity at each hierarchical level; (ii) the absence of a model to understand the AI complexity as an overall control system similar to the brain; (iii) the absence of a generic behavioural model. The vision formulated by this paper offers feasible approaches to each of the missing links, albeit in qualitative terms, by applying evolutionary/systemic thinking (Khatibi, 2011, 2012a) to the AI complexity and to some extent towards explaining intelligence.

The periphery of AI is culturally rich with an age-old debate among philosophers on intelligence and consciousness, which predates AI, already confused due to conflicting views. Neuroscience has resolved some of the conflicts but consciousness and intelligence are still open questions as functional brains are not normal laboratory materials. Owing to the missing links, the *referencing* (i.e. calibration or training) of AI to natural intelligence and vice versa (e.g. Motah, 2006) is not a well-established subject since each seems to operate in different ways. Nilsson (1974, 1998) argues that knowledge about the structure and function of the neuron – or any other basic component of the brain – is irrelevant to AI. This may be true but the synergy between them is overlooked. Evidently, the vision, presented in this paper, is not indifferent to these peripheral cultures and will be duly reviewed.

The subtlety of natural intelligence is that each individual of each species makes sense of its environment in its own way ultimately driven by natural selection. Intelligence is hierarchical with different implications in different contexts covering: (i) *reflex-based actions* induced by instinct displaying basic intelligence; (ii) *causative processes* acquired by learning and trial-and-error and some sort of thinking; (iii) *reactive intelligence* giving rise to innovation, design and proactive initiatives; and (iv) *anticipative intelligence* giving rise to strategic thinking. Similar to the existence of hierarchy and interconnectivity in the natural world, their neural systems comprise many interconnected internal subdivisions to make sense of infinite possibilities. This may be contrasted with AI applications, which are "islands" of mutually exclusive capabilities without any interconnectivity among different brands. Descriptive comparisons of AI with natural intelligence (see Sokolowski, R. 1988, Ratsch, *et al* 1998 and Motah, 2006) do not render a critical understanding of the differences.

The focus of the paper is that all faculties of natural intelligence are systemic and interconnected (or inclusive) but the AI applications are mutually exclusive. Even the ongoing Artificial General Intelligence (AGI) activities seeking “human level intelligence” are just technological solutions. This paper argues that without inclusivity, the AI-products may be intelligent but ought to be “dumb.” Thus, a scientific methodology is required to *reference* AI with natural intelligence to understand interconnectivity and hierarchy.

This paper reviews the challenges associated with studies of intelligence and consciousness and highlights conflicts within conventional schools of thoughts. The basic shortfall is that scientific methodologies have not been generalised to treat such problems and therefore not widely applied outside of traditional science. The paper attributes this shortfall to science being preoccupied with solving problems of physical and social complexity but overlooking its own product of countless “areas of complexity,” examples in AI include: games, diagnosing diseases, theorem proving, language/image understanding and robotics. This paper uses the term complexity to refer to the apparent lack of a structure in the sense of hierarchy and interconnectivity. Each area of complexity appears monolithic with no apparent interconnectivity among different areas and this has resulted in *ad hoc* methods of study specific to each area, for example, areas of AI bear little resemblance with one another and these in turn bear no resemblance to risk management, mathematical modelling or design.It is remarkable that biological systems, driven by natural selection, have given rise to hierarchical complexity over billions of years and it is equally remarkable that man has been able to articulate this hierarchy through science (physics, chemistry, biology, and so on) in the course of the last three centuries. It is not remarkable that science is overlooking the structure within its own products referred to as areas of complexity. This paper attributes the absence of a structure to missing links in science and outlines a model to unearth the structure of hierarchy and interconnectivity within each hierarchy. This model is analogous to the hierarchy in science from *physics*, *chemistry*, *biology*, *psychology*, *sociology* to *anthropology* but beyond their normal definitions, these terms are used in any potential area of complexity as: *physicalistic*, *chemicalistic*, *biologistic*, *psychologistic*, *sociologistic* and *anthropologistic*. Note that *physicalistic-anthropologistic* refers to the full hierarchy.

2. Literature Review and Gaps

The debate on AI/AGI often crosses borders with the wider disciplines including natural intelligence, neuroscience, psychology, cognitive science, sociology, linguistics, philosophy and engineering. This review suffices to some of these disciplines towards highlighting gaps.

2.1. Review of Philosophical Debates on Intelligence

Scientific discourse is driven by data and evidence and undergoes continual refinements and evolutionary transitions.

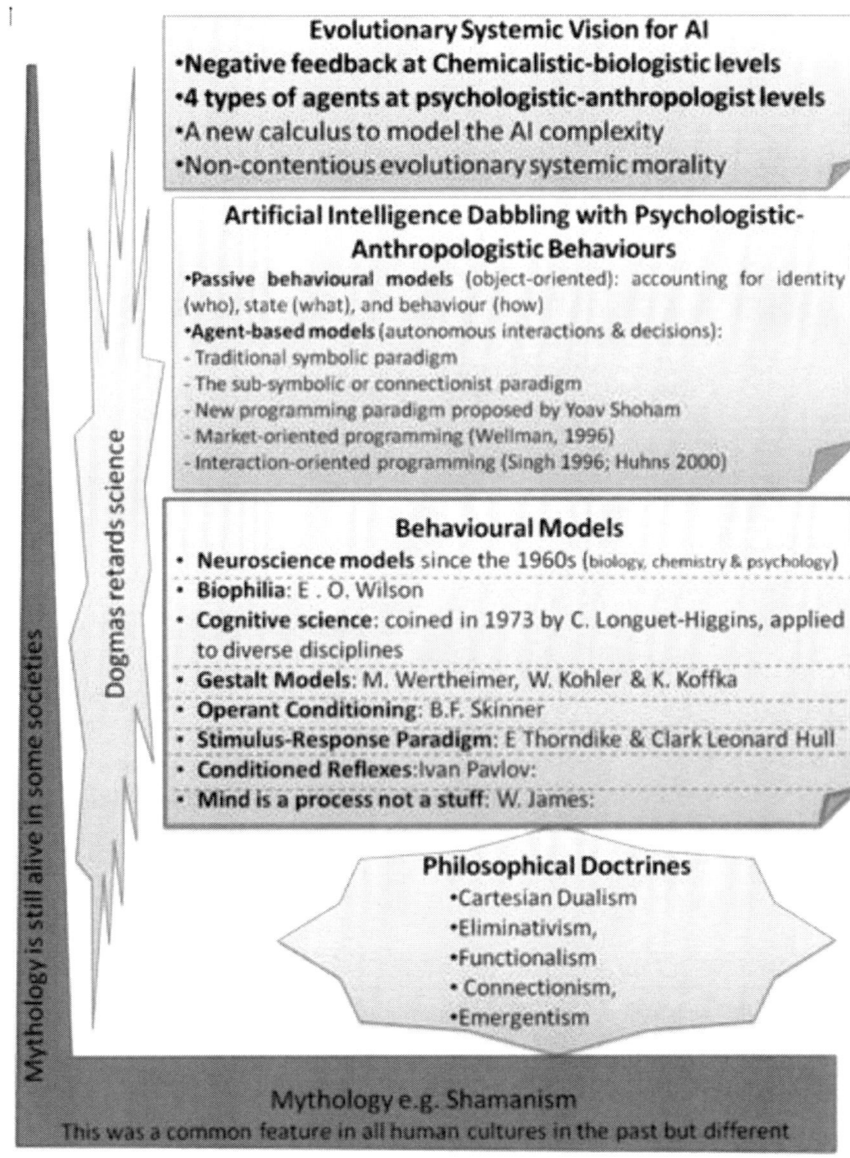

Figure 1. Diverse Streams of Thinking on Explaining Intelligence.

Neuroscience has a determining role as the most direct way to address the processing mechanisms underpinning natural intelligence. However, there are fundamental problems yet to be solved and as remarked by Hundert (1990, p189) and others, neuroscience is still in its infancy. Whilst science stimulates philosophical discourse, philosophy remains speculative and retards science. For instance, the metaphor of *thinking machine* is powerful but without its scientific definition, it encourages unscientific perceptions retarding AI development or sets unrealistic expectations.

The stimulation of the debate on natural intelligence predates that of AI driven by discourses on the philosophy of mind. A brief overview of the subject is given in Appendix I and depicted in Figure 1. Philosophies of mind are not able to offer an analytical approach to consciousness but instead map out a set of ideas or propositions supported by clever arguments. If there are undiscovered aspects about the mind (and this is highly likely), these doctrines will have to start all over again. This also explains why philosophical approaches fall into and out of vogue. Arguably, there are fundamental problems yet to be solved for the development of an analytical solution.

2.2. Review of AI Capabilities

The advent of computers in the 1940s–1950s was a technological breakthrough overcoming the major barrier of slow manual computations or slow information processing, which constricted scientific and technological capabilities prior to 1950.

This selective advantage of the first and early generations of computers was fragile in the sense that they implemented problem-specific programs and the first generation even required hand setting of the programs. In general, these early computer capabilities would not closely resemble modern ones but they transformed the knowledge capital of reductive science of the 19th century into working tools (Khatibi et al., 2003) and opened up new opportunities, e.g. artificial intelligence.

The emergence and subsequent proliferations of AI technologies have not followed any intelligence strategy but have largely been driven by opportunism. Works on machine intelligence are associated with the development of the electronic computer since 1941 and the term AI is the legacy of the Dartmouth conference, USA, in 1956.

The conference provided an impetus for the emergence of a new AI culture, including theories, principles, computer programs, computer languages and dedicated researchers. The lateral influence of the emerging cybernetics and systems approaches also provided some impetus to AI. A comprehensive background on AI is given by many authors, e.g. Jones (2008).

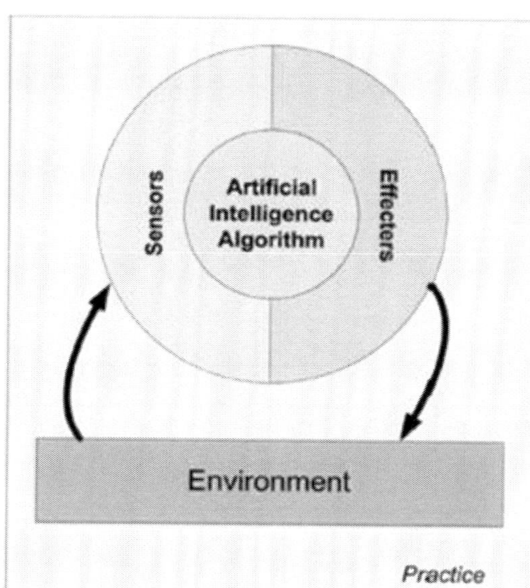

Figure 2. AI as a Version of a Control System, Jones (2008).

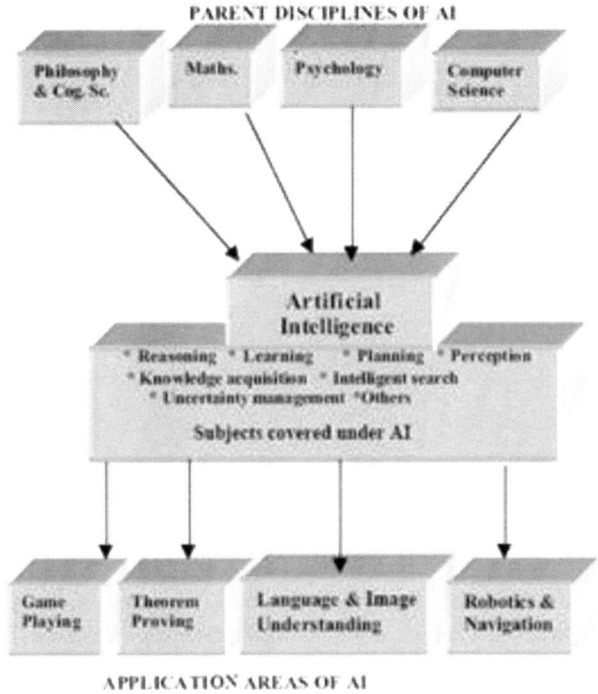

Figure 3. Disciplines of AI (http://www.learnartificialneuralnetworks.com/ ai.html#disciplines).

The *General Problem Solver* (GPS) program was one of the first AI products developed by Herbert Simon, J.C. Shaw, and Allen Newell and was released in 1957.

It was based on the *symbolic paradigm*, as discussed in Section 3.2, which treated any formalized symbolic problem. The basis of this paradigm also served the development of the Logic Theorist (a computer program written in 1955 and 1956 by Allen Newell, Herbert Simon and J. C. Shaw), artificially emulating the problem solving skills of a human being and in time it proved 38 of the first 52 theorems in Whitehead and Russell's *Principia Mathematica*, (Whitehead et al. 1910) and found new and more elegant proofs for some. This led to the LISP language (LISt Processing), developed by McCarthy in 1958 and released as the first AI language, still used today. Figures 2 and 3 show the roles of AI and its capabilities.

Without aiming to describe the various generations of computers or AI, each software feature taken for granted now emerged in the past in a simple form but was selected for a selective advantage through an evolutionary process.

Of these, information systems (IS) are outlined as they are most relevant background to AI. IS includes computer programs, programming languages and software systems – both software architecture and design. Today some 2500 programming languages are in use, many of which have generations and dialects. Their developments did not emerge according to a strategic plan but each facility was developed and selected for a variety of reasons and often as an outcome of exploring niche opportunities. There have been waves of improvements but the presence of many subtle barriers should not be surprising. For instance, the central theme of this paper is that hierarchy and interconnectivity are overlooked in AI and this is a barrier reducing the efficacy of the AI technologies.

The realisation of any AI initiative depends on investments but the outcomes are normally bespoke products and just another opportunistic addition to complexity and without strategy. Alternatively, steps can be taken both to produce the required product and to solve fundamental problems of hierarchy and interconnectivity, as discussed in Sections 5 and 6.

2.3. Background to Agents

Algorithmic languages, such as Fortran developed in 1957, have played a powerful role in transforming the vast array of scientific and mathematical equations into working tools. Although the models based on these languages have been used to simulate the states of the behaviour of complex systems, the complete description of behavioural systems needs to account for physicalistic-anthropologistic processes and effects. AI models do not represent the real system through physicalistic-anthropologistic hierarchies but instead they are a development of another category of computer programs, referred to as object-oriented models, and Agent-Based Models (ABM) to be detailed in Section 3.

Agents introduce capabilities simulating autonomous, proactive, responsive and adaptive acts. According to Guilfoyle (1998, p91) "Intelligent agents sprang into view in the early 1990s. Suddenly the business press and other specialist publications discovered that agents made good stories, and were potentially big business. But varieties of this supposedly new species had been living for years in different habitats." Nwana and Ndumu (1996, p29) attributes the roots of agent-based systems to Carl Hewitt's (1977) concurrent actor model.

The logic, rationale or rules for the mindset of agents are normally referred to as a paradigm, implying the emergence of a common approach among a particular group of practitioners.

For instance, examples of paradigms on ABM include: (i) Traditional symbolic paradigm is based on symbols, as a mathematically oriented way of abstractly describing processes leading to intelligent behaviour by applying information processing metaphor. (ii) The sub-symbolic or connectionist paradigm posits that intelligent behaviour is performed in a non-symbolic way and adopts the behaviourist approaches. (iii) Yoav Shoham has proposed a 'new programming paradigm,' based on a societal view of computation (Shoham, 1990). The key idea is to build computer systems as societies of *agents* and their key features are:

- Agents are autonomous, concurrently executing computer processes;
- Agents are *cognitive systems*, programmed in terms of beliefs, goals, and so on;
- Agents are *reasoning systems*, specified in terms of logic; and
- Agents communicate via *speech acts*.

Weiß (2002) outlines three paradigms for implementing agent systems: *agent-oriented* programming (Shoham, 1993), *market-oriented* programming (Wellman, 1996) and *interaction-oriented* programming (Singh, 1996; Huhns, 2000).

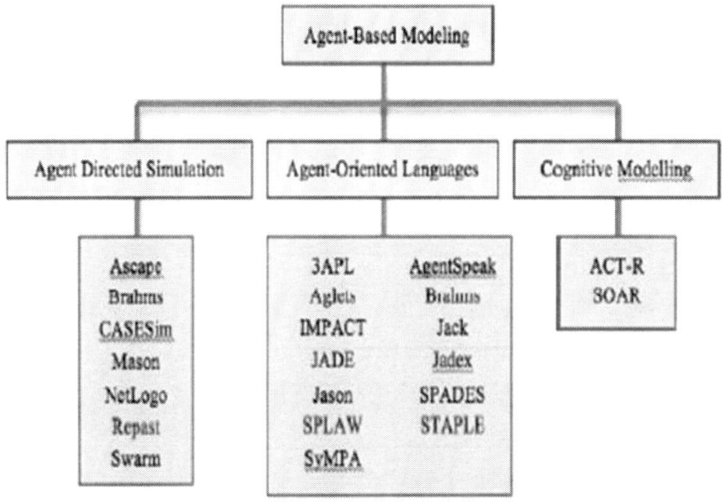

Figure 4. Diversification of Languages used in Agent-based Modelling.

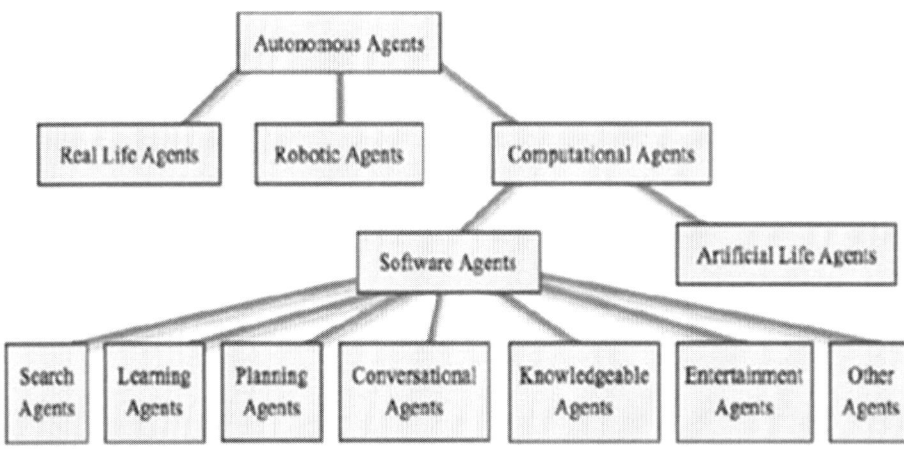

Figure 5. Autonomous Agent (cited on Franklin and Graesser, 1997).

The basic idea behind market-oriented programming is to view, design and implement agent systems according to economic principles of markets and market price systems. Against that, interaction-oriented programming is based on the idea that a designer's and programmer's focus should be on the events and processes occurring between (rather than within) agents.

Agent systems are also diversifying by the computer language used to develop them. For instance, Java and C/C^{++} were used in their development but several prototype languages have also emerged to better realize agent-specific conceptions and as an indicator of underlying challenges. The diversification of languages to produce agent-based models is shown in Figure 4. The application of agents has diversified, as indicated in Figure 5.

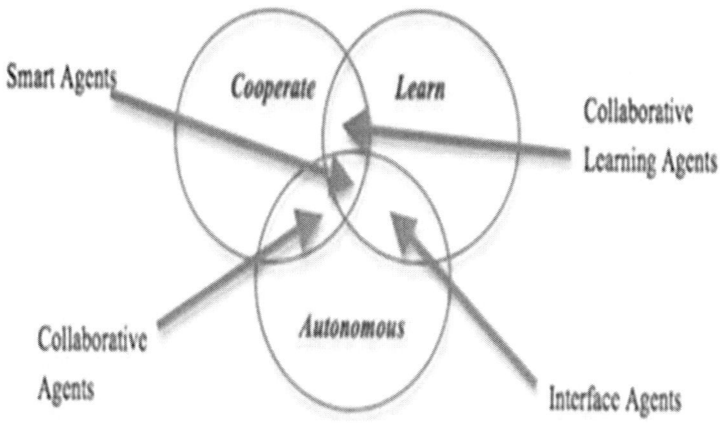

Figure 6. Various Attributes of Agents – Nwana and (Ndumu (1996).

Agents have been classified in various ways and Nwana and Ndumu (1996, p30) argue that they may be classified according to: (i) *mobility*: static or mobile agents depend if they move around their network system; (ii) *deliberative or reactive agents*: these hold internal symbolic reasoning model and engage in planning and negotiation with other agents in order to achieve their goals; (iii) *agents with several attributes*: these agents have at least three attributes of autonomy, learning, and cooperation. Agents with reactive capabilities would be able to take initiatives and smart agents would cope with at least three attributes, see Figure 6; (iv) *role-playing agents*: examples include Internet agents; (v) *hybrid agents.*

The authors did not find evidence for paradigms designed to cater for hierarchies and interconnectivities with the AI complexity and the AI product offering facilities for catering to hierarchies and interconnectivities within psychologistic-anthropologistic hierarchies. This is therefore identified as a gap in existing agent paradigms.

2.4. Evolutionary Systemic Modelling

An analytical approach to the science of complexity is inherent in a set of papers by Khatibi (2011a, 2012a) and a rather older generation of his ideas as in Khatibi (2003) and Khatibi et al (2003) and Khatibi (2006). The core idea is that evolution by the law of natural selection is the architect of natural and social complexity and systemic feedback loops are their architecture, where their integration provides the evolutionary systemic methodology. The rationale of this approach is given by Khatibi (2012a). The metaphors of architect and architecture used here are reminiscent of RNA and DNA but there are considerable differences too, for instance, the coding in RNA/DNA acts largely on the basis of well preserved patterns but natural selection adds favourable features to complexity as a blind architect.

Evolutionary systemic methodology builds on the pluralism of individuals, community and institutions as contexts of evolutionand operates at three dynamic, evolutionary and long timescales. The architecture for evolutionary transitions in institutions or systems can be explained by the four types of feedback complexes: (i) *zero^{+} feedback complexes:* catering for the process of "priming" of an entity towards offering a selective advantage; (ii) *positive* *feedback complexes:* maps the routemap of an entity after its selection, conjugated with diversification, adaptation and entropy, where the latter continually reduces efficacy and triggers a selective advantage for an evolutionary transition to the next loop; (iii) *negative feedback complexes:* this is the addition of a control mechanism to the entity as it was in its positive feedback loop; but after the evolutionary transition it can be trapped forever and hence suffer a loss of efficacy triggering the selective advantage for the next stage of evolutionary transitions; (iv) *feedforward complexes:* to escape the above trap. A complex refers to a system at its dynamic timescale plus its mindset or hereditary codes at the evolutionary timescale (see Khatibi 2012). Notably, as higher complexes are primed, they coexist pluralistically with the lower ones within the broader environment and can give rise to colourful dynamic situations. These complexes are also in evidence in the development of individuals but communities seem to display complex patterns, possibly analogous to the concept of the Rankine's passive and active soil pressures.

The problem with the above is that they are a qualitative account. Although the law of natural selection shows that even qualitative accounts can be very powerful in explaining very complex situations, nonetheless evolutionary systemic modelling can be mathematised by offering a new calculus to explains "complexity-generating objects" (outlined later).

A quantitative evolutionary systemic modelling capability is a missing tool in the science of complexity to be generally applicable even to manmade areas of complexity. It is essentially a building block approach, which explains all the "physicalistic" manifestations of an area of complexity in terms of chemicalistic, biologistic, psychologistic and sociologistic hierarchies. The idea has already been applied to explain qualitatively risk, mathematics and mathematical modelling by Khatibi (2011, 2012b, 2012c). This paper presents its application to AI and further work will address other aspects of manmade complexity.

2.5.Gap Analysis

This review identifies diversification in the various areas of the AI complexity and paper attributes it to the role of opportunism, where complexity is escalated without advancing science. Thus, AI capabilities are widespread but their implementations are extremely cumbersome. Jennings (2000) argues that there are three shortfalls in agent-based modelling: (i) the basic building blocks are too fine grained; (ii) the interactions are too rigidly defined; or (iii) insufficient mechanisms are available for dealing with organisational structure. Janca and Gilbert (1996) are explicit on some of the limitations of agent capabilities. Such systems apply: (i) to only a single application niche and therefore no currently available commercial agent system can work in more than one such application; (ii) to every system in which agent capability is built from zero, no building block approach is used; and (iii) "if I want to add agents to an existing mail system today, tough luck. I must get rid of the system in favour of a new mail system which includes agents."

The paper concurs with the above and recast them in the context of complexity in the following terms: (i) *monolithic products:* AI products are monolithic in the sense that each capability is built from zero for each application and without transferring relevant building blocks from other areas of application; (ii) *closed architecture*: each AI product is only applicable to a single niche and is bespoke; (iii) *absence of an insight and foresight:* each AI application is developed without reducing the underling complexity into hierarchies, and without identifying building blocks (or modularity) and interconnectivity among them at each hierarchy, all of which are prerequisite for the emergence of open architecture. Thus, AI neglects chemicalistic structures but instead focuses only on physicalistic manifestations.

The paper also presents a glimpse of discourses on natural intelligence, consciousness and AI. Their philosophical accounts are infested with contentions driven by presuppositions for the universal truth but the experience defies consistently such a quest. Also, their scientific account based on the review of agent-based modelling research works shows a procedure for retrofitting a modelling capability to a behavioural situation. Groundwork is yet to be carried out on transforming a particular problem to hierarchies, building blocks, feedback loops, and identify the roles of individuals, communities and institutions, map out the roles of selection, proliferation, diversification, adaptation, entropy, control, anticipation and pluralism and their subsequent dynamic behaviours. These will be captured in this paper by suggesting the application of the evolutionary systemic rationale to formulating scenarios.

3. Behavioural Modelling Capabilities

Behavioural models may be posed in two contexts: (i) scientific models applied to psychological, sociological and anthropological problems; (ii) AI models, which often overlook hierarchies and interconnectivities in physicalistic-anthropologistic hierarchies of the real system and replace them with a set of program facilities. There are two categories of AI behavioural models: (i) object-oriented behavioural models (OOBM) and (ii) agent-oriented behavioural models or agent-based models (ABM), catered for by *agents*, both of which are reviewed below. Notably, the concept of agent plays a key role in Minsky's (1986) *The Society of Mind.*

Generally, ABMs require knowledge of the state of the agents and this is derived from either *distributed models* such as those based on laws of nature/ empirical rules or *local models* such as those based on time series analysis. These models have also their equivalence in the study of the overall behaviour of human beings. For example, the term neuroscience is a branch of biology but now it represents a very complex interdisciplinary science to explain both human behaviours and the working of the brain/mind. Neuroscience is intrusive, which constrains the development of models hence their implementation needs dedicated resources. Local models are normally based on externally measured data, without process descriptions. Psychological models of behaviour are local models that study behaviour in terms of various measurements without involving the brain processes. Psychology emerged in the second half of the 19th century following the works of Wilhelm Wundt (1832–1920) and William James (1842–1910) and according to Edelman (1989, p37) the latter was the first to point out that consciousness is a process and not a substance, thus refuting Descartes' 17th century philosophy of mind. Figure 1 maps out models according to their resolutions.

3.1. Object-Oriented Behavioural Models

Object-oriented behavioural models are a category of computer programs, such as the Unified Modelling Language (UML), which provides facilities for behavioural models, more details may be obtained from textbooks, e.g. Schmuller (2002). Basically, passive behavioural models, incapable of spontaneous acts, are a way of thinking to reformulate an area of complexity in terms of computer facilities available, possibly in terms of identity (who – class models), state (what – functional models), and behaviour (how, if invoked or passive behavioural models).

Class models account for static structure of the classes within a given system in terms of identity, relationships, attributes and methods. As discussed by McNeile and Simons (2004), the activities relating to a class model are commonly known as structural modelling and they work in terms of *class diagrams*. Functional models refer to the computational structure of an object-oriented model using activity diagrams and Use Cases (a collection of scenarios describing the usage of a system). Behavioural models describe the control structure of a system by sequence of operations, identifying the states of objects or their interactions and therefore these models are dynamical. McNeile and Simons (2004) also refer to other types of behavioural models.

3.2. Review of Agent-Based Behavioural Modelling

Dawson *et al.* (2011) developed a proof-of-concept agent-based model for a successful operation of complex, interacting human and technological systems under incident management scenarios. Their work is summarised here for convenience to reflect the various facets behind the development of such AI models.

Human systems may be regarded as reflexive due to their adaptive nature. Thus, the challenges include modelling spatial and temporal behavioural variations; accounting for small perturbations when they can cause a large knock-on effect (as in nonlinear dynamic behaviours); and simulating interactive individuals and organisations with conflicting values and objectives and with many feedbacks inducing dampening and/or amplification effects. Agent-based modelling can support flood risk management capabilities in which agents (i) represent explicit interactions and feedback between floods and human responses by using dynamic updating of simulation engines; and (ii) cater for uncertainties about human behaviour and incomplete data.

Individual agents need to be capable of autonomous interactions and decision-making by (i) assessing the effects of the agents on the system as a whole and (ii) assessing its situation for making decisions on the basis of a set of rules. Agents may execute various behaviours appropriate for the system components they represent. At the simplest level, an agent-based model consists of a system of agents and the relationships between them.

Even a simple agent-based model can exhibit complex behaviour patterns because a series of simple interactions between individuals may result in more complex system-scale outcomes that could not have been predicted just by aggregating individual agent behaviours. One language used for developing agent-based model is NetLogo and the architecture of the system developed by Dawson et al. (2011) is given in Figure 7.

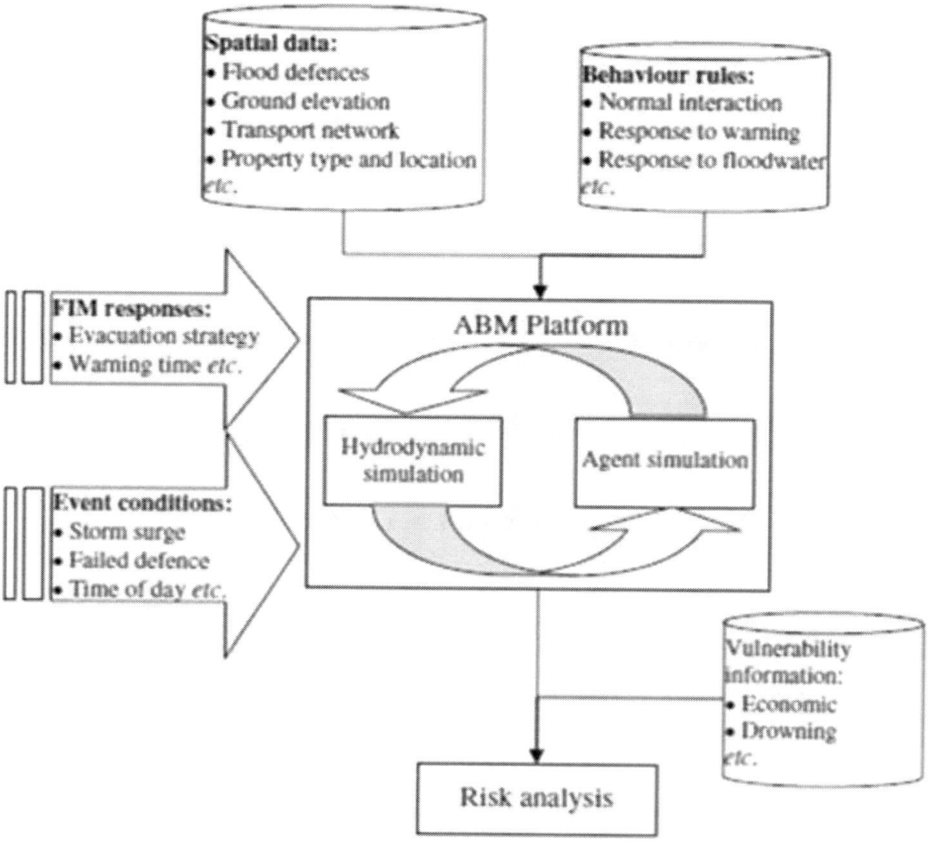

Figure 7. An Implementation of Agent-based Modelling System, Dawson *et al.* (2011).

Developing an ABM by Dawson *et al.* (2011) was their core development, interacting with the following components:

State Modelling: hydrodynamic models simulating flood waves using modelling engines and interacting directly with the agents and the built environment.

Spatial Modelling: describing features such as the flood protection structures, property types and their locations, floodplain elevations and the transport network.

Behaviour Modelling: A suite of behaviour rules defining the behaviour of agents and flood incident organisations.

Scenario manager: this, developed in the Python programming language, facilitates batch running of multiple simulations and sets up the model to test different types including: exploring the impacts of different storms, locations of flood defence failure and flood incident responses such as the amount of warning time given to people.

Damage functions: it describes, for example, the relationship between loss of life or some other measures of harm and the properties of the flood to quantify flood risk.

There are two components to determining agent behaviour: (i) determining the state of the agent and its behaviour under normal conditions at the beginning of an event; and (ii) the agent's respond to a flood incident thereafter.

Each agent is provided with a probabilistic finite state machine to describe its possible states, actions to take and the transitions between states. Agents are therefore able to make decisions and act autonomously by accessing and using considerable information such as the probability of being in a given state (e.g. home or work), as parameterised by using data from the National Travel Survey (ONS and DfT 2008) which gives:

- sample diaries of travel patterns for households
- proportion of journeys at each hour of the day
- proportion of journeys for a purpose (commuting, business, shopping, recreation)
- proportion of journeys driven by gender, household size, age and employment status

3.3. Gap Analysis

The above reviews reflect the state-of-the-art on behavioural modelling using AI and the current issues on agent-based modelling. Evidently, hierarchy and interconnectivity within each hierarchy is not influencing the ongoing thinking.

This paper is critical of overlooking hierarchy and interconnectivity and Section 5 presents a model to solve the problem.

4. Artificial Intelligence Appraised by Evolutionary Systemic

Science makes sense of the world by well established concepts of statics and dynamics.

Prior to the emergence of science, the prototype of these concepts in philosophy were known as ontology (what exists, beings, often regarded as permanently fixed) and epistemology (what can be known). Biology discovered the evolutionary timescale, where complexity increases in a very slow rate of change through the evolutionary mindset (hereditary codes).

However, the success of evolutionary thinking is yet to penetrate the wider sciences, where the dynamic and evolutionary scales of time are yet to be integrated. Without their integration, scientific thinking is blinkered and one is bound to suffer from *tunnel vision*, for which the only escape is through evolutionary thinking.

The rationale for evolutionary systemic modelling is presented by Khatibi (2012a), which is applied now to explain AI as an area of complexity. The study from the roots is necessary to eliminate blinkered thinking by showing the culture before an evolutionary transition was rich enough to generate the building block for the transition.

4.1. Zero$^+$ Feedback Stage of AI

The roots of AI go back to prehistory and these cannot be overlooked when presenting a vision on AI. Thus, the background information is outlined in Appendix II. However, as Clark (1994) states: "Until the rise of a more critical approach in the final decades of the eighteenth century people were happy to accept their history and the world they knew as the work of God without seeking to account for them in rational terms." So the past world was static (ontological) and not cumulative as they are now. There was not much evidence or critical thinking and any features selected in the makeup of past cultures were conserved rather zealously, which were often derived from coercion, perception, shaky interpretations or faith.

By the emergence of AI in the 20th century, a great deal had changed, as the societies were implicated by science, evidence, data and critical thinking were parts-and-parcels of most of the cultures. Building blocks of AI include a culture rife for AI, demand for intelligent approaches, mechanical/electronic components, control components and rules or logic. These were already part of the culture prior to the emergence of AI, as discussed in Appendix II.

4.2. Positive Feedback Stage

By 1940, control engineering/cybernetics had also become mature disciplines of science, each growing out of a hazy past. Furthermore, the role and rules of logic were also topical then emerging out of the past mindsets. These are not reviewed here but it is possible to trace them to their mythological origins at their $zero^{+}$ and positive feedback stages. However, each of these served as components to prototype AI products. The selective advantage for an AI prototype emerged during the WWII due to the need for code breaking. Electronic approaches became feasible to code breakers at Bletchley Park, England, using an *Enigma machine* (invented previously by German engineer Arthur Scherbius) operated by electromechanical rotor ciphers. By the 1950s, AI prototypes had emerged but were without an identity and this was realised in 1956 at the *Dartmouth Conference*.

Many predicted that a machine as intelligent as a human being would exist in no more than a generation, but this vision did not come true even with large research projects. Over the years, AI fortunes have gone through booms and busts (for more details, see Jones, 2008). This outcome is a characteristic feature of positive feedback loops. The gap analysis on agents outlined in Sections 2.5 and 3.3 reflects chasms in AI, which cannot be bridged. The comparison below with natural intelligence highlights the problem.

When AI is compared with natural intelligence, there are definite differences, as natural intelligence pertains to an interconnected faculty with different functional areas and capabilities, but this architecture is absent in AI. The scientific knowledge of natural intelligence increasingly shows that the functional areas in the brain are communicating with one another. The outcome is that all the higher species make sense of their environment, ostensibly, proportional to their faculties. The process of making sense is: (i) hierarchical, (ii) systemic by the virtue of feedback loops, and (iii) it is not automatic for each individual in each species to acquire the loops but a matter of development and learning. For an observer studying species, the hierarchies may be described as follows:

Physicalistic hierarchy focuses on a species with respect to such external properties as shape, colour, size, weight, intake, output, waste, interactions with the environment.

Chemicalistic hierarchy focuses on the chemical compounds from some 92 basic elements used in the makeup of a species and their functions and interactions with one another; within this hierarchy, the species are inanimate and may be considered as a *closed system* in the sense of the General Systems Theory (GST) given by von Bertalanffy (1956).

Biologistic hierarchy focuses on species in terms of reproducible individuals, each individual as a collection of interoperating cells oriented towards the common goal of survival; thus, they would all have a selection of negative feedback loops to reduce their internal entropy but at this level, there is no awareness of the environment.

Psychologistic, sociologistic and *anthropologistic* hierarchies involve the selection of more negative feedback loops by natural selection to provide selective advantages: for coping better with its environment (psychologistic traits) by an individual; for interacting individuals (sociologistic values), and for interacting societies (anthropologistic cultures). Species with hierarchies above that of biologistic can be considered as *open systems* in the sense of GST. These hierarchies should not imply that the brain is also organised hierarchically but can be trained to discern between traits of the psychologistic hierarchy and the value-laden sociologistic hierarchy.

It is not difficult to see that natural intelligence builds on inner and outer worlds, which are not the same as dualism. When an individual is subjected to significant changes, the inner world (chemicalistic-biologistic hierarchies) has the faculty of the brain, which provides active control (negative feedback loops) to coordinate the individual to function according to the needs of the environment or other stimuli. The workings of the brain have not been properly understood but it is generally agreed that it is a highly interconnected body of functional areas.

AI is a world on its own, currently subservient to human interests. Its developments are only focussed to a very specific area of human interests, where human activities are scrutinised at their physicalistic level by "if-then" rules. In current practices of AI, there is no interconnectivity among these AI rules from one AI capability to another and therefore each AI application area appears as an island. This is one way of saying that AI ventures have instigated entropy and if this carries on, economical investments may not render their due returns. Also, knowledge mining scenarios for coding behaviours are through "what-if" rules and overlook inherent hierarchies and interconnectivities. This makes AI susceptible to ever more sophisticated rules and susceptible to severe inaccuracies.

4.3. Negative Feedback Stage

AI capabilities aim to decipher the working of man, explain it by a set of rules and build in learning algorithms to respond to changes in these. Although learning is through appropriate negative feedback loops in AI/AGI, these are the loops of the learning process as invented by AI/AGI but not as an emulation of those loops in human actors. As soon as the circumstances change slightly, the AI learning process has to restart all over again. True negative feedback loops have to build on building blocks of a lower hierarchy.

4.4. Feedforward Stage

Nwana and Ndumu (1996, p30) state that "Truly smart agents do not yet exist." This is because over the past few decades the majority of AI researchers have focused on "narrow AI" in terms of producing AI systems displaying intelligence regarding specific, highly constrained tasks. Over recent years, however, there has been a resurgence of interests in developing "human level intelligence" and more broadly *Artificial General Intelligence* (AGI). This resurgence is by the virtue of improved computer speeds, but without the necessary conceptual leaps. Without conceptual developments, any progress is likely to be a wave in the boom-and-bust cycles. This paper argues that the problem of complexity defined in terms of hierarchy, interconnectivity among building block is the foremost problem of science in general and of AI/AGI in particular, without which AI/AGI is unlikely to emerge.

5. An Evolutionary Systemic Vision for AI

Science offers a potential model of hierarchy in terms of vertical steps from subatomic physics to anthropologic levels, where all of its premises use data/evidence to disaggregate natural/social complexity. The building of this knowledge base is a process, which started in the 17th century involved both unselecting past mindsets and a journey of discovery. Khatibi (2012a) models evolutionary transitions of science as: (i) since its emergence in the 17th century until 1950, science was reductive and the theory of evolution was one of its pinnacles; (ii) systems science emerged in 1950 by acquiring goal-seeking feedforward loops around reductive science with systems science as one of its pinnacles; and (iii) more recently science embodied the principle of sustainability as its foresight. Since its inception, science has also instigated countless new *areas of complexity* as by-products. Thus, life is no longer an outcome of natural/social complexity but is also affected by wider complexity.

As extant species and their intelligence have been selected by natural processes, there is no reason to regard the deciphering of intelligence to lie beyond human understanding, though current methodological and cultural problems mean that its resolution may take considerable time. Notably, evolutionary processes, themselves not intelligent, have given rise to intelligent organisms through a slow and ostensibly purposeless process of selection. The promise to explain intelligence is undertaken by neuroscience as a body of converging empirical approaches and knowledge. Neuroscience has not yet converged into an analytical framework and hence intelligent man is not yet near to deciphering this intelligence.

AI/AGI is in the grip of "physicalistic thinking" in the sense that it overlooks both hierarchy and interconnectivity and strips away all hierarchical features and lumps the remaining together. The paper suggests that this is not because of ignorance but possibly, there are *missing links* in the science of complexity, as follows: (i) *the absence of a model of hierarchy*: the hierarchical structure of science is used here to serve as a model of hierarchy for the AI complexity; (ii) *absence of a model of interconnectivity:* interconnectivities of the brain functional areas can serve as a model to understand the AI complexity at the chemicalistic level and the diversity of the brain among the species can serve to find a model for the diversification of the AI complexity at the biologistic level; and (iii) *absence of characterising agents:* the four feedback loops are suggested as a model to study interconnectivity at psychologistic-anthropologistic levels; and (iv) *absence of a quantitative model:* the new mathematics by Khatibi (2012a) suggests a possible approach to account for interconnectivity at each hierarchical level.

These four aspects are discussed below towards forming a vision for AI/AGI based on the evolutionary systemic rationale.

5.1. Modeling Hierarchical Architecture of AI Complexity

The first set of possible missing links in the science of hierarchy is the lack of a model for hierarchy. Khatibi (d, unpublished paper) discusses the hierarchies in science, as shown in Table 1.

Table 1. Hierarchical Structure of Science

	Scientific Disciplines	Emergent Properties	Resolution	Framing Theories
1	Subatomic Physics	The principle of uncertainty, Matter and energy	Below Å (1×10^{-10}m)	Quantum mechanics
2	Physics: studying external properties of matter/energy	Properties, Phase State	*Ad hoc* scales	Various laws of nature Empirical approaches Pertains to closed systems
3	Chemistry: Atoms, molecule and compound	Compounds have emergent properties	Atomic scale	Periodic Table Pertains to closed systems
4	Biology	Reproduction, Survival of the individual	Individuals of species, Transparency at food web	Cladistics, Pertains to open systems
5	Psychology	Individuals make sense of their environments, Traits	Individuals	childhood, young middle age old age
6	Sociology: interacting individuals of each species	Morality Values	-Individuals, -Communities, - Institutions, (human beings)	-Exclusion of species -Kinship -Relationships -Community, society
7	Anthropology: Study of cultures	- Ethics - Interacting societies - Politics - Economics	-Clans -Tribes -Nations, -Countries	Position of strength Ethics, law, governance United Nations Principle of Sustainability

This model is concerned with the dynamic natural and social complexity operating at various resolutions. Notably, science at the level of chemistry makes sense of diverse compounds of matter by understanding interconnectivity among the elements as captured through the Periodic Table; likewise, the interconnectivity among all the species is captured by the cladistics at the biological level. Notably, both the Periodic Table and the cladistics are two pinnacles of science, the equivalents of which are absent in psychological-anthropological levels. At these higher hierarchies, science is largely empirical and lacks a framing theory (a paradigm) for understanding inherent interconnectivities at each of these levels. This shortfall of science is even more pronounced at each area of complexity, e.g. AI.

It is a bold step to assume that both hierarchy and interconnectivity prevail in each area of complexity but the authors are not the first on this assumption. Checkland (1981, Chap 3) argues that "The idea that the architecture of complexity is hierarchical and that different languages of description are required at different levels has in recent years led to a developing interest in hierarchy theory as such. ... the structure of the planet we inhabit could only be hierarchical." It should be noted that the various forms of hierarchies at the sociologistic level customary in societies with class, caste, racism or sexism are normally unjust formations, selected by man in some societies. The authors do not support such social constructs and hold them not be relevant to the concept of hierarchy discussed here.

5.2. Interconnectivity at Chemicalistic-Biologistic Hierarchies

The second set of possible missing links is an apparent absence of the building blocks to define the AI complexity at its chemicalistic and biologistic levels.

Obviously, atoms and species are more appropriate to chemistry and biology and not meaningful at the chemicalistic and biologistic level of the AI complexity. Thus, in every chemicalistic and biologistic levels of complexity, the building blocks have to be identified, e.g. neurons in natural intelligence. Since the brain is a control mechanism using fuzzy logic as its fact engine, AI can also be regarded as a control/regulatory system with a set of computational engines or information processing engines serving the function of fuzzy logic.

It is understood that the collective workings of hundreds of brain functional areas constitute the intelligence of each species. This consideration invokes the thought that perhaps modular knowledge bases are needed to make sense of our environment in an artificial environment. This is not practiced at the present, as the common approach is to rush to the drawing board for each area of complexity in isolation to reinvent the wheel.

The brain also makes sense of the environment at psychologistic, sociologistic and anthropologistic levels through a whole range of sensors, fuzzy logic and actuators, as well as modulating the actions continuously by a diverse range of negative feedback loops. Table 2 compares attributes of natural intelligence with those of conventional and Evolutionary Systemic AI.

Thus, an understanding of natural intelligence is of the paramount importance to be emulated in AI. Instead, conventional AI/AGI churns out unique products, without establishing their interconnectivities and hierarchy. The paper just expresses an optimistic opinion that it is not impossible to modularise AI to emulate natural intelligence by establishing interconnectivities and hierarchies.

6. Interconnectivity at Psychologistic-Anthropologistic Levels

The third possible missing link is a model of interconnectivity to study AI at psychologistic, sociologistic and anthropologistic hierarchical levels. This section suggests a possible model to study intelligence of social systems towards applying it to the AI complexity.

6.1.Interconnectivity at Psychology – Anthropology Levels

The paper suggests that (i) the nature of intelligence itself depends on each of hierarchical levels and (ii) feedback loops can serve as a frameing theory at each of the levels to study intelligence as follows: (a) reflex-based actions associated with zero^{+} feedback loops, (b) causative processes (based on cause-and-effect rules) associated with positive feedback loops; (c) reactive intelligence driven by negative feedback loops; and (d) anticipative intelligence driven by feedforward loops. Some of attributes of the hierarchy are given in Table 3 and they are further discussed below.

Reflex-based actions: Reflex-based capabilities stem from basic needs to cope with the environment and as such, they are normally inherited by the species, as a whole. These capabilities often lead to safe strategies at the expense of opportunities, e.g. running away from danger as the expense of missing out any opportunity therein.

Causative processes: Knowledge retention is through passive community instruments and cultures. Inherent discrepancies actively drive diversity through trial and error procedure. These are all natural processes involving opportunism or some sort of thinking but the process is not intelligent.

Reactive intelligence: The individuals of the species in the animal kingdom capable of reactive intelligence (specially the individuals of human beings) develop their reactive intelligence on social and cultural matters on their own and enrich their culture as if reactive intelligence is passed on through generations. At the sociologistic level, the survival of the primed negative feedback loops is reinforced by a package of training or education to pass it on culturally through generations. The argument put forward by the paper is not Lamarckian, as Jean-Baptiste Lamarck (1744–1829) held that characteristics acquired by an organism passed on during its lifetime to its offspring (see Rice, 2007, p132).

In the approach outlined in Table 3, intelligence becomes a tangible emergent property and definable in terms of negative feedback and feedforward loops. Brain structure is now known to be an interconnected network of neural cells in their functional areas and not hierarchical. A hierarchy of consciousness seemingly emerges from their collective working.

6.2. Psychologistic–Anthropologistic Levels Modeling Scenarios

The above ideas on natural intelligence are transferrable to AI and the focus here is on agent-based modelling emulating behavioural systems. Thus, agents can be developed to emulate human behavioural (or animal behavioural) systems through their actions in the manner of reflex-based, causative, reactive and anticipative intelligence.

Without these capabilities, intelligent software systems are not being referenced to cultural systems driven by the mind and are not defensible, i.e. they are sophisticated tools following their program but not necessarily emulating human behavioural systems.

A systematic way to study human behaviours is through formulating scenarios and such methodologies have emerged on a pragmatic basis and supported by empirical observations. However, they are not systematic and their rationale remains unexplained. The challenge is to devise a systematic approach with a pool of scenarios and select intelligently from the pool based on the specific characteristic of a particular study. The following is suggested by Khatibi (d) based on the four feedback loops as follows:

- Establish baseline models without any change
- Identify areas of complexity relevant to a particular study

- Identify hierarchies appropriate to each area of complexity
- Identify building blocks at each hierarchical level, at which the four feedback loops would apply (although the aim is not to catalogue them), as outlined in Table 4.

Note that negative feedback loops are only possible through a subsystem composed of at least four components of (i) transmitters measuring the state of the main system; (ii) a fact engine deciding on the amount of change to maintain the main system within the prescribed set-points; (iii) actuators to modulate the changes at the interface; and (iv) a flexible system that can accommodate changes.

6.3. A Possible Alternative to ANNs

The fourth set of possible missing links identified by the paper is the absence of a quantitative approach to account for hierarchy and interconnectivity. It is intriguing such a mathematical capability to be absent in wider sciences to account for interconnectivity and hierarchy. This paper attributes this to the absence of an appropriate mathematical tool on the subject, which may be remedied by a new analytical calculus developed by Khatibi (unpublished) but outlined in this section.

In general, the paper invokes the idea that the current approaches to the mathematical formulation of neural networks are possibly wrong-footed but a different approach is feasible based on a new calculus by Khatibi (unpublished manuscript), although this calculus is still to be fully presented. For instance, the working of the brain has been emulated in Artificial Neural Networks (ANN) but neurons in AI are treated as discrete individual neurons connected to one another, each associated with a weighting and therefore their interactions or outputs are accounted for by sophisticated systems of equations and optimisation techniques (for one such textbook see, Hassoun 1995, pp 57-133). This is plausible for AI products but it is doubtful (and probably almost certain) that Dawkins' (1986) 'Blind Watchmaker' is capable of following such mathematics. So is there a possible alternative approach? The new calculus suggests clues towards a potential for an understanding of the working of neurons and brain processes, including intelligence.

Seeking solutions by emulating neural processes in AI and ANN is not taking a potential benefit of a mathematical world that is out there waiting to surface at least for three centuries. The mathematics of the new calculus is concerned with the sequences of natural numbers but this is counterintuitive, as there does not seem any apparent relationship between neurons and sequences of natural numbers. However, both conventional and new calculi of sequences of natural numbers are in fact the calculus of object selection. The conventional calculus of number sequences is a physicalistic treatment and ignores chemicalistic and biologistic properties in number sequences. However, (Khatibi, unpublished) builds on their chemicalistic and biologistic interconnectivity within the sequences of natural numbers, unearthing equivalents of their Periodic Table and tree of life.

The new calculus is very simple and well within the capability of students of secondary schools but, for the sake of lay readers to AI, it is outlined in Appendix III. This section just outlines qualitatively its relevance to AI. The core of the new calculus is that any sequence based on natural numbers, N, can be factorised into two sequences, $N=I*C$: (i) the sequence of invariants, *I*, the architecture and (ii) the sequence of counters, *C*, the architect. Note that the operation of multiplication represented by "*" represents a number of different such operators, some of which are new, although they are not discussed here.

I-sequences are generic signatures characterising both the whole sequence (*N*) and each term of the sequence. Thus, each term is simply a specific configuration of *I*-sequences and a higher order terms contains composites of the terms in lower order terms (see II.9.a and II.9.b). When the structure of a sequence is reduced, everything about it is clearly known quantitatively and this makes the new calculus very powerful and flexible.

The new calculus unearths interconnectivity and interoperability among a large family of sequences of natural numbers by deriving their sequences of invariants and operating on them. Each sequence is associated with: (i) a structure representing bondage or configuration of its building blocks, where the building blocks can represent numbers, atoms, cells or neurons; (ii) the complexity of the structure from one term to the next either increases or the next term grows by having the lower terms nested within itself; (iii) each structure is unique and may be associated with emergent properties; and (iv) knowledge of the *I*-sequence and the number of terms in the sequence is all that is needed to identify everything about it.

The characteristic features outlined above point toward a *phase-driven* methodology rather than the conventional *state-driven* methodologies. In state-driven methodologies, the weights associated with neural networks are identified by solving systems of over-redundant equations using optimisation techniques; whereas, the new calculus suggests that specific configurations, or the phase of building blocks, can be used to predict outcomes of each network in response to an input. As Nature operates at a diversity of resolutions, there ought to be cogency between Nature and the emergent properties of objects at a right resolution for an association with a particular sequence. If so, the new calculus has the versatility to understand sequences in a new light that was not possible previously as conventional calculi conceal their inner structure.

Neuroscience may benefit from the new calculus by identifying the performances of a group of neurons. This is particularly relevant as the new calculus is able to predict their performance from inputs alone by using the knowledge of their inner structures. A long shot of this thought-provoking idea is that various groups of neurons may have new emergent properties, ultimately explaining a hierarchy of emergent properties in the makeup of consciousness. It is highly unlikely that the neural networks in the brain are capable of solving complex equations but very likely that different association of the brain cells make up different sequence of invariants, after which, the output can be determined precisely for the specified number of layers. This new mathematics may provide tools to directly understand the workings of brain cells through function evaluation rather than through solving equations.

6.4. The Problem of Morality

AI invokes fundamental issues related to morality ranging from the takeover of humanity by robots to undermining normal employment. The paper is not concerned with specific issues of morality raised by AI, other than encouraging debates on any such concern. However, the concern raised by the paper is that scientific cultures for ethics and morality are rather poor and this has to be remedied. In a planet with the cureent global population rise of one billion individuals every dozen years, there ought to be a gap between science-pushed encroachments and Nature-tolerated resilience. Any control in population rise touches on some sort of morality. Something is not right and population rise is just one indicator and others include booms-and-busts in the economy or the efficacy of aid and development programs.

The problem is that morality as a discipline is highly contentious and the past concepts of moral authorities are open to abuse. This paper is not defining morality but acknowledges that it exists as a dimension of human culture and without any sophistication, it is the mindset that individuals avoid voluntarily, say, to steal, even when can get away. The agenda for morality is a legacy of philosophical traditions but is infested with their dogmatic presuppositions/ predispositions and often driven by "universal truth" rather than setting out gaol-seeking core principles that are to be interpreted on a case-specific basis.

Contention in the philosophy of morality is not surprising, as it is the outcome of overlooking the evolution dimension. The human species did not emerge with its present codes of morality but has selected each of them one-by-one and conversely has unselected many of them for good (e.g. witch hunting in the Middle Ages; and the public outcry against death by stoning of adulterous woman in some countries or against the caste system of social formations). Now, the principle of sustainability can be seen as a new code of morality being selected by the human species out of necessity. In this regard, the goal of sustainability represents an ideal and one that can be defined in scientific terms, e.g. (i) The Natural Step model, which provides a set of principles based on the best scientific knowledge on the ethical and equitable behaviour of human societies to shape a sustainable future (see e.g Nattrass and Altomare, 1999, amog others); or (ii) the 'triple bottom line' measuring a corporation's ultimate success or health in terms of not just by the traditional financial bottom line, but also by its social/ethical and environmental performances (see e.g. Savitz and Weber, 2006, among others). These can be applied as high-level principles in a feed-forward cycle to guide decisions and actions leading cumulatively towards the global sustainability as a necessary condition for continued collective wellbeing.

The morality in the past and present can all be explained by the evolutionary systemic rationale. Negative feedback loops, serving as morality, operate in the following ways: (i) morality in the individuals trigger immediate control actions; (ii) such feedback loops are operated within the communal conscience as active and passive control mechanisms and are normally slow but actions are mobilised at the point of instability; (iii) in institutions, ethics often serve as a passive leverage but can be enforced as an active force.

The position of the paper towards morality of developing human intelligence AI capabilities is founded on the sentiment that if something is out there to be discovered, it will probably be discovered given a freedom of thought upon which forces of natural selection may act. Consequently, it is a vain act to suppress innovations, though moral constrains. Thus, there is a need for the world consciousness to be mobilised in order to prevent the experience of reductive science, which was oblivious to its impacts. As a result, there are many problems infesting today's world including the rapid loss of biodiversity. If artificial intelligence can emulate human intelligence and enhance it by computer speed, this emerging outcome can be channelled towards accelerating the process towards socially-derivable goals such as sustainability. The pace of change can be expected to quicken and inevitably, new orders are likely to emerge.

Misues should clearly be prevented but the suppression of curiosity and open thinking is the least effective approach to determine the future new orders. The paper argues that the global moral culture has to be enhanced now as a preparation towards a world with the human level AI capabilities. For instance, the pluralistic rights for each and all the species must be integral to everything that man does within the old natural framework, turning round the current economic paradigm which is based on individual profit (marketed goods) relative to the 'worth' of collective wellbeing generated by the very non-marketed ecosystem processes that have become progressively undermined by historic narrowly-framed commercial exploitation of natural resources (MEA, 2005).

7. The Vision

The missing links in formulating a vision for AI/AGI are presented in this section but suggestions are also made for their resolutions and these are the basis for the emergence of a vision for AI/AGI. The idea underlying the vision is that the purposeless Nature is unlikely to develop a computational brain capable of solving or optimizing complex equations as implied by AI. So there is gap between natural intelligence and AI to the extent that one cannot be referenced to the other. Constraining AI to only natural means is not advocated here but a vision is reflected that advocates learning from Nature.

The vision is founded on: (i) *the evolutionary systemic rationale* integrating evolutionary thinking with systemic thinking in terms of feedback loops; (ii) *a new calculus of number sequences,* which can allow building mathematical models of complexity; and (iii) *morality* that acts among all the individuals not to exploit the resources possibly found now to own interests in the way reductive science facilitated the plundering of natural resources, during which reductive science was oblivious to morality.

The paper is neither producing a tested solution nor is intending to offer a speculative vision but to provoke thought. Hence, it is argued that the major problems to AI/AGI or any areas of complexity are possibly stemming from physicalistic approaches but breakthroughs may stem from analytical thinking.

If the new calculus is the starting point to unearthing the existence of neural structures, the AI/AGI outlook can be a different ballgame altogether. The implication would be that any area of complexity can then be treated in a quasi-identical manner to science or the domain of science can instantaneously expand greatly to cover most of these complexity areas. Also, science is empirical even at the psychologistic to anthropologistic hierarchies, which may be retarding its advances. The application of feedback loops at each hierarchy and then the pluralism of the hierarchies are likely to provide a new impetus to each hierarchical level.

8. Application Areas

The emergence of a new species through adaptive natural selection by mutations involves the reuse of extant systems but without reverse engineering. This process, lacking design and future direction, is famously summarised by Dawkins (1986) as 'The Blind Watchmaker', expressing how unintelligent Nature, driven by interconnectivity, creates new possibilities by

Table 2. Comparison of Basic Attributes of natural intelligence, conventional AI and Evolutionary Systemic AI

	Natural Intelligence	Conventional AI	Evolutionary Systemic AI
Basic unit	Neurons and networked functional areas of neurons	Not uniquely defined but this can be defined in terms of information generating functions	Information generated by a hierarchy of function: 1. Simple data processing 2. 'What-if' enquiries 3. Decision trees 4. Function evaluation 5. Solving equations 6. Optimisation problems 7. Problem of diminishing returns 8. Hybrids of the above
Interconnectivity	Neurons are interconnected within functional areas, in turn are interconnected to one another	Exploiting anything that is technically feasible for required outcomes. In ANN, perceptrons are treated as discrete neurons.	The new calculus invokes the thought that natural neurons may have bondage that has been overlooked by treating each perceptron as a discrete unit.
Hierarchy	This is an open question and current research shows that the brain is a network rather than a hierarchically organised complexity.	Exploitation of anything that is technically feasible towards required outcomes	This is an open question.
Logic	Fuzzy logic	Fuzzy logic, as well as any other logic if useful	(A) AI referenced to natural intelligence The AI architecture emulating the brain (B) AI not referenced to natural intelligence The AI architecture driven by strict non-contentious morality.

Table 3. Broad Attributes of Natural Intelligence

Attributes	Reflex-based actions	Causative processes	Transition	Reactive intelligence	Anticipative intelligence
Definitions	• Driven by instinct in most of the animal kingdom, • Require no intelligence	• Primed by learning rules describing cause-and-effect processes often through trial-and-error and then by selection. • Found in many species of the animal kingdom and build on reflex-based capabilities	• Loss of efficacy by escalating discrepancies becomes overwhelming	• Negative feedback loops are at a higher level of complexity to reduce entropy • Negative feedback emerges only when entropy severely reduces efficacy	• Feedforward loops are at a higher level of complexity to overcome possible stagnations
Driving Loop	• $Zero^{+}$ Feedback Loop • Thriving culture	• Positive Feedback Loop	• Triggered by a selective advantage for reducing entropy	• Negative Feedback	• Feedforward loops
	• The mind is encapsulated with system components capable of understanding inputs or outputs but triggers reflex-based actions to carry out certain functions or to control the processes	• Involve some sort of "thinking;" • Kick in positive feedback loops • During this process knowledge selection is Darwinian and slow	• The period of rethinking when signals of escalating entropy is picked up and solutions are sought by developing negative feedback loops	• The brain in the animal kingdom is a system of negative feedback loops to control the functioning of the bodies of the animals	• The increased complexity is modulated by foresight of future environmental fluctuations similar to negative feedback loops

Attributes	Reflex-based actions	Causative processes	Transition	Reactive intelligence	Anticipative intelligence
Processes	• Under this condition, either cause is known or effect but not both	• A blind eye is turned to discrepancies by believing in the truth of the causative processes		• Ever since its emergence some 600-700 million years ago, the brain keeps evolving through various species	•
Entropy	• Spontaneity in the background culture and manifests as a cause or effect and this is a driver to understand the effects of a cause or vice versa	• Knowledge bases is infected with growing discrepancies (entropy) • Increasing entropy reduces the efficacy of the processes in the long run	• Past solutions ceases to be efficacious: • Destruction • Boom-and-bust, or • Evolutionary transition	• Goal-fixation or stagnation • Goal displacement • Too strong against variations give rise to highly dynamic process of 'hunting'	• The complexity becomes highly dynamic and may be this state is an statement of a new type of entropy
Selective Advantage	• Triggering spontaneity and autonomy	• Survival • Learning new facts of life to create more opportunities	N/A	• Steady performance • The emergence of an insight	• Emergence of goal-orientation • Emergence of foresight • Clear statements of the mission

Table 4. Broad Attributes of Agents in Psychologistic–anthropologistic Levels Modelling Scenarios

Attributes	Reflex-based actions	Causative processes	Reactive intelligence	Anticipative intelligence
Psychologistic	• To emulate change scenarios triggered spontaneously by individuals (inputs or outputs) without knowing the impacts. with switch between rationality or irrationality, e.g. hotly debated bonus cultures	• To emulate change scenarios triggered by pioneering individuals noticing causative information bearing the traits of positive feedback loops with facilities to emulate their entropy, e.g. the actions of the individuals using fossil fuel and those identifying global warming	• Emulating change scenarios, in which individuals attempt to reduce entropy by initiating negative feedback, e.g. individuals committing to recycle household rubbish, or efficient use of energy	• To emulate change scenarios to modulate proactively the contributions or impacts of individuals with visions
Sociologistic	• To emulate change scenarios triggered spontaneously by sections of a society (inputs or outputs) without knowing the impacts with switch between rationality or irrationality, e.g. profit motives of	• To emulate change scenarios triggered by communal, social or institutional consciousness escalating causative processes with entropy, e.g. the actions of the individuals using fossil fuel and those identifying global warming	• Emulating change scenarios, in which sections of the society attempt to reduce entropy by initiating negative feedback, e.g. organisations having Carbon Footprint policies	• To prevent goal fixation or goal displacement by the virtue of foresight, e.g. the Green Movements in the 196s-1990s, until their ideals were enshrined to the Principle of Sustainability

Attributes	Reflex-based actions	Causative processes	Reactive intelligence	Anticipative intelligence
Anthropologistic	• To emulate change scenarios triggered spontaneously by various cultures (inputs or outputs) without knowing the impacts with switch between rationality or irrationality	• To emulate change scenarios enshrined by governance or cultures escalating causative processes with entropy, e.g. the policies and philosophies of utilitarianism during the Industrial Revolution	• Emulating change scenarios of reducing entropy is enshrined to the governance and global cultures, e.g. policies on the principle of sustainability	• To prevent goal fixation or goal displacement by the virtue of foresight coded to them, e.g. the principle of sustainability as one such foresight
Pluralism	• Each of the above types of agents is a mindset and can act on its own or the newly emerging mindset can create a pluralistic culture with the older mindsets. This is the strength of evolutionary systemic modelling, which emulates reality, where there is the pluralism of these mindsets associated with individuals, communities and institutions. Each mindset can act as an attractor or repellent in the sense of chaos and/or catastrophe theories and therefore in principle, it is possible to study the dynamics of complexity			

Table 5. Systemic Modularisation of Professional Practices

<table>
<tr><th>Timescales</th><th>Zero^{+} Feedback Loops</th><th>Positive Feedback Loops</th><th>Negative Feedback Loops</th><th>Feedforward Loops</th><th>Examples</th></tr>
<tr><td rowspan="2">Incident Management</td><td colspan="4">• For relatively short decision times, a diversity of professional practices have emerged to cater for a range of activities including, for example, Emergency Management, Disaster Management, Civil Protection, Crisis Management and Incident Management. These activities, which formalise responses rather than relying on new thinking, have generic differences which may be used for modularisation.</td><td rowspan="2">• Best practice, e.g. USA guidance in emergency management on Nuclear Regulatory Commission & Federal Emergency Management Agency in the 1980s
• The Civil Contingencies Act 2004 in the UK</td></tr>
<tr><td>• Guidelines with legally delegated powers to acts for a set short time and maintain the capability for triggering action, as and when needed during emergencies, incidents or crises</td><td>• Various practices driven by guidelines, policies and directives through lifecycle models such as: mitigation, preparedness, response and recovery</td><td>• Post-incident analysis is often carried out to improve practices and possibly feedback to future operations, guidelines and policies</td><td>• A whole range of plans driven by various initiatives, best practice guidance, policies and directives, e.g. Civil Contingency plans, including adaptive management</td></tr>
<tr><td>Activities directly driven by Policies, Directives or Acts</td><td colspan="4">• Policies, Directives, Memoranda of Understanding, Ministerial Guidance are examples of formal drivers guiding towards goals or a set of objectives and trigger a lifecycle of activities often over a time horizon of the next 20 years or so
• Examples include: the EU Flood Directive requiring: (i) assessing risk from all watercourses and coastlines; (ii) requires Flood Risk Management Plans by a range of mapping activities and coordinated measures to reduce flood risk; or the EU Water Directive requiring: (i) the consideration of all the element of water quality; (ii) developing River Basin Management Plans by setting the objectives for the river basin; (iii) reviewing impacts of human activity on the status of waters in the basin; (iv) analyzing economic use of water within the basin towards rational discussions on the cost-effectiveness of possible solutions.</td><td>• The European Flood Directive (2007/60/EC)
• The European Water Directive (2000/60/EC): River Basin Management Plans (RBMP),</td></tr>
</table>

Timescales	Zero^{+} Feedback Loops	Positive Feedback Loops	Negative Feedback Loops	Feedforward Loops	Examples
	The emerging trend is to reinforce the rights of the public to access gained information and to have a say in the planning process. • Competent authorities are often empowered to take on board the policies goals through guided studies and develop plans.				Promoting whole system modelling for strategic policy and investment planning at long term and large scale through the following set of processes: Driver, Pressures, current and future State of the system, Options and Responses (DPSOR). • The EIA Directive (85/337/EEC) • Masterplanning by private developers and LA
	• The owners of policies and directives are often bodies of governmental organisations empowered to transform the emerging ideas and formulate requirements, often with a focus on 'delivery' rather than environmental and/or social outcome	Competent authorities uphold compliances, safeguard the, public, monitor policies & give feedback to minsters • Studies are initiated to produce working documents to expedite the discharge of duties • Each type of study is often characterised by a lifecycle of activities e.g. DPSOR, S-P-R-C	• Until a formal review, the available information is regarded as fit for purpose • In democratic countries, periodic reviews are operated on an *ad hoc* basis, but vested interest may limit the amount of scrutiny. • In undemocratic countries, policies often remain unchanged for a long time, with a tendency to bend rules	• A range of management plans are produced, which serve as the foresight for competent authorities to discharge their obligatory duties, e.g. CFMP, SMP, SWMP in the UK • Environmental Statement • Both adaptive management above modifies intervention provide forward feedback	

Table 5. (Continued)

Timescales	$Zero^{+}$ Feedback Loops	Positive Feedback Loops	Negative Feedback Loops	Feedforward Loops	Examples
Practices	• Practices refer to the development of skilful professionals engaged on repeated exercises or activities requiring skills or specialism • An example includes masterplanning initiated by local authorities or private developers involve a process to undertake, analysis and prepare strategies and a set of proposals to plan for major changes including buildings, spaces, movement strategy, land use; a masterplan is a sophisticated model showing ways of integrating communities into the built/natural environments				• Planning studies • Flood Risk Assessment (FRA) studies • Masterplanning • Environmental Impact Assessment (EIA)
	• Statutory requirements driving compliances with standards, acts, policies & sustainability.	• Authorities or private developers initiating large scale developments	• Consultants, who have to provide solutions meeting statutory requirements	• Principle driving practices, e.g. risk-based decision-making & sustainability	

augmenting and adapting rather than discarding past characteristics. The current thinking in AI/AGI demonstrates poor reuse of past AI products, as the 'reinvention of the wheel' is a norm.

To illustrate this, attention is given to a recent Call for Proposal (see: - by the RCUK, Research Councils for the UK: http://www.nerc.ac.ukresearch/programmes/resilience /events/ao3.asp) with the focus on increasing social and economic resilience in earthquake-prone and volcanic regions by reducing risks from multiple natural hazards.

The anticipated tasks included: improving hazard forecasting, risk mitigation and preparedness and recovery. Another focus of the CfP was the improvement of the uptake of and responses to scientific advice through risk-based approaches to natural hazards, including assessment of uncertainties and the context-specific communications, with respect to the communities at risk and with other stakeholders.

Below, an outline is given to distinguish between two hypothetical proposals for this CfP involving the development of AI capabilities. The first 'traditional approach' uses physicalistic methods to AI inevitably without reusing past capabilities. The contrasting 'evolutionary systemic approach' pools past modules supplemented by new modules.

8.1. Traditional Approaches

Application of AI into the operation of complex, interacting human and technological systems under incident management scenarios was outlined in Section 3.2.

To supplement the study of resilience in earthquake-prone and volcanic regions with AI capabilities, it is likely to follow a similar procedure to that outlined in Section 3.2. The outcome would be the addition of one more area of complexity to existing ones without advancing science.

8.2. A Possible Application of the Vision

The starting point is that the development of AI-solutions for resilience in earthquake-prone and volcanic regions by reducing risks from multiple natural hazards may be driven by a variety of working practices. These may be modularised according to the timescale of activities and within each timescale, the four types of feedback loops may be used to categorise each activity. The suggested timescale of activities are: (i) *incident management,* over time spans of several days, (ii) *project timescale,* over finite times from several days to several years; (iii) *practice timescale,* over long time spans during which projects are continually delivered but practice-wise changes are very slow; (iv) *Policy timescales,* over some 20-30 years of time span, over which planning is carried out but changes may be initiated from time-to-time; (v) *foresights,* over time spans of some 100 years.

Based on the vision presented in this paper, the first intention would be to advance science by modularising many aspects of developing AI solutions. Whilst it is appreciated that one proposal cannot solve all the problems of science, at least steps can be taken to offer preliminary modules. Activities for each of the above timescales can be modularised and common areas can be regarded as a module, which can then serves as a pool of resource to be called in whenever needed. As new CfPs would requires new streams of problem areas, e.g. recent volcanic eruption or tsunami incidents, their peculiarities can then be identified and formulated under new modules, which can then be made available to the public. Table 5 shows an attempt for a preliminary modularisation of three of the activities.

CONCLUSION

Artificial Intelligence (AI) and Artificial General Intelligence (AGI) are developing in symbiotic relationships with computers and are utilising their resources, particularly in terms of processing power and speed, to deliver intelligence. The achievements are formidable but the following problems are identified through a critical examination: (i) *monolithic products:* AI products are monolithic in the sense that each capability is built from zero and without transferring relevant building blocks from other areas of application; (ii) *closed architecture:* each AI product is only applicable to a single niche and is bespoke; (iii) *absence of an insight and foresight*: each AI application is developed without reducing the underling complexity into hierarchies, and without identifying building blocks (or modularity) and their interconnectivity among them at each hierarchy, all of which are prerequisite for the emergence of open architecture.

The paper argues that AI/AGI is in the grip of "physicalistic thinking" in the sense of neglecting inherent chemicalistic structures by focusing on physicalistic properties and by stripping away all hierarchical features and lumping together the remaining features. It is suggested that the hierarchical architecture of science (physics, chemistry, biology, psychology, sociology and anthropology) is well suited to serve as a model to unearth hierarchies in the wider complexity including the AI complexity. The paper refers to these as physicalistic, chemicalistic, biologistic, psychologistic, sociologistic and anthropologistic levels.

The paper suggests that there are *missing links* in the science of complexity including the AI complexity, which can be addressed as follows: (i) the hierarchical structure in science can serve as model of hierarchy for the AI complexity; (ii) the brain can serve as a model to understand the AI complexity at the chemicalistic level and the diversity of the brains among the species can serve to find a model for diversification of the AI complexity at the biologistic level; (iii) the four feedback loops are used as a model to study interconnectivity at psychologistic-anthropologistic levels; and (iv) the new mathematical approach by Khatibi (unpublished but summarized in Appendix III) is used to suggest a possible mathematical modelling technique to account for interconnectivity at each hierarchical level.

The paper invokes a possible dependence of intelligence on hierarchical levels and feedback loops, which can serve as a paradigm to study intelligence as follows: (i) *reflex-based* actions associated with zero$^+$ feedback loops, (ii) *causative processes* (based on cause-and-effect rules) associated with positive feedback loops; (iii) *reactive intelligence* driven by negative feedback loops; and (iv) *anticipative intelligence* driven by feedforward loops.

This paper uses evolutionary systemic language (Khatibi 2011 and 2012a) and formulates a vision by arguing that just like the brain functioning as a control system, AI can provide extension to man's control capabilities. The vision put forward is based on the following: (i) *the evolutionary systemic rationale* integrating evolutionary thinking with systemic thinking in terms of feedback loops; (ii) *a new calculus of number sequences*, which can allow building mathematical models of complexity; and (iii) *morality* that acts among all the individuals not to exploit the resources possibly found now for own interests in the way reductive science facilitated the plundering of natural resources, during which reductive science was oblivious to morality. When an overview is taken, it seems AI/AGI is an integral part of enterprises of complexity and by treating the problems through the vision presented in this paper, new mileages in science are feasible with the purpose that each species have pluralistic natural rights to own our common world and man has no right to overrule their rights – AI can provide us the tools to monitor these rights.

ACKNOWLEDGMENTS

This paper is a contribution to open science and to stimulate debate. The views shared here by the authors do not necessarily represent their affiliated organisation and they hope the benefits only remain in the service of open science. The authors take this opportunity to express their gratitude to Mr. John Little of New College, Swindon, UK, for thoroughly reviewing the paper and offering detailed comments and adding value to the paper.

APPENDIX I – PHILOSOPHY OF MIND

Philosophical discourses are "messy," as there is no reference point to regard a particular doctrine obsolete and therefore current thinking is burdened with those of the past. Searle (1997, P. ii) states that from the vantage point of the past and ongoing debate, "the greatest single philosophical obstacle to getting a satisfactory account of consciousness is our continued acceptance of a set of obsolete categories" and presuppositions "inherited from our religious and philosophical tradition." The paper categorises past traditions as follows:

Mythological Age: The past mythological narratives have lost their influence but of these shamanism remains very important to understand the perception of mind among primitive man and societies (see for example Eliade 1989 or Armstrong 2005);

Dogma: Religious dogmas emerged in the past and still persist to retard science.

Reason: Benjamin Franklin (1706-90) is credited as the first by defining the human being in terms of a "tool-making animal" arguing that "The making and using of tools is the first evidence of the human ability to use reason to solve problems," Guisepi (http://history-world.org/paleolithic1.htm). Zantonavitch (http://rebirthofreason.com/Articles/Zantonavitch/Sumerian_Philosophy.shtml) argues that both the Egyptians and the Sumerians did not have true philosophy but had *something* in the form of observations involving simple ethics or 'wisdom texts'.

The Hellenistic tradition of philosophy flourished 6th centuries B.C. out of past traditions of reasoning. It boosted rationalism and idealism but collapsed in circa the second century A.D. coinciding with the emergence of Christianity. The Greek philosophy was then revived by the traditions of Islamic philosophy in the 9th century to boost the discourse in Islamic cultures. However, the rationalistic traditions of Islamic philosophy also collapsed by the 14th

century. Nonetheless, worldly thinking also emerged in Chinese and possibly Indian philosophies. Traditional reasoning denies change or dynamism and oblivious of evolution and hence it is normally ontological.

Modern (European) philosophy: Appealing to arguments and abandoning faith re-emerged in the 17th century in Europe thanks to: (i) humanistic movements in the 15th-16th century; (ii) the emergence of science in the 17th century; and (iii) the revival of Greek philosophy. Ever since this revival, these doctrines have remained oblivious to evolution particularly since the publication of *On the Origin of Species* by Charles Darwin (1809–82) in1859 and hence they are ontological in essence. The emerging doctrines have been rich and stimulating and one application of modern philosophy is the philosophy of mind with various reviews and anthologies, e.g. Lyon (1995), Rey (1997), Heil (2004), Chalmers, (2002). The doctrines include: Cartesian dualism (which is discredited now), various forms of eliminativism (Feyerabend 1963, Rorty, 1965, Churchland and Churchland, 1998 and Stich 1983), functionalism (Putnam 1975), connectionism (http://plato.stanford.edu/entries /connectionism/), emergentism (http://plato.stanford.edu/entries/properties-emergent/) and more recently explaining the mind by computer analogies. The list is not exhaustive but these doctrines only feed to philosophical discourses with hardly any benefit to the general study of intelligence or AI.

Neuroscience-based philosophical debates: This tradition is relevant to this paper. The problem is that neuroscientists are busy with their scientific research using the usual scientific tools without aiming to communicate their thinking, while the philosophers of mind debate their points of view. This tradition is purported by a host of philosophers including Searle (1997), Dennett (1991), Edelman (1992), Churchland (1996). These recent traditions are either built on past traditions or run in parallel with new traditions stemming from psychological studies, e.g. 'biophilia' as presented by Wilson (1984), which are reflected in Figure 1.

Owing to the work of neuroscience-based philosophers of mind, it is possible to take stock of past thinking and to delineate between relevance and irrelevance. The line may be drawn in part by consensus.

The followings are suggested by this paper: (a) All conscious experiences are explained by the behaviours of neurons and are themselves emergent properties of the system of neurons, as argued by Searle (1997), which is deemed to be consensus. (b) There is a gap between the mind (observed physical and mental phenomena) and its extant explanations. (C) The essential functional element for consciousness is debated: Penrose (1994) attributes it to the quantum scale, Edelman to "neural groups," Churchland (1996) denies the existence of consciousness and liken it to the rejected phlogiston. This paper argues that the problem of complexity has been overlooked in the third item.

Neuroscience-based philosophies of mind enjoy healthy debate but, as reflected, these are all merely clever arguments.

This is because the dimensionality of the problem of consciousness has not been identified, if indeed there is such a thing as the dimensions of consciousness. Searle (1992, pp128-141) argues that "it is not possible to give a definition of "consciousness" in terms of necessary and sufficient conditions, nor is it possible to define it in the Aristotelian fashion by way of genus and differentia. He presents a dozen features delineating the structure of consciousness shown in Figure I.1.

Features Delineating the Structure of Consciousness (Searle 1992)

i. Finite modalities, e.g. the various senses;
ii. Non-pathological unity of multiple states, e.g. having a toothache, a visual experience, thinking about a subject, etc;
iii. Intentionality;
iv. Subjective feeling;
v. Connections between intentionality and consciousness;
vi. Gestalt structure of conscious experience;
vii. Familiarity;
viii. Overflow, in the sense that conscious states extend beyond the immediate context;
ix. Centre and periphery;
x. Boundary conditions;
xi. Mood; and
xii. Pleasure and un-pleasure.

Figure I.1. Features Delineating the Structure of Consciousness (Searle 1992).

APPENDIX II – ZERO$^+$ FEEDBACK LOOP FOR AI

The roots of AI go undoubtedly back to prehistory and its traces can be learned from mythology. Khatibi (2012b) reflects on current thinking that *model thinking* is instinctive and argues that language is ostensibly its first fruit. This emerged in prehistory, where modelling uses one entity as a surrogate representation for understanding another. The emergence of language did not make model thinking obsolete but they coevolved together giving rise to more cultural or technological constructs. Although there were not a diversity of artefacts in prehistory and therefore there was no sense of artificial intelligence then in its modern sense, dream and language were two prerequisite building blocks for the emergence of imaginative constructs akin to artificial intelligence.

It is now certain that man is just a species belonging to the mammalian class and therefore up to the Middle Palaeolithic Age, there were very few distinguishing features between the modern man and other species of the same class. Clark (1994) states that "Until the rise of a more critical approach in the final decades of the eighteenth century people were happy to accept their history and the world they knew as the work of God without seeking to account for them in rational terms. ... When Sir James Frazer reviewed the myths relating to the creation of our human race he reminded us that Prometheus created our first parents 'by fashioning them, like a potter, out of clay'. He went on to state that analogous myths occurred not only in the second chapter of Genesis but also in various forms among peoples as widely as the Pacific inlanders, the Australian aborigines, the negroes of Togoland, the Eskimo of Alaska and the American Indians of California, Louisiana and Peru. ... they served a purpose and met a common need, one felt as much among civilized societies as among the preliterate tribes encountered by ethnologists in the field."

The imaginative powers of primitive man culminated in creating powerful supernatural images, some of which still plague modern cultures. It is not difficult to argue that those imaginative creatures are the roots of AI but under the conditions prevailing then. Instances of AI are found in the various mythologies. McCorduck (2004) identifies two extreme views as: (i) the Hellenic point of view, which generally welcomes the idea of thinking machines (e.g. Homer's Iliad robots made by the Greek god Hephaestos, also known, in Roman mythology, as Vulcan, or the Chinese tales of human-like machines that could think); and (ii) the Hebraic view point, driven by the Second Commandment prohibiting the making of graven images and regards the whole idea of thinking machines wicked, even blasphemous and therefore forbids humans to take on the creative privileges of divinities. It is easy to dismiss the above but they are just used for mining information from the seemingly irrelevant past mindsets.

The Middle Ages are dotted with stories of mechanical intelligent machines in the Islamic cultures, Europe and China (see http://www.aaai.org/AITopics/pmwiki /pmwiki.php/AITopics/BriefHistory). McCorduck (2004) identifies this dual attitude to modern times and gives the case of Wolfgang von Kempelen, who created *The Turk*, the name given to a chess-playing automaton, created by in order to impress the Empress Maria Theresa of Austria-Hungary in 1770 and then began a series of performances, which continued for 85 years, throughout Europe and eventually in the United States. She also recites the Mary Shelly's *Frankenstein*, which is the pessimistic attitude towards emerging science during the industrial revolution. Evidently, the mindset prior to the Middle Ages mirrors that afterwards.

The building blocks of AI are: a culture rife for AI, demand for intelligent approaches, mechanical components, electronic components, control components and rules or logic. Prior to the spread of Arabic-Indian numerals, even addition and subtraction was a difficult task and this created for an artefact to carry out calculations.

It should not be surprising that the Sumerian abacus emerged some 5000 years ago and this probably emerged in other cultures, e.g. Egypt, China and Greece. Although calculation became simple after the Arabic-Indian numerals, the emergence of reductive science in the 17th century created demand for more involved calculations and this gave rise to calculating machines. One basic design of an Analytical machine was given by Charles Babbage (1792-1871) but he never finished it. This was one of the first steps towards AI in its modern sense.

Scientific and philosophical cultures for AI were rife even before the advent of computers. The discourse in these cultures have brought forth some of the attributes of intelligence, e.g. consciousness, intentionality, thinking, problem-solving.

Many of these cultural building blocks are semantic constructs and often tautological without any real significance. For instance, Searle (1992, http://users.ecs.soton.ac.uk /harnad/Papers/Py104/searle.prob. html#BFN1) discusses *unity* as one of attributes of consciousness concerning the recognition of non-pathological forms, e.g. a pain in the elbow, a feeling of warmth when occurring simultaneously. According to him, Kant called this feature "the transcendental unity of apperception" but it is referred to as "the binding problem" in neurobiology. Obviously transcendental categories are obsolete and have been deselected from science. Another example is *qualia*, where this philosophical notion has been unselected from science. The problem is that all these terms and constructs are lurching around with no catalogue for obsolete ones.

The past mindsets on AI were fashioned through mechanical thinking but this is now obsolete, although mechanical components still contribute to AI but only in a subsidiary capacity. The paper argues that even many of the doctrines in philosophical discourses are obsolete but there is no methodology to catalogue them. However, the past optimistic and pessimistic perceptions on AI should not be dismissed and equally relevant today. AI should not be disassociated from morality.

APPENDIX III

Outline of a New Calculus for Hierarchy and Interconnectivity

The main preoccupation in conventional calculus of number sequences is the basic operation of finding the sum of all consecutive terms using such methods as the Umbral method (Bucchianico *el al.*, 2000) or those based on Stirling numbers of second kind, e.g. see Brualdi (1992).

These conventional approaches provide no hint towards a possible connection between sequences of natural numbers and the workings of neurons but the account below on the new calculus by Khatibi (unpublished) suggests new possibilities. The basis of these ideas is introduced in Electronic Supplementary Material (ESM).

A sequence is a highly ordered set of numbers and in a large family of them: (i) there is a hidden structure; (ii) the structure can be defined by the first few terms of the sequence; and (iii) the structure is nested in each of the remaining terms in a predictably countable manner; (iv) the structure can be unearthed by transforming the sequence into the product of the sequence of invariants and the sequence of counters. Some examples of sequences include:

$$\{1^m \quad 2^m \quad \ldots \quad n^m\} \qquad \text{(III.1.a)}$$

$$\{1^m \quad 3^m \quad \ldots \quad (2n-1)^m\} \qquad \text{(III.1.b)}$$

$$\{1^m \quad 4^m \quad \ldots \quad (3n-2)^m\} \qquad \text{(III.1.c)}$$

$$\{(1\times 2)^m \quad (2\times 3)^m \quad \ldots \quad (n-1)^m \times (n)^m\} \qquad \text{(III.1.d)}$$

where *m* and *n* is any integer numbers. Consider the following examples:

$$\{1^2 \quad 2^2 \quad 3^2 \quad \ldots \quad n^2\}=\left\{1\binom{i+1}{i-1}+1\binom{i}{i-2}\right\}_{i=1,n}=$$

$$\{1 \quad 1\}\odot\left\{\binom{i+1}{i-1} \quad \binom{i}{i-2}\right\} \qquad \text{(III.2)}$$

Note that for this power sequence with its power *m*=2, the sequence of invariants is $\{1 \quad 1\}$ and the sequence of counters is: $\left\{\binom{i+1}{i-1} \quad \binom{i}{i-2}\right\}$; the first two terms are sufficient to derive the sequence and define its structure, where this is nested in all the other terms.

For the power, m=3, we have:

$$\{1^3 \quad 2^3 \quad 3^3 \quad \ldots \quad n^3\}=\left\{1\binom{i+2}{i-1}+4\binom{i+1}{i-2}+1\binom{i}{i-3}\right\}_{i=1,n} \quad \text{(III.3)}$$

In which the sequence of invariants is: $\{1 \quad 4 \quad 1\}$ and the sequence of counters is self-evident.

These sequences of invariants are interoperable but the rules, though simple, are not in the mainstream mathematics. Based on Khatibi (unpublished) and ESM, a glimpse of the process is presented by transforming $\{1 \quad 1\}$ into $\{1 \quad 4 \quad 1\}$ is as follows:

$$\{1 \quad 1\}\widehat{\odot}\left\{1 \quad {2 \atop 2} \quad 1\right\}=\{1\text{x}1 \quad 1\text{x}2+2\text{x}1 \quad 1\text{x}1\}=\{1 \quad 4 \quad 1\}=\{1 \quad 4 \quad 1\} \quad \text{(III.4)}$$

Although the above mathematics is very simple, their deeper understanding is not possible without the full knowledge of the rules. Nonetheless, the essentials are clear in the above examples. It is noted that: $2! = \{1+1\}$, and $\{3! = 1+4+1\}$. Using the above rule, it can be shown that: 4!=1+11+11+1 and 5!=1+26+66+26+1. These elegant expansions are indeed all needed to manipulate powers 4 and 5 sequences of natural numbers, and so on (i.e. m=4 and 5 in (1)). This new calculus is extremely simple and well within the capability of students of secondary schools. Without any trouble, power m of the sequence of natural numbers can be treated by the above transformations.

There is a structure in the number sequences, which can be unearthed by the new calculus and very much related to combining and associating objects. One example is presented just to illustrate the idea. This is not too easy and not too difficult.

The main idea is that the building blocks (the sequence of invariants) are replicated in a nested manner with certain logic. The logic is not described fully but it can be spotted with some attention from the presentations below, in which the ith-term will have i-sub-terms (ST) and sub-terms will have sub-sub-terms (sst) and sub-sub-terms have sub-sub-sub-term (ssst), as this is a third degree power sequence.

Consider the sequence:

$$\{1^3 \quad 2^3 \quad 3^3 \quad \ldots \quad n^3\} \quad \text{(III.5)}$$

The first element is $\{1^3 = 1\}$ (III.6)

$$2^3 = \overbrace{\overbrace{\overbrace{\overset{\frown}{1}}^{ssst1}}^{sst1}}^{ST1} + \overbrace{\overbrace{\overset{\frown}{1}}^{sst1} + \overbrace{\overbrace{\overset{\frown}{1}}^{ssst1} + \overbrace{1+4}^{ssst2}}^{sst2}}^{ST2} \quad \text{(III.7.a)}$$

(III.7.a) is summarised as:

$$2^3 = \binom{2+2}{2-1}\times\mathbf{1}+\binom{2+1}{2-2}\times\mathbf{4} = 4\times\mathbf{1}+1\times\mathbf{4} = 8 \quad \text{(III.7.b)}$$

$$3^3 = \overbrace{\overbrace{\overbrace{1}^{ssst1}}^{sst1}}^{ST1} + \overbrace{\overbrace{1}^{sst1} + \overbrace{\overbrace{1}^{ssst1} + \overbrace{1+4}^{ssst2}}^{sst2}}^{ST2} + \overbrace{\overbrace{1}^{sst1} + \overbrace{1+1+4}^{sst2} + \overbrace{\overbrace{1}^{ssst1} + \overbrace{1+4}^{ssst2} + \overbrace{1+4+\underbrace{1}_{TBB}}^{ssst3}}^{sst3}}^{ST3}$$

(III.8.a)

where TBB is the Trailing Building Block and in this way the two values of 1 are differentiated. (II.8.a) is summarised as:

$$3^3 = \binom{3+2}{3-1} \times \mathbf{1} + \binom{3+1}{3-2} \times \mathbf{4} + \binom{3}{3-3} \times \mathbf{1} = 10 \times \mathbf{1} + 4 \times \mathbf{4} + 1 \times \underbrace{\mathbf{1}}_{\boldsymbol{TBB}} = 27$$

(III.8.b)

The inner structure of the fourth term is worked out as follows:

$$4^3 = \overbrace{\overbrace{\overbrace{1}^{ssst1}}^{sst1}}^{ST1} + \overbrace{\overbrace{1}^{sst1} + \overbrace{\overbrace{1}^{ssst1} + \overbrace{1+4}^{ssst2}}^{sst2}}^{ST2} + \overbrace{\overbrace{1}^{sst1} + \overbrace{1+1+4}^{sst2} + \overbrace{\overbrace{1}^{ssst1} + \overbrace{1+4}^{ssst2} + \overbrace{1+4+\underbrace{1}_{TBB}}^{ssst3}}^{sst3}}^{ST3} +$$

(III.9.a)

$$\overbrace{\overbrace{1}^{sst1} + \overbrace{1+1+4}^{sst2} + \overbrace{\overbrace{1}^{ssst1} + \overbrace{1+4}^{ssst2} + \overbrace{1+4+\underbrace{1}_{TBB}}^{ssst3}}^{sst3} + \overbrace{\overbrace{1}^{ssst1} + \overbrace{1+4}^{ssst2} + \overbrace{1+4+\underbrace{1}_{TBB}}^{ssst3} + \overbrace{1+4+\underbrace{1}_{TBB}}^{ssst4}}^{sst4}}^{ST4}$$

(III.9.b)

(III.9.b) is summarised as:

$$4^3 = \binom{4+2}{4-1} \times \mathbf{1} + \binom{4+1}{4-2} \times \mathbf{4} + \binom{4}{4-3} \times \mathbf{1} = 20 \times \mathbf{1} + 10 \times \mathbf{4} + 4 \times \underbrace{\mathbf{1}}_{\boldsymbol{TBB}} = 64$$

(III.9.c)

The above may be generalised as follow:

$$i^3 = \mathbf{1} \times \binom{i+2}{i-1} + \mathbf{4} \times \binom{i+1}{i-2} + \mathbf{1} \times \binom{i}{i-3} \qquad \text{(III.10)}$$

The hidden strengths of this new calculus are as follows:

- The elements in a sequence are not what they seem to be (just numbers) but systematically nested (or structured) building blocks.
- These examples show that the sequence of counters (the rules) is intrinsic and within the element, which is not a *meta*-rule.

- Each sequence is composed of two main divisions: (i) the first few terms (in the above case the first three terms) define the structure; (ii) this structure is nested in each of the remaining terms in a specific countable manner.
- The terms outside those defining the structure (in the above case the 4th term, for example) are repetitions and the building blocks, i.e. no new building blocks are introduced thereafter.
- When a sequence is reduced and its structure is unearthed, e.g. (III.3), everything is then known about it by just knowing the number of the term.
- There are also growth layers not introduced here, where these layers do not add anything to the complexity of the sequence but the process is a matter of physicalistic process of arranging more structured units in a specific order.
- A sequence can be characterised by its sequence of invariants, whereas the sequence of counters is generic.
- Completely arbitrary sequences can be taken as the sequence of invariants, such as $\{1 \ -1 \ 10\}$, or $\{1 \ 9 \ -2\}$. When these are multiplied by the same sequence of counters, they are likely not to produce interesting sequences (i.e. they may not have any emergent properties). For instance, the first one of the above arbitrarily selected sequences of invariants: $\left\{1\binom{i+2}{i-1} - 1\binom{i+1}{i-2} + 10\binom{i}{i-3}\right\}_{i=1,n}$ produces: $\{1 \ 3 \ 16 \ 50 \ 115 \ 221 \ 378 \ 596 \ 885 \ ...\}$.

Possible Application in Artificial Neural Network

The new calculus offers a possible alternative approach to Artificial Neural Network (ANN).

Consider the simple building blocks and rules given as:

$\{1 \ 1 \ 1 \ 1 \ 1 \ 1 \ 1 \ 1 \ 1\}$ [illegible] $3 \ -3 \ 3 \ 1$ (III.11.a)

and the convolutionary rule given by:

$\{1 \ 1 \ 1 \ 1 \ 1 \ 1 \ 1 \ 1 \ 1\}$ [illegible] $1 \ 1 \ 1 \ 1 \ 1$ (III.11.b)

Generation 1 is given by:

$\{1 \ 2 \ 3 \ 4 \ 5 \ 6 \ 7 \ 8 \ 9 \ 8 \ 7 \ 6 \ 5 \ 4 \ 3 \ 2 \ 1\}$ [illegible] $4 \ 1 \ 4 \ 5 \ 5 \ 4 \ 1 \ 4 \ 1$ (III.11.c)

The building blocks may be imagined to be numerical values assigned to the skill areas to do a certain job and the rules may be imagined the minimum competence level. Dividing every element by the sum of the values of all the elements, (III.11.c) may be assumed to be a measure of characterisitics (or the integrity) of the product. Assuming that the rules are applied again and again through generations, the characteristics of the product can be assessed after many generations. Figure III.1 shows how the characteristic of the system smoothens after a few generations.

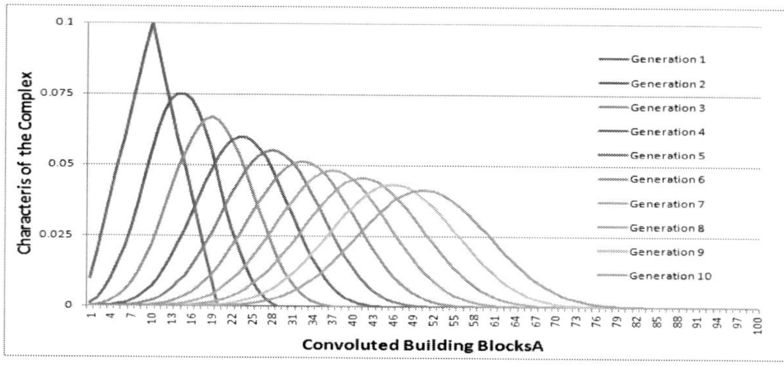

Figure III.1 Normalized Elements of a Product Sequence in Generations.

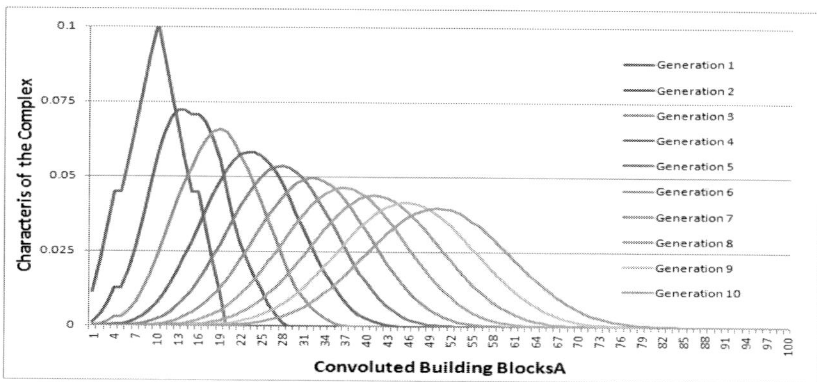

Figure III.2 – Same as III.1 but Fifth Element reduced from 1 to 0.

However, when the elements or the rules change, the response on the product can measure the integrity of the product. For instance, if the middle element corresponding to the rule is made zero, -1, and -2, the effects ca easily be measured and these are displayed in Figures III.2-III..4

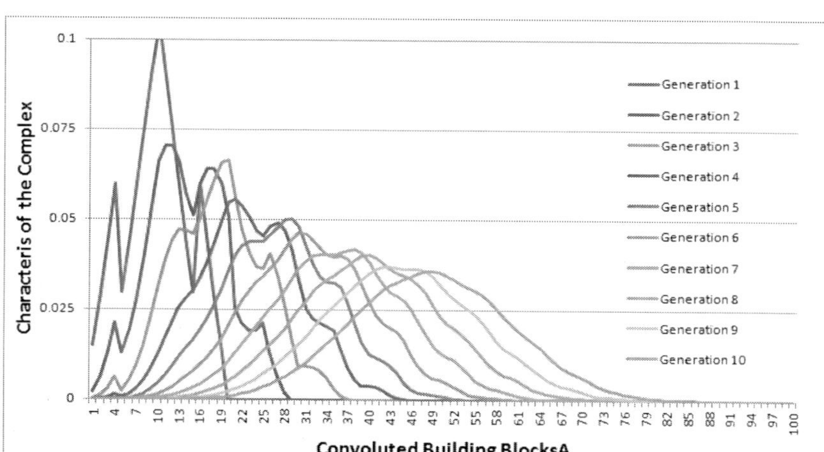

Figure III.3 – Same as III.1 but Fifth Element Reduced to -2 from 1

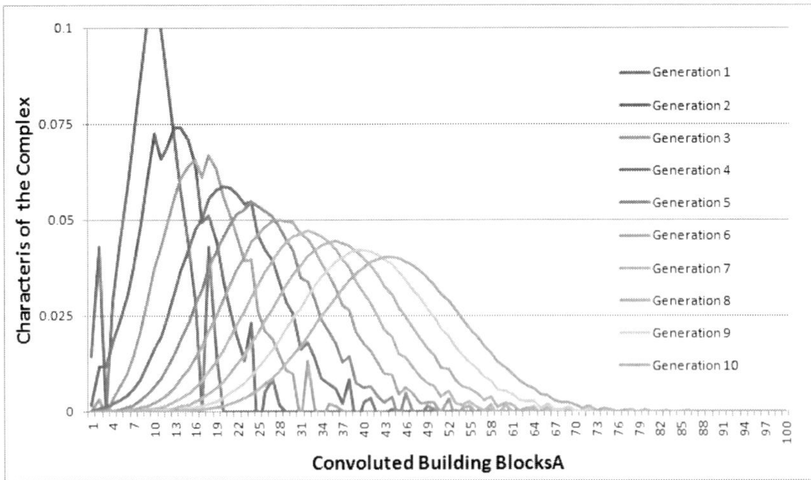

Figure III.4 – Same as III.1 but 3rd Element = 2, 4th Element =-2 and 5th Element =2

Conventional ANN models would employ polynomials by fitting equations to data, which are known to suffer from over-redundancy. However, the use of the new calculus provides new computational possibilities, where it is far easy to work out sums and other operations. So the task of solving over-redundant equations can be replaced by searching for the kernel of the problem for any given datasets to achieve best match between the model and observed data. It will be surprising that Nature has selected computational skills to solve these over-redundant equations but it is not difficult to see that each kernel is associated with a certain level of regeneration or convolution to produce selectable effects.

REFERENCES

Armstrong, K., 2005. A Short History of Mythology. Publ. By Canongate, Edinburgh.

Brualdi, R. A. (1992), Introductory Combinatorics, published by Pearson.

Bucchianico, A. Di., Loeb, D. (200), "A Selected Survey of Umbral Calculus," the Electronic *Journal of Combinatorics,* http://www.combinatorics.org/index.html.

Chalmers, D.J., (2002), *Philosophy* of Mind: Classical and Contemporary Readings, published by Oxford University.

Checkland, P., (1981), Systems Thinking, Systems Practice, John Wiley and Sons Ltd.

Churchland, PM and Churchland, P.S., (1998) On the Contrary: Critical Essays 1987-1997. Cambridge, Massachusetts: The MIT Press.

Churchland, P.M. (1996), The engine of reason, the seat of the soul, The Massachusetts Institute of Technology.

Clark G. (1994) Space, Time and Man, Published by Cambridge University Press, Canto Edition, Pp106-07.

Copeland (1993), Artificial intelligence, a philosophical introduction, Blackwell Publishing Ltd Daniel, D. (1993), Ai: The Tumultuous History of the Search for Artificial Intelligence.

Dawkins, R. (1986), The Blind Watchmaker, New York: W. W. Norton and Company, Inc.

Dawson, R., Pepper R. and Wang, M. (2011), "An agent-based model for risk-based flood incident" *Journal of Natural Hazards,* (DOI 10.1007/s11069-011-9745-4).

Dennett, D.C. (1991), Consciousness explained, The Penguin Press.

Edelman, G. (1992), Bright air, bright fire, on the matter of the mind, Penguin Group.

Eliade, M., (1989), Shamanism, archaic technique of ecstasy, Published by Arkana.

Feyerabend, P. (1963), "Mental events and the brain," *J. of Philosophy,* 60:295-296.

Franklin, S., and Graesser, A., (1997), Is it an Agent, or just a Program?: A Taxonomy for Autonomous Agents. Proceedings of the Third International Workshop on Agent Theories, Architectures, and Languages, published as Intelligent Agents III, *Springer-Verlag,* 1997, 21-35.

Frazer, J. G. (2010), Some Primitive Theories of the Origin of Man , http://www.fullbooks.com/Darwin-and-Modern-Science4.html, and cited in Clark G. (1994) *Space, Time and Man,* Published by Cambridge University Press, Canto Edition, Pp106-07.

Guilfoyle (1998, p91), "Vendors of intelligent agent technologies: a market overview," in Agent technology, function, applications, and markets, ed. Jennings, N.R. and Woolridge M.I., Springer, 1998.

Guisepi, R. A., (editor), Civilizations Past And Present, The Development Of The Human Race, The Civilizations Of The Ancient Near East, (http://history-world.org /paleolithic1.htm), access on 10 Octomer 2011.

Hassoun, M.H., (1995), Fundamentals of artificial neural networks, The Massachusetts Institute of Technology.

Heil, J., (2004), Philosophy of Mind: A Guide and Anthology, Oxford University Press.

Hewitt, Carl (1977),"Viewing control structures as patterns of passing messages," *Artificial intelligence*, 8(3), 323-364.

Huhns, M.N. (2001), "Interaction-oriented programming", Agent-oriented software engineering. Proceedings of the First International Workshop (AOSE-2000), Ed. P. Ciancarini and M.J. Wooldridge.

Hundert, E.M. (1990), Philosophy, psychiatry and neuroscience – three approaches to the mind, Clarendon Press, Oxford.

Janca P.C., and Gilbert, D. (1996), "Practical design of intelligent agent systems" in Agent technology, function, applications, and markets, ed. Jennings, N.R. and Woolridge M.I., *Springer,* 1998.

Jennings, N.R. (2000), "On agent-based software engineering," Artificial Intelligence, Volume 117, Issue 2, pp277-296, (http://www.sciencedirect.com/science/article /pii/ S0004370299001071).

Jones, M.T. (2008) Artificial Intelligence: A Systems Approach, Infinity Science Press LLC.

Khatibi, R. H., Jackson, D., Curtin, J., Whitlow, C., Verwey, and Samuels, P. (2003), "Vision statement on open architecture for hydraulic modelling software tools", *Journal of Hydroinformatics*., Vol. 6.1, PP 57-74 (http://www.iwaponline.com/jh /006/ jh0060057.htm).

Khatibi, R.H., (2003), "Systemic knowledge management in hydraulic systems: I. A postulate on paradigm shifts as a methodological tool", Journal of Hydroinformatics; Vol. 5.2, PP 127-140.

Khatibi, R., Lincoln, R., Jackson, D., Surendran, S., Whitlow, C., and Schellekens, J., (2004), "Systemic Data Management for Mathematical Modelling of Environmental Problems", *Journal of Management of Environmental Quality,* Vol. 15, No. 3, Pp. 318-330, (http://www.emeraldinsight.com/journals.htm?articleid=1454620).

Khatibi, R., (2006) Barriers inherent in Flood Forecasting and their Treatments, Chapter 29 of the book: River Basin Management for Flood Risk Mitigation, Ed. D.W. Knight and A.Y. Shamseldin; (http://www.crcnetbase.com/doi/abs/10.1201/9781439824702.ch29).

Khatibi, R., (2011), "Evolutionary Systemic Modelling for Flood Risk Management Practices," *Journal of Hydrology,* Vol. 401 Issue 1-2, Pp 36–52 (http://dx.doi.org /10.1016/j.jhydrol.2011.02.006).

Khatibi, R., (2012a), "Learning from Natural Selection in Biology: Reinventing Existing Science to Generalise Theory of Evolution – Evolutionary Systemic Modelling"

Khatibi, R., (2012b) "Axiomatisation of evolutionary transitions in mathematics by evolutionary and systemic thinking" – a chapter in this e-book.

Khatibi, R., (2012c) "A Vision for Mathematical Modelling by using Evolutionary Systemic Modelling" – a chapter in this e-book.

Khatibi, R., (unpublished) "Analytical Calculus of Sequences of Natural Numbers in Combinatorial Spaces," – summarized in this paper

Lyons, W. (1995), Modern philosophies of mind, Orion Publishing Group.

McCorduck, P. (2004), Machines who think: a personal inquiry into the history and prospects of artificial intelligence, A Peters Ltd, (http://www.pamelamc.com/html /machines_who_think.html and http://www.amazon.com/exec/obidos/ASIN /1568812051/ pamelamccordu-20).

McNeile, A., and Simons, N. (2004), Methods of behaviour Modelling – *A Commentary on Behaviour Modelling Techniques for MDA*, http://www.metamaxim.com/download /documents/Methods.pdf (accessed on 1 Oct 2011).

MEA, (2005), Millennium Ecosystem Assessment, *Ecosystems and Human Well-being: Synthesis*. Island Press, Washington DC.

Motah, M, (2006), "The Ontogeny of Memory and Learning: Natural Intelligence versus Artificial Intelligence in Information Technology Education," Proceedings of the 2006 Informing Science and IT Education Joint Conference, Salford, UK – June 25-28, Pp221-230.

Nattrass, B.F. and Altomare, M. (1999), The natural step for business: wealth, ecology and the evolutionary corporation, New Society Publishers.

Nilsson, N.J. (1974) "Artificial Intelligence," IFIP Congress 74, Stockholm, Sweden, August 5-10, 1974. Cited in Klopf, A.H. (1975), "A Comparison of Natural and Artificial Intelligence," SIGART Newsletter No. 52 June 1975.

Nilsson, N.J. (1998), Artificial Intelligence: a new synthesis, Morgan Kaufmann Publishers, Inc.

Nwana H.S. and Ndumu, D.T., (1996), "A brief introduction to software agent technology," in Agent technology, function, applications, and markets, ed. Jennings, N.R. and Woolridge M.I., Springer, 1998.

ONS and DfT (2008) Transport statistics bulletin: National Travel Survey 2008. Department for Transport, London.

Penrose, R. (1994), Shadows of the mind, Oxford University Press.

Penrose, R. (1997), The large, the small and the human mind, Cambridge University Press.

Putnam, H. (1975), Mind, Language and Reality. Philosophical Papers, vol. 2, Cambridge: Cambridge University Press.

Ratsch, U., Richter, M.M., and Stamatescu, I.-O. (1998) Intelligence and artificial intelligence: an interdisciplinary debate, Published by Springer-Verlaq.

Rey, G. (1997), Contemporary philosophies of mind, Blackwell Publishers Inc.

Rice, S.A., 2007. Encyclopaedia of Evolution, Facts on File, Inc. Infobase Publishing, New York, NY 10001.

Rorty, R. (1965), "Mind-body identity, privacy and categories," *Riveiew of metaphysics,* 29, 1:24-54.

Ryle, G., (1990), The concept of mind, Penguin Books.

Savitz, A.W. and Weber, K. (2006) The triple bottom line: how today's best-run companies are achieving, Jossey Bass, A Wiley Imprint.

Schmuller, J., (2002), Teach yourself UML in 24 hours, Sams Publishing.

Searle, J.R. (1992), The rediscovery of the mind, The Massachusetts Institute of Technology.

Searle, J.R. (1997, P. ii), The mystery of consciousness, Granata Publications.

Shoham, Y. (1990). Agent-Oriented Programming (Technical Report STAN-CS-90-1335). Stanford University: Computer Science Department.

Shoham, Y. (1993), "Agent-oriented programming". Artificial Intelligence, 60(1):51-92.

Singh. M.P. (1996), "Toward interaction-oriented programming", Proceedings of the 2nd International Conference on Multi-Agent Systems (ICMAS-96), page 457.

Sokolowski, R. (1988) "Natural and Artificial Intelligence" Vol. 117, No. 1, Artificial Intelligence (Winter, 1988), pp. 45-64; Published by: The MIT Press on behalf of American Academy of Arts and Sciences; Stable URL: http://www.jstor.org /stable/20025138.

Stich, S.P., (1987). "Review of Searle, J., Minds, Brains and Science." Philosophical Review XCVI, 1:129-133.

Strutt, J., and Lakey, J.R.A., Education, Training and research in emergency planning and management; Conference transactions of IMechE on *Emergency Planning and Management*; pp. 75-98; published by Mechanical Engineering Publications Limited for the Institute of Mechanical Engineering, London.

von Bertalanffy, L., 1956. General systems theory. *General Systems,* I, 1–10.

von Bertalanffy, L., 1962. General system theory. General Systems, VII, 1–20.

Weiß, G. (2002), "Agent Orientation in Software Engineering," Knowledge Engineering Review, January 2002.

Wellman, M.P. (1996), Market-oriented programming: Some early lessons. In S.H. Clearwater, editor, Market-based control, pages 74-95, World Scientific.

Whitehead, Alfred North, and Bertrand Russell (1910), *Principia Mathematica*, 3 vols, Cambridge University Press.

Wilson, E.O. (1984). *Biophilia*. Cambridge: Harvard University Press.

Yudkowsky, E. (2006), "Artificial Intelligence as a Positive and Negative Factor in Global Risk" Global Catastrophic Risks, eds. Nick Bostrom and Milan Cirkovic Draft of August 31, 2006.

Zantonavitch, K., Sumerian Philosophy, (http://rebirthofreason.com/Articles/Zantonavitch /Sumerian_Philosophy.shtml), accessed on 1 November 2011.

In: Natural Selection
Editors: J. R. Lynch and D. T. Williamson

ISBN: 978-1-62257-467-4

Chapter 3

AN EVOLUTIONARY COMPARATIVE VIEW FOR LONGEVITY-RELATED OXIDATIVE STRESS RESISTANCE MECHANISMS IN ANIMAL SPECIES

Reinald Pamplona[1,*] and Gustavo Barja[2]
[1]Department of Experimental Medicine,
University of Lleida-IRBLleida, Lleida, Spain
[2]Department of Animal Physiology II,
Complutense University, Madrid, Spain

ABSTRACT

The increase in oxygen concentration in the Earth's atmosphere represented an important selective pressure for living organisms and contributed to set up the pace of evolutionary changes in physiological and metabolic systems. Despite the evolution of molecular mechanisms to use oxygen for efficient energy production represented a relevant evolutionary innovation, the reduction-oxidation (or redox) reactions associated with its use have been responsible for the production of free radicals and, more in general, of reactive species (RS). These products of oxygen metabolism may have damaging effects on organism, in essence, due to oxidation of essential cellular components. Consequently, animals have evolved mechanisms to avoid or minimize the production of RS of oxidative metabolism as well as of antioxidant defense systems needed to limit the pro-oxidant activity of RS. The result was a high diversity in molecular and structural antioxidant defenses as well as in redox signaling pathways perfectly integrated in the cellular metabolic machinery. However, the antioxidant machinery is not 100% efficient in mopping RS up, especially when there is a strong imbalance between production of RS and antioxidant response. The consequence of such an imbalance is the generation of oxidative damage to biomolecules. This kind of biochemical stress of cells has been called oxidative stress, which may be defined as the

* Corresponding author: Dr. Reinald Pamplona, Department of Experimental Medicine, University of Lleida, Biomedical Research Institute of Lleida, Biomedicine Building I, Avenue Alcalde Rovira Roure 80, Lleida 25198, Catalonia, Spain. Phone: +34 973702442; Fax: +34 973702426.
E-mail: reinald.pamplona@mex.udl.cat.

rate at which oxidative damage is generated. Recent studies on animal species suggest that oxidative stress could also represent a universal constraint of life-history evolution, and a modulator of phenotypic development. Oxidative stress could therefore have worked as a relevant selective pressure, and that is why animal taxa have evolved a myriad of mechanisms to regulate and adjust the redox balance. This high diversity in antioxidant mechanisms also suggests that organisms have a great deal of flexibility in how they deal with RS challenges across time, conditions, and tissue types, as well as that different organisms may have evolved different strategies for dealing with similar oxidative challenges. In the following paragraphs, we review the high diversity of molecular and structural defenses against RS evolved by animals. In particular, it is our goal to describe physiological mechanisms underlying the production of RS and the biochemical regulation of defense mechanisms, to extend the traditional concept of antioxidant to other structural and functional factors affecting the 'whole' organism, and to review the state of the art about the link between oxidative stress and longevity. It is proposed that evolution of long-lived animals tends to decrease the rate of generation of endogenous damage and to increase the molecular resistance to the oxidative damage in long-lived species.

Keywords: aging; birds; complex I; electron transport chain; free radicals; mammals; membrane unsaturation; mitochondria; mitochondrial DNA; oxidative damage; peroxidizability index; phylogenomics; protein methionine content; reactive carbonyl species; invertebrates; vertebrates

INTRODUCTION

Oxygen appeared in significant amounts (about 1%) in the Earth's atmosphere some 2,700 million years ago, and geological evidence suggests that this was due to the photosynthetic activity of the cyanobacteria. As they split water to obtain their essential requirement for hydrogen atoms, thus gaining reducing power to drive their metabolism, microorganisms released tonnes of oxygen into the atmosphere [Falkowski and Godfrey 2008]. First eukaryotes with mitochondria, resulting via endosymbiotic events and lateral gene transfer to eukaryotic host cells, appeared between 2,300 and 2,000 million years ago when rise in oxygen levels to about 5-18% took place [Embley and Martin 2006]. When the Earth's atmosphere changed to the oxygen-rich state that we know today, the anaerobic life forms existing at that time adapted, died, or retreated to places with little or no oxygen was present [Lane 2002].

In this scenario of evolutionary change, living cells emerged as open systems, exchanging matter and energy with their surroundings, extracting and channeling energy to maintain them in a dynamic steady state distant from equilibrium. One of the distinctive properties of living organisms is a high degree of chemical complexity and molecular organization, defined functions for each of their components and tightly regulated interactions among them. The overall network of interconnected reaction sequences that interconvert cellular metabolites constitutes cellular metabolism [Nelson and Cox 2005]. Metabolism is adapted and optimized to achieve balance and economy [Melendez-Hevia et al. 2008]. So, one of the attractors involved in biomolecular evolution is the minimizing of unwanted side reactions. However, uncontrolled and potentially deleterious endogenous reactions occur, even under physiological conditions.

Life on this planet first evolved in a reducing atmosphere. The shift from a reducing environment to an oxidizing one certainly resulted in an evolutionary pressure [McCord 2000]. Consequently, aerobic life demanded metabolic innovations. Organisms that evolved the capacity to cope with oxygen did best, because they could simultaneously evolve to use oxygen for efficient energy production and for other oxidative reactions. In fact, reduction-oxidation (or redox) reactions (redox reactions involve the transfer of electrons or hydrogen atoms from one reactant to another) are at the core of the metabolic machinery. With the exception of some anaerobic and aerotolerant species, all organisms require oxygen for efficient production of energy by the use of electron transport chains that eventually donate electrons to oxygen (oxygen is an ideal terminal electron acceptor for biological systems), such as those in the mitochondria of eukaryotic cells and the cell membranes of many bacteria. Organisms that tolerated oxygen could also evolve to use it for metabolic transformations catalysed by oxidase, oxygenase and hydroxylase enzymes. Evolution of efficient energy production allowed the emergence of complex multicellular organisms [Falkowski and Godfrey 2008].

As oxygen levels rose in the atmosphere, living organisms began, however, to experience oxygen toxicity. The damaging effects of oxygen and other reactive molecular species on cells are, in essence, due to oxidation of essential cellular components [Halliwell and Gutteridge 2007]. Consequently, in addition to evolutionary attempts to avoid or minimize the production of reactive byproducts of oxidative metabolism, another very important direction was the induction of antioxidant defence systems [Halliwell 1999]. The result was a huge diversity of such antioxidant molecules, as well as redox signaling pathways perfectly integrated in the cellular metabolic machinery in order to regulate and to minimize the impact of the resulting oxidative-derived molecular damage (oxidative stress). So, reactive species production becomes a primary and universal constraint in life-history evolution [Dowling and Simmons 2009].

The involvement of oxidative stress in determining the animal longevity is intrinsically related to its key role in the origin and evolution of life. Oxidative stress is not only a natural consequence of aerobic life, but it has also been a determinant factor which demanded structural and functional adaptations to the cells and organisms that, in turn, determined their longevities [Pamplona and Barja 2007].

THE LONGEVITY-RELATED OXIDATIVE STRESS RESISTANCE MECHANISMS IN ANIMAL SPECIES

Rate of Generation of Mitochondrial Reactive Oxygen Species and Animal Longevity

Generation of Mitochondrial Reactive Oxygen Species (ROS)

Inside mitochondrial respiratory chain electrons from reduced substrates are passed from complexes I and II of the electron transport chain (ETC) through complexes III and IV to oxygen, forming water and causing protons to be pumped across the mitochondrial inner membrane. The proton motive force set up by proton pumping drives protons back through the ATP synthase in the inner membrane, forming ATP from ADP and phosphate [Scheffler

1999; Treberg and Brand 2011]. In this context, a major side reaction is that electrons may leak from the respiratory chain and react with oxygen to form the free radical superoxide. Superoxide anion, the product of a one-electron reduction of oxygen, is the precursor of most ROS and a mediator in oxidative chain reactions [Loschen et al. 1971; Boveris and Chance 1973; Loschen et al. 1974; Forman and Kennedy 1974; Chance et al. 1979]. There is, however, a lack of stoichiometric coupling of reactive oxygen species (ROS) production to oxygen consumption [Barja 2007]. Currently, it is well known that mitochondrial ROS (mtROS) generation occurs at complex I [Herrero and Barja 1997; Barja and Herrero 1998; Genova et al. 2001; Kushnareva et al. 2002; Lambert and Brand 2004; Kudin et al. 2004; Ohnishi et al. 2005; Chen et al. 2005] and at complex III [Boveris et al. 1976; St-Pierre et al. 2002]. Concerning the electron transport component responsible for mtROS generation within complex I, flavin mononucleotide, ubisemiquinone species, or iron-sulphur clusters have been proposed [Pamplona and Barja 2003; Brand et al. 2004; Lenaz et al. 2006; Pamplona and Barja 2006; Murphy 2009; Koopman et al. 2010; Brand 2010; Stefanatos and Sanz 2011].

Are there biological adaptation mechanisms with ability to modulate the rate of mitochondrial free radical generation? Available evidence seems to suggest that this is the case. Among these adaptations the following can be discerned: i/ to adapt the amount of the respiratory complex/es responsible for ROS generation. Thus, a decrease in the amount of complex I protein can lead to a decreased rate of ROS generation [Pamplona et al. 2005; Lambert et al. 2010]; ii/ to adjust the amount of uncoupling proteins. The mitochondrial superoxide production is very sensitive to the proton motive force, so it can be strongly decreased by mild uncoupling [Echtay et al. 2003; Azzu et al. 2010; Toime and Brand 2010]; iii/ to regulate the degree of electronic reduction of these generators: the higher their degree of reduction, the higher will be their rate of ROS production [Kushnareva et al. 2002; Barja 2007]; and iv/ to modify by enzymatic and non-enzymatic pathways such as S-nitrosation [Foster and Stamler 2004; Bruckdorfer 2005; Dahm et al.2006], acetylation [Janssen et al. 2006], and glutathionylation [Han et al. 2003; Taylor et al. 2003] specific peptides of the ROS generators.

Mitochondrial ROS Production and Animal Longevity

Several works have documented the relationship between mitochondrial ROS production and animal longevity. Early studies [Sohal et al. 1990; Ku and Sohal 1993; Ku et al. 1993], however, only included animal species following the "rate of living theory" *–the inverse relationship between longevity and metabolic rate-* [Pearl 1928], so the results obtained could also be interpreted as a correlate of that phenomenon [for a critical review of the rate of living theory see Lints 1989; Olshansky and Rattan 2005; Olshansky and Rattan 2005; Hulbert et al. 2007; Demetrius et al. 2009]. Species with short longevity could show high mitochondrial ROS production simply because their rates of mitochondrial oxygen consumption are higher. Fortunately, comparative physiology offers more possibilities to test the hypothesis that mitochondrial ROS production is involved in the rate of aging and, consequently, in determining longevity. Three groups of homoeothermic vertebrates have an extraordinarily high longevity in relation to their body size and metabolic rate: birds, bats and primates [Holmes et al. 2001]. Birds can be a strong test because if a low rate of ROS production contributes to slow the aging rate, the mitochondria of birds should show a low rate of ROS generation in spite of the high rate of oxygen consumption of these animals [Holmes et al. 2001]. Thus, in spite of their high rates of whole body O_2 consumption, parakeets, canaries and pigeons (with longevities of 21, 24 and 35 years, respectively) have lower rates of mtROS generation than mice and rats (3.5 and 4 years of longevity, respectively), even though their metabolic rates and body sizes are of a similar magnitude [Ku and Sohal 1993; López-Torres et al. 1993; Barja et al. 1994; Herrero and Barja 1998; St Pierre et al. 2002; Lambert et al. 2010]. Hence, all the comparative investigations performed in mammals and birds show that mitochondrial ROS generation is lower in long- than in short-lived animals. In addition, the difference in ROS production between these species occurred at only one of the electron transport complexes: complex I. The low rate of mtROS production of birds, in spite of their high rates of oxygen consumption, is possible because a/ the percent of total electron flow in the respiratory chain directed to ROS production (*the % free radical leak, %FRL*) is lower [Barja 2004], and b/ the total amount of mitochondrial complex I is also lower [Pamplona et al. 2005; Lambert et al. 2010]. This means that their respiratory chain transports electrons more efficiently avoiding univalent electron leaks to oxygen upstream of cytochrome oxidase.

Other interspecies comparisons between rodents with widely different longevities but similar body size [see Barja 2004], or between rodents and bats [Brunet-Rossinni 2004] also showed that the longest-lived species had lowest mitochondrial ROS production or leak. Recent studies have also found much lower rates of ROS generation in human (*primates -and especially humans- also live much longer than expected for their body size and metabolic rate*) than in rat brain mitochondria [Kudin et al. 2004]. Finally, if mitochondrial oxygen radical production controls aging rate, the rate of ROS production of animals with different metabolic rates and body sizes but similar longevity should be equal. This is exactly what happened in another interspecies comparison when mitochondrial ROS production of mouse (longevity, 3.5 years) and rat (longevity, 4 years) were compared [Herrero and Barja 1997; Herrero and Barja 1998]. The longevity of these species are similar (3.5 and 4 years), whereas metabolic rate is threefold higher and body size is around 15-fold lower in the mouse than in the rat.

Globally, these studies found that the rate of mtROS production is lower in the tissues of long-lived than in those of short-lived animal species. However, some comparative studies on longevity and physiology have not taken into account the possible confounding effects of body size and phylogeny [Speakman 2005]. Indeed, the significant relationship between longevity and ROS production by heart mitochondria reported by Ku et al. (1993) became non-significant ($P = 0.62$) when the published data were subjected to residual analysis. In other words, in that particular dataset, the changes in ROS production rates and longevity could be explained by their dependence on body size, confounding any interpretation of the effects of ROS production on longevity. Similarly, published correlations between longevity and ROS production in mitochondria from kidney [Ku et al. 1993] and liver [Sohal et al. 1990] became non-significant when re-analyzed in the same way (P values were 0.28 and 0.15 for kidney and liver, respectively). For this reason, it is particularly relevant the results obtained in a recent study where 12 animal species (final dataset comprised mouse, rat, white-footed mouse, naked mole-rat, Damara mole-rat, guinea pig, baboon, little brown bat, Brazilian free-tailed bat, ox, pigeon and quail) were analyzed [Lambert et al. 2007]. In this work, the reported results suggest that, in general, longevity associates negatively with mtROS production across species over two classes of vertebrate homeotherms, independently of body size or phylogeny. However, further studies with greater numbers of species were required to test the generality of the hypothesis, particularly because after the phylogenetically independent contrast analysis one outlier was found: the naked mole-rat.

The naked mole-rat is the longest-lived rodent known [Sherman and Jarvis, 2002; Buffenstein, 2005] with a longevity higher than 28 years and with a body size of approximately 30 g. The possible reasons for this exception are not known, but naked mole-rats are 'unusual' in several other ways, being highly eusocial, having poor endothermy and exhibiting nonavoidance of inbreeding depression [Buffenstein, 2005]. They have virtually undetectable cellular glutathione peroxidase activity [Andziak et al., 2005] and apparently high levels of some oxidative damage markers in certain tissues [Andziak et al., 2006]. Consequently, for this particular case, it is plausible to speculate that other(s) longevity-related oxidative stress resistance mechanisms could play a more significant role than mtROS production (see later). Alternatively, it may be that some of the underlying assumptions involved in the analysis are invalid. The method of phylogenetically independent contrasts assumes equal rates of evolution along all branches of the tree for both ROS production and longevity. If this assumption is invalid for either trait in naked mole-rats, then the calculated branch length will be incorrect resulting in a spurious contrast value.

In agreement with the low mitochondrial free radical production of long-lived animal species, there is an adaptive response concerning endogenous cellular antioxidant systems. Thus, long-lived animal species, including vertebrates and invertebrates, constitutively have lower (instead of higher) tissue levels of antioxidant enzymes and low molecular weight endogenous antioxidants [Pérez-Campo et al. 1998; Sanz et al. 2010], as well as repair systems [Salway et al. 2010], than short-lived ones [reviewed in Pamplona and Barja 2007; Sanz et al. 2009]. That characteristic can thus explain why endogenous tissue antioxidants correlate negatively with longevity across species: long-lived animals have constitutively low levels of antioxidants because they produce ROS at a low rate. If long-lived animals had high rates of ROS production together with their very low levels of endogenous antioxidants, their tissue cells would not be able to maintain oxidative stress homeostasis. With these premises, it is suggested that decreasing mitochondrial ROS production instead of increasing antioxidants or repair systems makes sense when considered from the point of view of evolution of longevity among species. It would be very inefficient to generate large amounts of ROS and, afterwards, try to intercept them before they reach cellular components like DNA, or even worse, try to repair DNA after heavily damaging it. This makes even more sense taking into account that cells possess the ability to regulate and to adjust the gene expression of cellular antioxidants – cell exposition to oxidants can regulates the expression of a large number of genes, tipically between 1% and 5% of the genome [Giles 2009]- and the potentially high energetic cost of continuously maintaining high levels of antioxidant and repair molecules in cells (to our knowledge, however, no data are available about energetic costs to maintain oxidative stress homeostasis for a given animal species).

Findings from a Phylogenomic Approach

In accordance with this interpretation, a recent study [Jobson et al. 2010] with a phylogenomic approach to identify the genetic targets of evolution for increased longevity in mammals shows, by comparing the nonsynonymous and synonymous evolution of 5.7 million codon sites across 25 species, that genes involved in antioxidation have played no detectable role in the evolution of longevity in mammals. So, in this context, it is reasonable to suppose that lowering the rate of mtROS production could be a highly conserved mechanism developed during evolution across species. It would allow long-lived species to decrease steady-state oxidative damage to cellular constituents and, consequently, the aging rate.

Mitochondrial DNA and Animal Longevity

Physiological Mitochondrial DNA (mtDNA) Oxidation

Among different kinds of molecular damage caused by ROS, that to mtDNA is important for longevity because it can lead to irreversible loss or alteration of its coded information [Marnett and Plastaras 2001], which will be especially deleterious in post-mitotic tissues. Determinant mechanisms of the steady-state level of mtDNA damage include: i) the structural susceptibility of the mtDNA molecule to damage (*see next section*); and ii) the mtDNA location. Mitochondrial DNA is located very close to or even in contact with the site/s of mitochondrial ROS production [Scheffler 1999]. The main electron transport components responsible for ROS generation, the flavin and all the FeS clusters, are located in the hydrophilic complex I domain facing the mitochondrial matrix [Barja 2007; Murphy 2009; Treberg et al. 2011], which would facilitate close proximity, or even the contact between mtDNA and the ROS generator. In contrast, complex III produces ROS directed to the cytosolic side of the inner membrane, although recent studies suggest that part of the production could also occur towards the matrix side [Brand 2010]. These facts can justify that, in post-mitotic tissues, oxidative damage is around 5-9 fold higher in mtDNA than in nDNA [Richter et al. 1988; Hamilton et al. 2001], as described in the brain and heart of all the 11 mammalian and bird species studied [Herrero and Barja 1999; Barja and Herrero 2000].

mtDNA Oxidation and Animal Longevity

Because long-lived animal species have low rates of mtROS generation, this should be reflected in the steady-state level of oxidative damage and the accumulation of somatic mutations in their mtDNA. In agreement with their low rates of mitochondrial ROS production, most long-lived animal species studied have lower steady-state levels of 8-oxodG (*a biomarker of DNA oxidation*) in their mtDNA [Barja and Herrero 2000; Sanz et al. 2009] and lower rates of urinary excretion of 8-oxo-7,8-dihydroguanine [Foksinski et al. 2004] than short-lived ones. Long-lived birds also usually show lower oxidative damage to mtDNA in their tissues than short-lived rodents of similar body size [Herrero and Barja 1999]. The naked mole-rat -as exceptionally long-lived animal model– exhibit, in a contradictory way, greater levels of mtDNA damage in liver and kidney, but lower levels of urinary excretion than physiologically age-matched mice, and equal to that of same-aged mice [Andziak et al. 2006]. Finally, it has been also described that the rate of accumulation of mtDNA mutations with age is also much slower in humans than in mice [Wang et al. 1997; Gredilla et al. 2010].

In order to maintain mitochondrial genomic integrity and to avoid the gradual accumulation of oxidative DNA base lesions, mtDNA damage is repaired by a number of mechanisms [for review see Bohr 2002]. In general, oxidized mtDNA bases are removed by base excision repair (BER) mechanisms [Bohr 2002; Gredilla et al. 2010] and although these correcting mechanisms show high fidelity, DNA lesions can still accumulate with age. DNA polymerase enzymes (as component of the BER machinery) have a proof-reading error-correcting capacity, and when mitochondrial DNA polymerase proof-reading activity was eliminated in mice there was a 3-5 fold increase in mitochondrial mutations and a decreased longevity in such mice [Trifunovic et al. 2004; Kujoth et al. 2005]. In support for a relevant role of mtDNA mutations in the rate of aging, in a recent study has been demonstrated that mutations in mtDNA profoundly affect the assembly of fully functional ETC complexes, leading to mitochondrial dysfunction and ultimately, to the activation of a mitochondrial-mediated apoptotic program, as mechanism underlying the accelerated aging process of this experimental model [Hiona et al. 2010].

All that might suggest that the flux rates of both ROS attack on and repair of DNA are much higher in the mtDNA of short-lived than in that of long-lived animals, and are also much higher in the mtDNA than in the nDNA of all species [Barja 2000]. In addition, the higher rate of mtROS production and mtDNA susceptibility to oxidative damage of short-lived animals may be an important cause of their much faster rate of accumulation of mtDNA mutations during aging [Wang et al. 1997; Melov et al. 1999].

Mitochondrial DNA and Longevity from a Phylogenomic Approach

The susceptibility of mtDNA to damage could be related to the simplest property of a DNA sequence, the proportion of A, C, G and T in the mtDNA molecule. In a recent study [Samuels 2005], analyzing the 'light strand' mtDNA sequence of 94 animals species (including invertebrates, birds and mammals) it was found that short-lived species have higher A and T abundances and lower C abundances than the long-lived species, while G is almost uniformly low in all species. The light strand was given this name because it is unusually deficient in G, and data from this study shows that this characteristic low G abundance carries through to all of these species. A conclusion is that there is some strong evolutionary pressure to keep the G abundance of the light strand low. The standard hypothesis is that the asymmetric mode of replication of the mtDNA molecule causes an increase in the G to A transition mutations on the light strand, depleting it of G nucleotides. In addition to this finding, it was verified that this pattern of nucleotides affects to a DNA sequence property, the free energy (*a physical property of the double-stranded DNA molecule related to the binding energy between the two DNA strands. The more negative the free energy is, the less likely is the spontaneous separation of the two strands through thermal fluctuations, conferring to the mtDNA a greater structural stability and lesser susceptibility to damage*) [Samuels 2005]. The data showed a strong relationship between the mtDNA free energy and longevity. Thus, the longer the maximum longevity of a species, the lower is the free energy of mtDNA.

Considering that among the four nucleobases, guanine has the lowest oxidation potential and is thus generally most easily oxidized [Bjelland and Seeberg 2003], it is proposed that these mtDNA base sequence patterns and derived properties represent an evolutionary adaptation of long-lived species to increase mtDNA resistance to endogenous damage by chemical side-reactions and so to decrease the susceptibility of mtDNA to damage and mutation. A similar evolutionary adaptation has been demonstrated for proteins (changing the methionine content) and membrane unsaturation (changing the polyunsaturated fatty acids content) in long-lived species as detailed in a later section of this chapter.

In this line, in a recent study [Nabholz et al. 2008], analyzing the lineage-specific mitochondrial mutation rate across 1,696 mammalian species and comparing it with the nuclear rate, a selected decrease of nuclear base substitution rate was reported in long-lived species, in agreement with the evidence for a causal role of mtDNA mutations in aging

[Kujoth et al. 2007]. This suggests that natural selection tends to decrease the mitochondrial mutation rate in long-lived species. In accordance with this interpretation, the study from Jobson et al [2010] with a phylogenomic approach to identify the genetic targets of natural selection for increased longevity in mammals shows, by comparing the nonsynonymous and synonymous evolution of 5.7 million codon sites across 25 species, that genes involved in DNA replication/repair or antioxidation have played no detectable role in the evolution of longevity in mammals, reinforcing the relevance of the association mtROS production-mtDNA oxidation-mtDNA mutations.

So, it is plausible to suppose that neither 8-oxodG repair nor endogenous antioxidants should be continuously maintained at levels higher than required because that would be energetically costly. Both the antioxidants and the mtDNA repair systems are inducible [Giles 2009; Gredilla et al. 2010]. Thus, they can be transitorily increased at moments of higher than normal oxidative damage to quickly return to their basal levels when the situation is normalized. The lower steady-state oxidative damage in the mtDNA of long-lived animals is much better obtained by decreasing the rate of mtROS generation than by increasing antioxidants and repair because this is much less costly and much more efficient and simple.

Protein Methionine Content and Animal Longevity

Physiological Role of Methionine and Methionine-Derived Compounds

Methionine (Met) is an essential amino acid that plays a relevant role in protein structure and function and metabolism [Stipanuk 2004; Brosnan and Brosnan 2006], as well in the synthesis of eukaryotic proteins as the initiating amino acid (N-formylmethionine serves the same function in prokaryotic organisms) [Drabkin and Rajbhandary 1998; Brosnan and Brosnan 2006]. The role of Met in protein structure and function is basically related with the physico-chemical properties conferred by the presence of the sulfur atom and the terminal methyl group. Thus, the electronegative character of the sulfur atom and the presence of the terminal methyl group determine that Met is among the most hydrophobic of the amino acids [Brosnan and Brosnan 2006]. This means that most of the Met residues in globular proteins are found inside hydrophobic core, being a small fraction of the Met residues exposed at the protein surface; in membrane protein domains, methionine is often found to interact with the lipid bilayer. A key aspect of the metabolic functions of methionine is serving, via S-adenosylmethionine (SAM) [Catoni 1953], as the methyl donor for essentially all known biological methylation reactions [Clarke and Banfield 2001; Stipanuk 2004; Brosnan and Brosnan 2006]. Indeed, SAM, regarded as the essential biological methyl donor, is one of the most versatile compounds in life, second perhaps only to ATP, in a myriad of biochemical processes, many of which date back to the early origin of life.

The high-energy sulfonium compound SAM serves primarily as a methyl donor via reactions catalyzed by a variety of methyltransferases and involving a variety of acceptors. These SAM-dependent methylations are essential for metabolism of a variety of cellular components including creatine, epinephrine, carnitine, phospholipids, proteins, DNA and RNA [Clarke and Banfield 2001; Stipanuk 2004; Brosnan and Brosnan 2006]. In addition to its role as a methyl donor, it serves as a source of methylene groups for the synthesis of cyclopropyl fatty acids, amino groups in biotin synthesis, and aminopropyl groups for synthesis of the polyamines spermidine and spermine [Fontecave et al. 2004].

Finally, a compound generated in the methionine transsulfuration pathway deserves a special mention: cysteine [Banerjee et al. 2003; Finkelstein 2000]. Cysteine has a critical function in protein structure by virtue of its ability to form both interchain and intrachain disulfite bonds with other cysteine residues. Most disulfite linkages are found in proteins destined for export or residence on the plasma membrane. In addition with its structural role, cysteine is required for glutathione synthesis, besides taurine, which plays a central role as antioxidant defense against oxidative stress [Halliwell and Gutteridge 2007].

Methionine and Animal Longevity

All amino acid residues of proteins are susceptible to oxidative modification by one or more forms of ROS [Halliwell and Gutteridge 2007]. Although most amino acids in proteins can be modified by diverse endogenous oxidants, residues commonly modified include tyrosine, cysteine, and methionine [Dean et al. 1997]. Oxidation of methionine and cysteine residues generates methionine sulfoxide (MetSO) and disulfides in proteins, which may lead to loss of their biological activity [Chao et al. 1997], and deprives Met of its function as methyl donor. Methionine residues in proteins are particularly susceptible to oxidation by various reactive intermediates and the sensitivity of proteins to oxidative stress increases as a function of their number of methionine residues [Stadtman et al. 2003]. Consequently, the lower the protein methionine content, the higher should be the protein resistance to oxidative damage. In accordance with this, it has been recently found that methionine is the amino acid present in the whole tissue proteome from heart, as post-mitotic tissue, that strongly correlates with mammalian longevity, and that this correlation is negative [Ruiz et al. 2005]. In addition, protein methionine content is also lower in tissues (brain and skeletal muscle) of long-lived birds than in short-lived mammals of similar body size [Portero-Otin et al. 2004; Pamplona et al. 2005]. So, available evidences suggest that the longer the longevity of a species, the lower is its tissue protein methionine content. In these studies, however, a correlation for cysteine cannot be rule out because this amino acid was not measured by a constraint inherent to the analytical method used.

Most of the protein oxidative modifications are irreversible, i.e., the only known way to eliminate the protein adduct is to catabolize the protein. However, some oxidative modifications are reversible, e.g., methionine sulfoxide (MetSO) and disulfides. This kind of modification can be repaired by methionine sulfoxide reductase in a thioredoxin-dependent reaction [Moskovitz et al., 2001]. Most cells contain two different reductases (MsrA and MsrB) that catalyze the thioredoxin [Th(SH)2]-dependent reduction of MetSO back to Met. Significantly, the oxidized form of thioredoxin (Th-S-S) produced in the reduction of MetSO can be converted back to Th(SH)2 by the enzyme thioredoxin reductase (Thr) in an NADPH-dependent reaction. The disulfides can be reduced to cysteine residues by thiol-disulfide exchange reactions. These reactions are carried out by thiol transferases (glutathione S-transferase) that catalyze reactions between glutathione and thioredoxin to regenerate the protein sulfhydryl groups. S-thiolated formation of disulfide bond in protein with glutathione (e.g., protein S-SG disulfide) is largely reduced to the sulfhydryl group by glutaredoxin [Gravina and Mieyal 1993; Jung and Thomas 1996]. In this context, it is very illustrative that knocking out MsrA lowers longevity and increases protein carbonyls and sensitivity to hyperoxia in mice [Moskovitz et al. 2001]. The opposite manipulation, overexpression of Msr A, increases longevity and delays the aging process in *Drosophila* [Ruan et al. 2002]. On the other hand, the oxidized form of thioredoxin, produced during the reduction of methionine sulfoxide, can be converted back to reduced thioredoxin by the enzyme thioredoxin reductase. In agreement with a methionine oxidation role in aging, it has been reported that overexpression of thioredoxin reductase increases longevity in mice [Nakamura et al. 2002; Mitsui et al. 2002].

So, because organisms contain methionine sulfoxide reductases that can catalyze the thioredoxin-dependent reduction of the sulfoxides back to methionine, it was proposed that the cyclic oxidation/reduction of methionine residues might serve as antioxidants to scavenge ROS [Stadtman et al. 2002]. Consequently, it can be inferred from the available data that whole proteomes from aerobic long-lived animal species also evolved by maintaining the minor abundance of structural components highly susceptible to oxidative damage, obtaining a lower susceptibility to oxidative stress and, consequently, a higher structural stability.

Protein Methionine Content and Longevity from a Phylogenomic Approach

From an evolutionary comparative approach, a recent study [Aledo et al. 2010] addresses the association between species longevity and methionine usage in mitochondrial proteins using a meta-examination of mitochondrial genomes from 168 species of mammals, encompassing 24 different orders. Results showed that methionine usage and longevity exhibited a significant negative correlation. However, species are part of a hierarchically structured phylogeny, raising the possibility that data from different species may not necessarily be statistically independent from one another. Therefore, to correct for non-independence of the individual-species data, phylogenetically independent contrasts were calculated. Interestingly, the association between methionine content and longevity became more significant after phylogenetic correction. The association of methionine enrichment with decreasing longevity also remained significant after mathematical correction for body mass, a potentially confounding factor known to correlate with longevity [de Magalhães et al. 2007; Speakman, 2005]. The independence of the methionine-longevity correlation from the variable body mass indicates that even in species with identical size, the shorter-lived animals exhibit an enrichment of methionine. In fact, the correlation using mass-corrected variables became much stronger. Extending these findings, Aledo et al [2010] studied the relationship between longevity and methionine content within those eutherian orders with 10 or more species present in their compilation. All the analyzed orders, Carnivora ($n = 41$), Primates ($n = 23$), Artiodactyla ($n = 20$), Cetacea ($n = 18$) and Rodentia ($n = 10$), showed a negative covariance between longevity and methionine content. However, after correcting for the phylogenetic inertia only Carnivora, Primates and Rodentia exhibited statistically significant negative correlations between the variables of interest. Furthermore, within these orders, the link between high longevity and low methionine content remained significant after controlling for body mass.

So, mtDNA-encoded proteins from short-lived mammals are enriched in methionyl residues when compared to those from long-lived mammals. Interestingly, the results suggest that methionine addition to proteins in short-lived species, rather than methionine loss from proteins in long-lived species is responsible for the observed negative correlation between methionine content and longevity. Furthermore, methionine abundance in the general group of nuclear DNA encoded mitochondrial proteins did not significantly correlate with longevity, suggesting that the protein methionine content is circumscribed to a small subset of mitochondrial proteins localized in the inner mitochondrial membrane where ROS are generated. By contrast, when using this dataset, cysteine fails to show a link with longevity within the Mammalia class after correction for the effects of phylogeny.

However, when the meta-examination of mitochondrial genome sequences is extended to 248 animal species with known longevities, including mammals, birds, reptiles, amphibians, fish, insects, crustaceans, arachnids, and helminths, a slightly different and varied picture seems to appear [Moosmann and Behl 2008; Bender et al. 2008; Moosmann 2011]. So, the analysis reveals that the frequency with which cysteine is encoded by mitochondrial DNA is a specific and phylogenetically ubiquitous molecular indicator of aerobic longevity: long-lived species synthesize respiratory chain complexes which are depleted of cysteine. Cysteine depletion was also found on a proteome-wide scale in aerobic versus anaerobic bacteria, archaea, and unicellular eukaryotes; in mitochondrial versus hydrogenosomal sequences; and in the mitochondria of free-living, aerobic versus anaerobic-parasitic worms. The association of longevity with mitochondrial cysteine depletion persisted after correction for body mass and phylogenetic interdependence, but it was uncoupled in helminthic species with predominantly anaerobic lifestyle [Moosmann and Behl 2008]. By contrast, methionine usage in mitochondrially encoded proteins is strikingly higher than in nuclear-encoded proteins in the majority of animals (which is the inverse behavior compared to cysteine), and there is no significant correlation between mitochondrial methionine usage and longevity [Moosmann 2011].

The dichotomy methionine accumulation vs cysteine depletion in mitochondrial respiratory chain complexes probably should be considered in the context of antioxidant protection (for methionine) vs avoidance of oxidation-derived effects (for cysteine). So, the topological selectivity of longevity-dependent cysteine depletion to the inner mitochondrial membrane might find its most probable explanation in the concurrence of three circumstances that are not to be found elsewhere in the cell: exceptionally high concentrations of multispan transmembrane proteins of outstanding hydrophobicity, their particularly high exposure to ROS generated *in situ*, and the unique chemical reactivity of thiols in nonpolar environments. Under these premises, cysteine is avoided in order to prevent cysteine–catalyzed chain–transfer reactions and thus the formation of extended networks of internally cross-linked hydrophobic proteins that might eventually lead to a mitochondrial dysfunction [Moosmann 2011].

Membrane Unsaturation and Animal Longevity

Biomembranes and Unsaturated Fatty Acids

Biological membranes are dynamic structures that generally consist of bilayers of amphipathic molecules held together by non-covalent bonds [Yeagle 1993; Vance and Vance 1996]. Phospholipids, the predominant membrane lipids, consist of a hydrophilic head group with attached hydrophobic acyl chains. The acyl chains are either saturated (SFA), monounsaturated (MUFA) or polyunsaturated (PUFA) hydrocarbon chains that normally vary from 14 to 22 carbons in length [Wallis et al. 2002]. In eukaryotic cells from vertebrate species, the average chain length of a biological membrane is strictly maintained around 18 carbon atoms, and the relative distribution between saturated and unsaturated fatty acids follows a ratio of 40:60 [Pamplona 2008]. Polyunsaturated fatty acids (PUFAs) are generally synthesized by the modification of saturated fatty acid precursors that are products of fatty acid synthase. The desaturase enzymes, which are conserved across kingdoms, insert double bonds at specific carbon atoms in the fatty acid chain and the fatty acid elongation system elongates the precursors in two-carbon increments [Nakamura and Nara 2004]. The fatty acid desaturation pathway and the deacylation-reacylation cycle are the mechanisms responsible for the particular fatty acid composition of cell membranes.

The susceptibility of membrane phospholipids to oxidative alterations is related to two inherent traits, the physico-chemical properties of the membrane bilayer and the chemical reactivity of the fatty acids composing the membrane [Hulbert et al. 2007; Pamplona 2008]. The first property is related with the fact that oxygen and free radicals are more soluble in the fluid lipid bilayer than in the aqueous solution. Thus, membranes contain an interior organic phase, in which oxygen may tend to concentrate. Therefore, these differences in solubility are important when considering the availability of oxygen/free radicals for chemical reactions inside living systems: organic regions may contain more free radicals than aqueous regions [Moller et al. 2005; Gamliel et al. 2008] and, consequently, membrane lipids become primary targets of oxidative damage.

The second property is related to the fact that PUFA residues of phospholipids are extremely sensitive to oxidation. Every membrane phospholipid contains an unsaturated fatty acid residue esterified to the 2-hydroxyl group of its glycerol moiety. Many of these are polyunsaturated and the presence of a methylene group between two double bonds renders the fatty acid sensitive to ROS-induced damage, their sensitivity to oxidation increasing exponentially as a function of the number of double bonds per fatty acid molecule [Holman 1954; Bielski et al. 1983]. Consequently, polyunsaturated fatty acid side chains (two or more double bonds) are much more easily attacked by radicals than are saturated (no double bonds) or monounsaturated (one double bond) side chains.

Combining the relative susceptibilities of different fatty acids with the fatty acid composition of membrane lipids it is possible to calculate a peroxidizability index (a measure of the susceptibility to peroxidation; Peroxidizability Index (PI) = 0.025 x (% monoenoics) + 1 x (% dienoics) + 2 x (% trienoics) + 4 x (% tetraenoics) + 6 x (% pentaenoics) + 8 x (% hexaenoics)) for any particular membrane. The peroxidation index of a membrane is not the same as its unsaturation index (sometimes also called its "double bond index" (DBI) = 1 x (% monoenoics) + 2 x (% dienoics) + 3 x (% trienoics) + 4 x (% tetraenoics) + 5 x (% pentaenoics) + 6 x (% hexaenoics)]) which is a measure of the density of double bonds in the membrane. For example, a membrane bilayer consisting solely of MUFA will have an unsaturation index of 100 and a peroxidation index of 2.5, while a membrane bilayer consisting of 95% SFA and 5% DHA will have an unsaturation index of 30 and a peroxidation index of 40. This means that although the 5% DHA-containing membrane has only 30% the density of double bonds of the monounsaturated bilayer, it is 16-times more susceptible to peroxidative damage.

Lipid peroxidation generates hydroperoxides and endoperoxides which undergo fragmentation to produce a broad range of reactive intermediates called reactive carbonyl species (RCS) with three to nine carbons in length, the most reactive being α,β-unsaturated aldehydes [4-hydroxy-trans-2-nonenal (HNE) and acrolein], di-aldehydes [malondialdehyde (MDA) and glyoxal], and keto-aldehydes [4-oxo-trans-2-nonenal (ONE) and isoketals] [Esterbauer et al. 1991; Thorpe and Baynes 2003]. 2-Hydroxyheptanal (2-HH) is another major aldehydic product of lipid peroxidation of PUFAn-6, while 4-hydroxyhexenal (4-HHE) is generated in a lower yield. Additionally, a number of other short chain aldehydes are produced during lipid peroxidation through poorly understood mechanisms.

These carbonyl compounds, ubiquitously generated in biological systems, have unique properties contrasted with free radicals. For instance, compared with ROS, reactive aldehydes have a much longer half-life (i.e., minutes to hours instead of microseconds to nanoseconds for most free radicals). Further, the non-charged structure of aldehydes allows them to migrate with relative ease through hydrophobic membranes and hydrophilic cytosolic media, thereby extending the migration distance far from the production site. Based on these features alone, these carbonyl compounds can be more destructive than ROS and may have far-reaching damaging effects on target sites within or outside membranes [Pamplona 2011].

Carbonyl compounds react with nucleophilic groups in macromolecules (lipoxidation reactions) like proteins, DNA, and aminophospholipids, among others, resulting in their chemical, nonenzymatic, and irreversible modification and formation of a variety of adducts and crosslinks collectively named Advanced Lipoxidation Endproducts (ALEs) [Thorpe and Baynes 2003; Pamplona 2011].

These aldehydes were believed to produce only "cytotoxic" effects associated with oxidative stress, but evidence is increasing that these compounds can also have specific signaling roles inducing adaptive responses driven to decrease oxidative damage and improve antioxidant defenses [Pamplona 2008; Pamplona 2011]. Two of these mechanisms deserve a special mention: the regulation of uncoupling protein activity [Echtay et al. 2003; Brand et al. 2004], and the activation of the antioxidant response signaling pathway [Wakabayashi et al. 2004; Copple et al. 2008; Giles 2009], both involved in the prevention of oxidative damage effects.

With these premises, and maintaining other physiological properties, it is plausible to postulate that a low degree of fatty acid unsaturation in cellular membranes could be advantageous by decreasing their sensitivity to lipid peroxidation. This would also protect other molecules against lipoxidation-derived damage. So, it is proposed that membrane unsaturation acts as a structural antioxidant system and it is adapted to the animal longevity.

Membrane Unsaturation and Animal Longevity

The relationship between the degree of membrane unsaturation and longevity has been evaluated in mammals –including humans- and birds, strains of mice that vary in longevity and animal species that are exceptionally long-living, differences in longevity between queen and worker honeybees, strains of *D. melanogaster* that vary in longevity, and from experimental extensions of longevity in rodents, flyes and worms. Unfortunately, not data are currently available for exceptional models such as bivalves, bats, and non-human primates.

Thus, it has been found that long-lived animals (birds and mammals, including humans) have a lower degree of total tissue and mitochondrial fatty acid unsaturation and PI than short-lived ones (Table 1). In agreement with this, it has been demonstrated that in long-lived animal species a low degree of total tissue and mitochondrial fatty acid unsaturation is accompanied by a low sensitivity to *in vivo* and *in vitro* lipid peroxidation and a low steady-state level of lipoxidation-derived adducts in both tissue and mitochondrial proteins from organs like skeletal muscle, heart, liver, and brain [Pamplona 2008; Pamplona et al. 2002]. In this sense, lipofuscin, often considered a hallmark of aging also shows an accumulation rate that inversely correlates with longevity [Terman and Brunk 2004].

Table 1. Comparative studies between membrane unsaturation (peroxidizability index, PI) and longevity in animal species (by chronological order)

Species compared	Organ	PI long-lived species	References
Rat-Pigeon-Human	Liver mitochondria	Lower	[Pamplona et al. 1996]
SAM-R/1 vs SAM-P/1 mice	Liver	Lower	[Park et al. 1996]
8 mammalian species	Liver mitochondria	Lower	[Pamplona et al. 1998]
Rat vs pigeon	Heart mitochondria	Lower	[Pamplona et al. 1999]
Mouse vs canary	Heart	Lower	[Pamplona et al. 1999]
Mouse vs parakeet	Heart	Lower	[Pamplona et al. 1999]
Rat vs pigeon	Liver mitochondria	Lower	[Gutierrez et al. 2000]
Rat vs pigeon	Heart mitochondria and microsomes	Lower	[Gutierrez et al. 2000]
8 mammalian species	Heart	Lower	[Pamplona et al. 2000]
7 mammalian species	Liver	Lower	[Pamplona et al. 2000]
8 mammalian species	Liver mitochondria	Lower	[Portero-Otin et al. 2001]
Rat vs pigeon	Skeletal muscle	Lower	[Portero-Otin et al. 2004]
Mouse, parakeet, canary	Brain	Lower	[Pamplona et al. 2005]
8 mammalian species	Heart	Lower	[Ruiz et al. 2005]
Strains of mice (Idaho, Majuro and WT)	Skeletal muscle and liver	Lower	[Hulbert et al. 2006]
Naked-mole rat vs mouse	Skeletal muscle mitochondria and Liver mitochondria	Lower	[Hulbert et al. 2006; Mitchell et al. 2007]
12 mammalian species and 9 bird species	Skeletal muscle	Lower	[Hulbert et al. 2007]
10 mammalian species and 8 bird species	Liver mitochondria	Lower	[Hulbert et al. 2007]
Queen honey bees vs workers	Head, thorax, abdomen	Lower	[Haddad et al. 2007]
13 bird species	Heart	Lower	[Buttemer et al. 2008]
Echidna vs mammals	Liver, liver mitochondria, and Skeletal muscle	Lower	[Hulbert et al. 2008]
Humans (long-lived vs control group)	Erythrocytes	Lower	[Puca et al. 2008]
D. melanogaster (long-lived mutant strains)	Whole organism and mitochondria	Lower	[Sanz et al. 2010, #]
C. elegans (long-lived mutant strains)	Whole organism	Lower	[Shmookler Reis et al. 2011]
Rat vs pigeon	Erythrocytes, heart, kidney, liver, skeletal muscle (whole tissue and mitochondria)	Lower	[Montgomery et al. 2011]
D. melanogaster (wild-type strains)	Whole organism and mitochondria	Lower	#

R. Pamplona, unpublished results.

These findings were consistent with the negative correlation observed between longevity and the sensitivity to lipid autoxidation of mammalian kidney and brain homogenates [Cutler 1985].

Interestingly, the findings from a recent work [Puca et al. 2008] centered in offspring of long-lived individuals again seem to reinforce the association between membrane unsaturation and longevity. Thus, the fatty acid composition of erythrocyte membranes from 41 nonagenarian offspring were compared with 30 matched controls. Genetic loci were also tested in 280 centenarians and 280 controls to verify a potential genetic predisposition in determining unique lipid profile (allele and genotyping frequencies at endothelial-nitric oxide synthase and delta-5/delta-6 and delta-9 desaturase loci were considered). The results of this study demonstrated significant differences in the lipid composition of erythrocyte membranes derived from nonagenarian offspring versus matched controls. This is indicative of reduced oxidative stress and increased membrane integrity at the cellular level for nonagenarian offspring compared with the general population under investigation. In this context, it is plausible to infer that lipid composition of erythrocyte membranes could represent a useful biomarker of longevity.

While longevity can differ dramatically between mammal and bird species, there can also be significant longevity differences within a species. For example, populations of two wild-derived strains of mice display extended longevity (both mean and maximum longevity) compared to genetically heterogenous laboratory mice when kept under identical conditions [Miller et al. 2002]. The PI of both skeletal muscle and liver phospholipids of the two wild-type mice strains with the extended longevity was significantly smaller than that of the laboratory mice [Hulbert et al. 2006]. This is interesting because, since the different mice strains were fed the same diet, it shows that the differences in membrane composition between species are not determined by dietary differences but is genetically controlled. In this line, it is also interesting that in the senescence-accelerated mouse (SAM) strain, the SAM-prone mice have greater levels of the highly polyunsaturated peroxidation-prone fatty acids 22:6 n-3 and 20:4 n-6 and lower levels of the less peroxidation-prone 18:2 n-6 PUFA in their membranes, and consequently have a greater PI than the SAM-resistant mice [Park et al. 1996]. SAM-prone mice also show greater degrees of lipid peroxides in their tissues than do SAM-resistant mice.

Two exceptionally long-living mammalian species (naked mole-rats and echidnas) also have membrane fatty acid profiles that are resistant to lipid peroxidation as one would predict from their longevities. Thus, when membrane fatty acid composition was measured in tissues from naked mole-rats, the longest-living rodents known with a recorded longevity exceeding 28 years [Buffenstein 2005] it was found that they have very low levels of 22:6n-3 in their tissue phospholipids compared to mice. Although both mice and naked mole-rats have similar levels of total unsaturated fatty acids in their tissue phospholipids, the low 22:6n-3 levels of the naked mole-rats result in lower PI and more peroxidation-resistant membranes in skeletal muscle and liver mitochondria [Hulbert et al. 2006; Mitchell et al. 2007]. In a similar way, the echidna *Tachyglossus aculeatus*, a monotreme mammal from Australia that is exceptionally long-living and with a documented longevity of 50 years, also have a membrane composition resistant to lipid peroxidation [Hulbert et al. 2008] Accordingly, membrane lipids of echidna tissues (skeletal muscle, liver and liver mitochondria) were found to have a lower content of polyunsaturates and a higher content of monounsaturates, resulting in a low PI and, consequently, indicating that the cellular membranes of echidnas is peroxidation-resistant.

Interestingly, when the calculated peroxidation index was plotted against maximum longevity, the echidna values conformed to the relationship for mammals in general [Hulbert et al. 2008].

Honeybees (*Apis mellifera*) and flyes (*D. melanogaster*) provide another example of variation in longevity within a species. In the honey bee, depending on what they are fed, female eggs become either workers or queens [Winston 1987]. So, queens and workers share a common genome. However, the longevity of queens is an order-of-magnitude longer than workers. In order to test if differences in membrane composition could be involved it was compared the fatty acid composition of phospholipids of queen and worker honey bees [Haddad et al. 2007]. The cell membranes of both young and old honey bee queens were highly monounsaturated with very low content of PUFAs. Newly emerged workers shown a similar membrane fatty acid composition to queens but within the first week of hive life, they increase the polyunsaturate content and decrease the monounsaturate content of their membranes, probably as a result of pollen consumption. This means their membranes likely become more susceptible to lipid peroxidation in this first week of hive life. So, the results again support the suggestion that membrane composition might be an important factor in the determination of longevity. Assuming the same slope of the relationship between membrane PI and longevity as previously observed for mammal and bird species, it was proposed [Haddad et al. 2007] that the 3-fold difference in PI of phospholipids of queens and workers is large enough to account for the order-of-magnitude difference in their longevity. In another approach, these predictions have also been tested in a comparison among three wild type strains of *D. melanogaster* differing in their longevities (a long-lived strain: Oregon R, and two short-lived strains: Canton S and Dahomey). The results also confirm the present of an inverse correlation between membrane unsaturation and lipoxidation-derived molecular damage and longevity. So, the longer the longevity of the *Drosophila* strain, the lower is the membrane unsaturation [R. Pamplona, unpublished results].

In order to clarify whether the low membrane unsaturation of long-lived animals protects their cellular components from lipid oxidation and lipoxidation-derived molecular damage, studies of experimental dietary modification of *in vivo* membrane fatty acid unsaturation have been performed [Herrero et al. 2001; Portero-Otin et al. 2003; Pamplona et al. 2004]. These studies were specially designed to partially circumvent the homeostatic system of compensation of dietary-induced changes in membrane unstauration which operates at tissue level. The obtained findings demonstrate that lowering the membrane unsaturation of cellular membranes protects post-mitotic tissues against lipid peroxidation and lipoxidation-derived macromolecular damage.

The relevance of membrane unsaturation in determining longevity has also been recently reinforced by using *Drosophila* as experimental model [Sanz et al. 2010] In this study, it was created transgenic strains of *Drosophila* that express yeast NDI1 (in yeast, the single-subunit NADH dehydrogenase Ndi1 serves as a non-proton-translocating alternative enzyme that replaces complex I, bringing about the reoxidation of intramitochondrial NADH) ubiquitously. NDI1 expression mitigated the aging associated decline in respiratory capacity and the accompanying increase in mitochondrial free radical production, and resulted in decreased accumulation of markers derived from a lipoxidative stress in aged flies, resulting in an increased longevity. This lower lipoxidation-derived damage in the long-lived strains is also linked to an adaptive response with a low degree in membrane unsaturation [R. Pamplona, unpublished results].

In another approach with *C. elegans* as experimental model [Shmookler Reis et al. 2011; Hulbert 2011] it was analyzed the fatty acid profile of lipids extracted from strains of *C. elegans* that vary in longevity by ~10-fold, and report several significant log-linear correlations between longevity and fatty acid composition. The results are strongly influenced by two mutant strains (daf-2 and age-1, both long-lived mutant strains linked to a disruption of the insulin like-signaling pathway) that show the greatest longevities. Their findings can be summarized thus; comparing the shortest-living to longest-living strains total monounsaturates increased from 34% to 48%, total polyunsaturates decreased from 37% to 26%, and PI decreased from 141 to 81. Interestingly, these changes were remarkably similar to correlations between PI and longevity reported previously for mammals and birds. Furthermore, it was also found a reduction in *C. elegans* longevity with the addition of PUFAs to their diet. All together, these finding clearly support a role for membrane unsaturation in the determination of longevity.

Globally, all these comparisons a) support an important role for membrane fatty acid composition in the determination of longevity, b) reinforce the idea that the connection between membrane unsaturation and longevity is not restricted to vertebrates, and c) suggest that membrane composition is regulated in a species-specific way.

Mechanism Responsible for the Longevity-Related Differences in Fatty Acid Profile. Findings from a Phylogenomic Approach

The membrane acyl composition of the animal species studied indicates that their biological membranes maintain a similar a) fatty acid average chain length (18 carbon atoms), b) ratio of saturated versus unsaturated fatty acids (ratio 40:60), and c) phospholipid distribution irrespective of animal longevity [Pamplona 2008]. The low PI observed in long-lived species are due to changes in the type of unsaturated fatty acid that participates in membrane composition. So, there is a systematic redistribution between the types of PUFAs present from highly unsaturated fatty acids such as 22:6n-3, 20:5n-3, and 20:4n-6 in short-lived animals to the less unsaturated 18:3n-3, 18:2n-6, and 18:1 in the long-lived ones, at mitochondrial and tissue level [Pamplona et al. 2002; Hulbert et al. 2007]. Furthermore, the PI of the respective diets did not correlate with longevity. This indicates again that the contribution of the variations in the degree of unsaturation of dietary fats to the inter-species differences is, if any, very modest.

The mechanisms responsible for the longevity-related differences in fatty acid profile can be related, in principle, to the fatty acid desaturation pathway, and the deacylation-reacylation cycle. The available estimates of delta-5 and delta-6 desaturase activities indicate that they are several folds lower in long-lived species than in short-lived ones [Hulbert et al. 2007; Shmookler reis et al. 2011; Pamplona et al. 2002]. This can explain why e.g. 22:6n-3 and 20:4n-6 decreases and 18:2n-6 and 18:3n-3 increases, from short- to long-lived animals, since desaturases are the rate-limiting enzymes of the n-3 and n-6 pathways synthesizing the highly unsaturated PUFAs 20:4n-6 and 22:6n-3 from their dietary precursors, 18:2n-6 and 18:3n-3, respectively. Thus, desaturation pathways would make available *in situ* the n-6 and n-3 fatty acids to phospholipid acyltransferases in order to remodel the phospholipid acyl groups. The fact that acyltransferase/n-6 desaturase activity ratio is about 10:1 in tissues [Ivanetich et al. 1996] reinforces the idea that regulation of desaturases can be the main limiting factor responsible for the observed membrane unsaturation-longevity relationship. However, a role for a phospholipid-specific deacylation-reacylation system cannot be discarded since it has been observed that the longevity-related redistribution particularly affects the phosphatedylcholine and phosphatidyletanolamine fractions in liver mitochondria, and does not modify cardiolipin [Portero-Otin et al. 2001; Mitchell et al. 2007]. In accordance with this interpretation, a recent and already mentioned study [Jobson et al. 2010] with a phylogenomic approach to identify the genetic targets of evolution for increased longevity in mammals has been published. In this work, comparing the nonsynonymous and synonymous evolution of 5.7 million codon sites across 25 species, shows that genes involved in lipid composition (and particularly desaturation system) have collectively undergone increased selective pressure in long-lived species. So, cellular membrane has apparently been the optimized feature.

Animals with a high longevity have a low degree of membrane fatty acid desaturation based in the redistribution between types of PUFAs without any alteration in the total (%) PUFA content, average chain length and phospholipid distribution. This may be viewed as an elegant evolutionary strategy, because it decreases the sensitivity to lipid peroxidation and lipoxidation-derived damage to cellular macromolecules without strongly altering fluidity/microviscosity, a fundamental property of cellular membranes for the proper function of receptors, ion pumps, and transport of metabolites. This would occur because membrane fluidity increases acutely with the introduction of the first and less with the second double bond (due to their introduction of "kinks" in the fatty acid molecule), whereas additional (the third and following) double bonds cause few further variations in fluidity [Brenner 1984]. This is so because the kink has a larger impact on fluidity when the double bond is situated near the center of the fatty acid chain (first double bond) than when it is situated progressively nearer to its extremes (next double bond additions). In the case of the sensitivity to lipid peroxidation, however, double bonds increase it irrespective of their location at the center or laterally on the fatty acids [Holman 1954].

Table 2. Summary of longevity-related oxidative stress resistance mechanisms in long-lived species

	Long-lived vs short-lived animal species		Long-lived vs short-lived animal species
Mitochondrial free radical production	↓	Methionine Content in whole proteome	↓
mtComplex where ROS production decreases	Complex I	Membrane Unsaturation (b)	↓
mtComplex I content	↓	In vitro lipid peroxidation	↓
% Free Radical Leak	↓	Lipoxidation-derived mtDNA Damage	n.d.
Oxidation-derived mtDNA Damage (a)	↓	Lipoxidation-derived Protein Damage (c)	↓
DNA methylation	n.d.		

↑, increase; ↓, decrease; n.d.: not determined; =, unchanged.

(a) measured as 8-oxo-7,8-dihydro-2'deoxyguanosine; (b) Measured as Double Bond Index and Peroxidizability Index; (c) measured as Advanced Lipoxidation Endproducts.

Note: For specific References see text.

Thus, by substituting fatty acids with four or six double bonds by those having only two (or sometimes three) double bonds, the sensitivity to lipid peroxidation is strongly decreased in long-lived animals, whereas the fluidity of the membrane would be essentially maintained. This hypothesis, reminiscent of membrane acclimation to different environments at PUFA level in poikilotherms and bacteria, has been denominated homeoviscous longevity adaptation [Pamplona et al. 2002].

Conclusion

The increase in oxygen concentration in the Earth's atmosphere represented an important selective pressure and contributed to set up the pace of evolutionary changes in physiological and metabolic systems. Despite the evolution of the capacity to use oxygen for efficient energy production represented a relevant evolutionary innovation, the redox reactions associated with its use were responsible for the production of free radicals and, more in general, of reactive species (RS). The damaging effects of oxygen and other reactive molecular species on cells are, in essence, due to oxidation of essential cellular components. Consequently, in addition to evolutionary attempts to avoid or minimize the production of reactive byproducts of oxidative metabolism, another very important feature was the generation and selection of antioxidant defense systems. The result was a high diversity in molecular and structural antioxidant defenses, as well as in redox signaling pathways perfectly integrated in the cellular metabolic machinery. So, oxidative stress resulting from biomolecular oxidative damage due to the imbalance between RS production and antioxidant response has become a universal constraint of life-history evolution in animals and modulator of phenotypic development and trade-offs.

Redox balance is an important selective pressure faced by most organisms, and a myriad of mechanisms have evolved to regulate and adjust this balance. This diversity of mechanisms means that organisms have a great deal of flexibility in how they manage with RS challenges across time, conditions, and tissue types, as well as that different organisms may evolve different strategies for dealing with similar challenges. Available evidences suggest that aerobic life, and long-lived species, evolved by reducing the relative abundance of those structural components that are highly susceptible to oxidative damage, but without renouncing them, thus conferring to the cellular constituents a higher structural stability and lower susceptibility to oxidative damage. In addition, long-lived species also show low rates of mitochondrial reactive species generation and oxidative molecular damage. So, longevity-related oxidative stress resistance mechanisms have apparently been the optimized features (Table 2).

Acknowledgments

Investigations of the author of this review have been supported in part by I+D grants from the Spanish Ministry of Science and innovation (BFU2008- 00335/BFI; BFU2011-23888/BFI), and BSCH-UCM (2009-2010) to G.B; and grants from the Spanish Ministry of Science and Innovation (BFU2009-11879/BFI; BFU2012/39496/BFI), and the Generalitat of Catalonia (2009SGR735) to R.P. No competing financial interests exist.

REFERENCES

[1] Aledo JC, Li Y, de Magalhães JP, Ruíz-Camacho M, Pérez-Claros JA (2010) Mitochondrially encoded methionine is inversely related to longevity in mammals. *Aging Cell* 10: 198-207.

[2] Andziak B, O'Connor TP, Buffenstein R (2005) Antioxidants do not explain the disparate longevity between mice and the longest-living rodent, the naked mole-rat. *Mech. Ageing Dev.* 126: 1206–1212.

[3] Andziak B, O'Connor TP, Qi W, Dewaal EM, Pierce A, Chaudhuri AR, Van Remmen H, Buffenstein R (2006) High oxidative damage levels in the longest-living rodent, the naked mole-rat. *Aging Cell* 5: 463–471.

[4] Azzu V, Jastroch M, Divakaruni AS, Brand MD (2010) The regulation and turnover of mitochondrial uncoupling proteins. *Biochim. Biophys. Acta* 1797: 785-791.

[5] Banerjee R, Evande R, Kabil O, Ojha S, Taoka S (2003) Reaction mechanism and regulation of cystathionine beta-synthase. *Biochim. Biophys. Acta* 1647: 30-35.

[6] Barja G (2000) The flux of free radical attack through mitochondrial DNA is related to aging rate. *Aging Clin. Exp. Res.* 12: 342-355.

[7] Barja G (2004) Aging in vertebrates and the effect of caloric restriction: a mitochondrial free radical production-DNA damage mechanism?. *Biol. Rev.* 79: 235-251.

[8] Barja G (2007) Mitochondrial oxygen consumption and reactive oxygen species production are independently modulated: implications for aging studies. *Rejuvenation Res.* 10: 215-224.

[9] Barja G, Cadenas S, Rojas C, Pérez-Campo R, López-Torres M (1994) Low mitochondrial free radical production per unit O2 consumption can explain the simultaneous presence of high longevity and high aerobic metabolic rate in birds. *Free Radic. Res.* 21: 317-327.

[10] Barja G, Herrero A (1998) Localization at complex I and mechanism of the higher free radical production of brain non-synaptic mitochondria in the short-lived rat than in the longevous pigeon. *J. Bioenerg. Biomembr.* 30: 235-243.

[11] Barja G, Herrero A (2000) Oxidative damage to mitochondrial DNA is inversely related to maximum life span in the heart and brain of mammals. *FASEB J.* 14: 312-318.

[12] Bender A, Hajieva P, Moosmann B (2008) Adaptive antioxidant methionine accumulation in respiratory chain complexes explains the use of a deviant genetic code in mitochondria. *Proc. Natl. Acad. Sci. USA* 105: 16496-16501.

[13] Bielski BH, Arudi RL, Sutherland MW (1983) A study of the reactivity of HO_2/O_2- with unsaturated fatty acids. *J. Biol. Chem.* 258: 4759-4761.

[14] Bjelland S, Seeberg E (2003) Mutagenicity, toxicity and repair of DNA base damage induced by oxidation. *Mut. Res.* 531: 37–80.

[15] Bohr VA (2002) Repair of oxidative DNA damage in nuclear and mitochondrial DNA, and some changes with aging in mammalian cells. *Free Rad. Biol. Med.* 32: 804-812.

[16] Boveris A, Cadenas E, Stoppani AOM (1976) Role of ubiquinone in the mitochondrial generation of hydrogen peroxide. *Biochem. J.* 156: 435-444.

[17] Boveris A, Chance B (1973) The mitochondrial generation of hydrogen peroxide: general properties and effect of hyperbaric oxygen. *Biochem. J.* 134: 707–716.

[18] Brand MD (2010) The sites and topology of mitochondrial superoxide production. *Exp. Gerontol.* 45: 466-472.
[19] Brand MD, Affourtit C, Esteves TC, Green K, Lambert AJ, Miwa S, Pakay JL, Parker N (2004) Mitochondrial superoxide: production, biological effects, and activation of uncoupling proteins. *Free Radic. Biol. Med.* 37: 755-767.
[20] Brenner RR (1984) Effect of unsaturated fatty acids on membrane structure and enzyme kinetics. *Progr. Lipid Res.* 23: 69-96.
[21] Brosnan JT, Brosnan ME (2006) The sulfur-containing amino acids: an overview. *J. Nutr.* 136: 1636S-1640S.
[22] Bruckdorfer R (2005) The basics about nitric oxide. *Mol. Aspects Med.* 26: 3–31.
[23] Brunet-Rossinni AK (2004) Reduced free-radical production and extreme longevity in the little brown bat (Myotis lucifugus) versus two non-flying mammals. *Mech. Ageing Dev.* 125: 11-20.
[24] Buffenstein R (2005) The naked mole-rat: A new long-living model for human aging research? *J. Gerontol.* 60: 1369-1377.
[25] Buttemer WA, Battam H, Hulbert AJ (2008) Fowl play and the price of petrel: long-living Procellariiformes have peroxidation-resistant membrane composition compared with short-living Galliformes. *Biol. Lett.* 4: 351-354.
[26] Catoni GL (1953) S-Adenosylmethionine: a new intermediate formed enzymatically from L-methionine and adenosine triphosphate. *J. Biol. Chem.* 204: 403–416.
[27] Chance B, Sies H, Boveris A (1979) Hydroperoxide metabolism in mammalian organs. *Physiol. Rev.* 59: 527–605.
[28] Chao CC, Ma YS, Stadtman ER (1997) Modification of protein surface hydrophobicity and methionine oxidation by oxidative systems. *Proc. Natl. Acad. Sci. USA* 94: 2969-2974.
[29] Chen Q, Chen YR, Chen CL, Zhang L, Green-Church KB, Zweier JL (2005) Superoxide production by NADH dehydrogenase induces self-inactivation with specific protein radical formation. *J. Biol. Chem.* 280: 37339-37348.
[30] Clarke S, Banfield K (2001) S-Adenosylmethionine-dependent methyltransferases. In: Carmel R, Jacobsen DW, editors. Homocysteine in health and disease. Cambridge: Cambridge University Press; p. 63–78.
[31] Copple IM, Goldring CE, Kitteringham NR, Park BK (2008) The Nrf2-Keap1 defence pathway: role in protection against drug-induced toxicity. *Toxicology* 246: 24–33.
[32] Cutler RG (1985) Peroxide-producing potential of tissues: inverse correlation with longevity of mammalian species. *Proc. Natl. Acad. Sci. USA* 82: 4798-4802.
[33] Dahm CC, Moore K, Murphy MP (2006) Persistent S-nitrosation of complex I and other mitochondrial membrane proteins by S-nitrosothiols but not nitric oxide or peroxynitrite. *J. Biol. Chem.* 281: 10056-10065.
[34] de Magalhães JP, Costa J, Church G (2007) An analysis of the relationship between metabolism, developmental schedules, and longevity using phylogenetic independent contrasts. *J. Gerontol. A Biol. Sci. Med. Sci.* 62: 149-160.
[35] Dean RT, Fu S, Stocker R, Davies MJ (1997) Biochemistry and pathology of radical-mediated protein oxidation. *Biochem. J.* 324: 1-18.
[36] Demetrius L, Legendre S, Harremöes P (2009) Evolutionary entropy: a predictor of body size, metabolic rate and maximal life span. *Bull. Math. Biol.* 71: 800-818.

[37] Dowling DK, Simmons LW (2009) Reactive oxygen species as universal constraints in life-history evolution. *Proc. R. Soc. B*. 276: 1737-1745.
[38] Drabkin HJ, Rajbhandary UL (1998) Initiation of protein synthesis with codons other than AUG and amino acids other than methionine. *Mol. Cell Biol*. 15: 5140-5147.
[39] Echtay KS, Esteves TC, Pakay JL, Jekabsons MB, Lambert AJ, Portero-Otín M, Pamplona R, Vidal-Puig AJ, Wang S, Roebuck SJ, Brand MD (2003) A signalling role for 4-hydroxy-2-nonenal in regulation of mitochondrial uncoupling. *EMBO J*. 22: 4103-4110.
[40] Embley TM, Martin W (2006) Eukaryotic evolution, changes and challenges. *Nature* 440: 623-630.
[41] Esterbauer H, Schaur RJ, Zollner H (1991) Chemistry and biochemistry of 4-hydroxynonenal, malonaldehyde and related aldehydes. *Free Radic. Biol. Med*. 11: 81-128.
[42] Falkowski PG, Godfrey LV (2008) Electrons, life and the evolution of Earth's oxygen cycle. *Phil. Trans. R. Soc. B*. 363: 2705-2716.
[43] Finkelstein JD (2000) Pathways and regulation of homocysteine metabolism in mammals. *Semin. Thromb. Hemost*. 26: 219-225.
[44] Foksinski M, Rozalski R, Guz J, Ruszowska B, Sztukowska P, Ptwowarski M, Klungland A, Olinski R (2004) Urinary excretion of DNA repair products correlates with metabolic rates as well as with maximum life pans of different mammalian species. *Free Rad. Biol. Med*. 37: 1449-1454.
[45] Fontecave M, Atta M, Mulliez E (2004) S-Adenosyl methionine: nothing goes to waste. *Trends Biochem. Sci*. 29: 243–249.
[46] Forman HJ, Kennedy JA (1974) Role of superoxide radical in mitochondrial dehydrogenase reactions. *Biochem. Biophys. Res. Commun*. 60: 1044–1050.
[47] Foster MW, Stamler JS (2004) New insights into protein S-nitrosylation. Mitochondria as a model system. *J. Biol. Chem*. 279: 25891–25897.
[48] Gamliel A, Afri M, Frimer AA (2008) Determining radical penetration of lipid bilayers with new lipophilic spin traps. *Free Radic. Biol. Med*. 44: 1394-1405.
[49] Genova ML, Ventura B, Giulano G, Bovina C, Formiggini G, Parenti Castelli G, Lenaz G (2001) The site of production of superoxide radical in mitochondrial Complex I is not a bound ubisemiquinone but presumably iron-sulphur cluster N2. *FEBS Lett*. 505: 364-368.
[50] Giles GI (2009) Redox-controlled transcription factors and gene expression. In: Jacob C, Winyard PG, eds. Redox signaling and regulation in biology and medicine. Wiley-VCH Verlag GmbHandCo. KGaA, Weinheim, pp. 245-270.
[51] Gravina SA, Mieyal JJ (1993) Thioltransferase is a specific glutathionyl mixed disulfide oxidoreductase. *Biochemistry* 32: 3368-3376.
[52] Gredilla R, Bohr VA, Stevnsner T (2010) Mitochondrial DNA repair and association with aging-an update. *Exp. Gerontol*. 45: 478-488.
[53] Gutiérrez AM, Reboredo GR, Arcemis CJ, Catalá A (2000) Non-enzymatic lipid peroxidation of microsomes and mitochondria isolated from liver and heart of pigeon and rat. *Int. J. Biochem. Cell Biol*. 32: 73-79.
[54] Haddad LS, Kelbert L, Hulbert AJ (2007) Extended longevity of queen honey bees compared to workers is associated with peroxidation-resistant membranes. *Exp. Gerontol*. 42: 601-609.

[55] Halliwell B (1999) Antioxidant defence mechanisms: from the beginning to the end (of the beginning). *Free Radic. Res.* 31: 261-272.
[56] Halliwell B, Gutteridge JMC (2007) Free radicals in biology and medicine. Oxford University Press, UK.
[57] Hamilton ML, Guo Z, Fuller CD, Van Remmen H, Ward WF, Austad SN, Troyer DA, Thompson I, Richardson A (2001) A reliable assessment of 8-oxo-2-deoxyguanosine levels in nuclear and mitochondrial DNA using the sodium iodide method to isolate DNA. *Nucleic Acids Res.* 29: 2117-2126.
[58] Han D, Canali E, Rettori D, Kaplowitz N (2003) Effect of glutathione depletion on sites and topology of superoxide and hydrogen peroxide production in mitochondria. *Mol. Pharmacol.* 64: 1136-1144.
[59] Herrero A, Barja G (1997) Sites and mechanisms responsible for the low rate of free radical production of heart mitochondria in the long-lived pigeon. *Mech. Ageing Dev.* 98: 95-111.
[60] Herrero A, Barja G (1998) H_2O_2 production of heart mitochondria and aging rate are slower in canaries and parakeets than in mice: sites of free radical generation and mechanisms involved. *Mech. Ageing Dev.* 103: 133-146.
[61] Herrero A, Barja G (1999) 8-oxodeoxyguanosine levels in heart and brain mitochondrial and nuclear DNA of two mammals and three birds in relation to their different rates of aging. *Aging Clin. Exp. Res.* 11: 294-300.
[62] Herrero A, Portero-Otín M, Bellmunt MJ, Pamplona R, Barja G (2001) Effect of the degree of fatty acid unsaturation of rat heart mitochondria on their rates of H2O2 production and lipid and protein oxidative damage. *Mech. Ageing Dev.* 122: 427-443.
[63] Hiona A, Sanz A, Kujoth GC, Pamplona R, Seo AY, Hofer T, Someya S, Miyagawa T, Nakayama C, Samhan-Arias AK, Servais S, Barger JL, Portero-Otin M, Tanokura M, Prolla TA, Leeuwenburgh C (2010) Mitochondrial DNA mutations induce mitochondrial dysfunction, apoptosis and sarcopenia in skeletal muscle of mitochondrial DNA mutator mice. *PLoS One* 5: e11468.
[64] Holman RT (1954) Autoxidation of fats and related substances. In: Holman RT, Lundberg WO, Malkin T (eds) Progress in Chemistry of Fats and Other Lipids, Pergamon Press, London, 51-98.
[65] Holmes DJ, Flückiger R, Austad SN (2001) Comparative biology of aging in birds: an update. *Exp. Gerontol.* 36: 869-883.
[66] Hulbert AJ (2011) Longevity, lipids and C. elegans. *Aging* 3: 81-82.
[67] Hulbert AJ, Beard LA, Grigg GC (2008) The exceptional longevity of an egg-laying mammal, the short-beaked echidna (Tachyglossus aculeatus) is associated with peroxidation-resistant membrane composition. *Exp. Gerontol.* 43: 729-733.
[68] Hulbert AJ, Faulks SC, Buffenstein R (2006) Oxidation-resistant membrane phospholipids can explain longevity differences among the longest-living rodents and similarly-sized mice. *J. Gerontol. A Biol. Sci. Med. Sci.* 61: 1009-1018.
[69] Hulbert AJ, Faulks SC, Harper JM, Miller RA, Buffenstein R (2006) Extended longevity of wild-derived mice is associated with peroxidation-resistant membranes. *Mech. Ageing Dev.* 127: 653-657.
[70] Hulbert AJ, Pamplona R, Buffenstein R, Buttemer WA (2007) Life and death: metabolic rate, membrane composition, and life span of animals. *Physiol. Rev.* 87: 1175-1213.

[71] Ivanetich KM, Bradshaw JJ, Ziman MR (1996) Delta 6-desaturase: improved methodology and analysis of the kinetics in a multi-enzyme system. *Biochim. Biophys. Acta* 1292: 120-132.

[72] Janssen RJRJ, Nijtmans LG, Van der Heuvel LP, Smeitink JAM (2006) Mitochondrial complex I: structure, function and pathology. *J. Inherit. Metab. Dis*. 29: 499-515.

[73] Jobson RW, Nabholz B, Galtier N (2010) An evolutionary genome scan for longevity-related natural selection in mammals. *Mol. Biol. Evol*. 27: 840-847.

[74] Jung CH, Thomas JA (1996) S-glutathiolated hepatocyte proteins and insulin disulfides as substrates for reduction by glutaredoxin, thioredoxin, protein disulfide isomerase, and glutathione. *Arch. Biochem. Biophys*. 335: 61-72.

[75] Koopman WJ, Nijtmans LG, Dieteren CE, Roestenberg P, Valsecchi F, Smeitink JA, Willems PH (2010) Mammalian mitochondrial complex I: biogenesis, regulation, and reactive oxygen species generation. *Antioxid. Redox. Signal* 12: 1431-1470.

[76] Ku HH, Brunk UT, Sohal RS (1993) Relationship between mitochondrial superoxide and hydrogen peroxide production and longevity of mammalian species. *Free Radic. Biol. Med.* 15: 621-627.

[77] Ku HH, Sohal RS (1993) Comparison of mitochondrial pro-oxidant generation and anti-oxidant defenses between rat and pigeon: possible basis of variation in longevity and metabolic potential. *Mech. Ageing Dev*. 72: 67-76.

[78] Kudin AP, Bimpong-Buta NY, Vielhaber S, Elger CE, Kunz WS (2004) Characterization of superoxide producing sites in isolated brain mitochondria. *J. Biol. Chem*. 279: 4127-4135.

[79] Kujoth GC, Hiona A, Pugh TD, Someya S, Panzer K, Wohlgemuth SE, Hofer T, Seo AY, Sullivan R, Jobling WA, Morrow JD, Van Remmen H, Sedivy JM, Yamasoba T, Tanokura M, Weindruch R, Leeuwenburgh C, Prolla TA (2005) Mitochondrial DNA mutations, oxidative stress, and apoptosis in mammalian aging. *Science* 309: 481-484.

[80] Kushnareva Y, Murphy AN, Andreyev A (2002) Complex I-mediated reactive oxygen species generation: modulation by cytochrome c and NAD(P)+ oxidation-reduction state. *Biochem. J.* 368: 545-553.

[81] Lambert AJ, Boysen HM, Buckingham JA, Yang T, Podlutsky A, Austad SN, Kunz TH, Buffenstein R, Brand MD (2007) Low rates of hydrogen peroxide production by isolated heart mitochondria associate with long maximum lifespan in vertebrate homeotherms. *Aging Cell* 6: 607-618.

[82] Lambert AJ, Brand MD (2004) Inhibitors of the quinone-binding site allow rapid superoxide production from mitochondrial NADH:ubiquinone oxidoreductase (complex I). *J. Biol. Chem.* 279: 39414-39420.

[83] Lambert AJ, Buckingham JA, Boysen HM, Brand MD (2010) Low complex I content explains the low hydrogen peroxide production rate of heart mitochondria from the long-lived pigeon, Columba livia. *Aging Cell* 9: 78-91.

[84] Lane N (2002) Oxygen. The molecule that made the world. Oxford University Press, UK.

[85] Lenaz G, Fato R, Genova ML, Bergamini C, Bianchi C, Biondi A (2006) Mitochondrial Complex I: structural and functional aspects. *Biochim. Biophys. Acta* 1757: 1406-1420.

[86] Lints FA (1989) The rate of living theory revisited. *Gerontology* 35: 36-57.

[87] Lopez-Torres M, Perez-Campo R, Rojas C, Cadenas S, Barja G (1993) Maximum life span in vertebrates: relationship with liver antioxidant enzymes, glutathione system,

ascorbate, urate, sensitivity to peroxidation, true malondialdehyde, in vivo H2O2, and basal and maximum aerobic capacity. *Mech. Ageing Dev.* 70: 177-199.

[88] Loschen G, Azzi A, Richter C, Flohe L (1974) Superoxide radicals as precursors of mitochondrial hydrogen peroxide. *FEBS Lett.* 42: 68–72.

[89] Loschen G, Flohe L, Chance B (1971) Respiratory chain linked H2O2 production in pigeon heart mitochondria. *FEBS Lett.* 18: 261–264.

[90] Marnett LJ, Plastaras JP (2001) Endogenous DNA damage and mutation. *Trends Genet.* 17: 214-221.

[91] McCord JM (2000) The evolution of free radicals and oxidative stress. *Am. J. Med.* 108: 652-659.

[92] Meléndez-Hevia E, Montero-Gómez N, Montero F (2008) From prebiotic chemistry to cellular metabolism-the chemical evolution of metabolism before Darwinian natural selection. *J. Theor. Biol.* 252: 505-519.

[93] Melov S, Coskun PE, Wallace DC (1999) Mouse models of mitochondrial disease, oxidative stress, and senescence. *Mut. Res.* 434: 233-242.

[94] Miller RA, Harper JM, Dysko RC, Durkee SJ, Austad SN (2002) Longer life spans and delayed maturation in wild-derived mice. *Exp. Biol. Med.* 227: 500-508.

[95] Mitchell TW, Buffenstein R, Hulbert AJ (2007) Membrane phospholipid composition may contribute to exceptional longevity of the naked mole-rat (Heterocephalus glaber): a comparative study using shotgun lipidomics. *Exp. Gerontol.* 42: 1053-1062.

[96] Mitsui A, Hamuro J, Nakamura H, Kondo N, Hirabayashi Y, Ishizaki-Koizumi S, Hirakawa T, Inoue T, Yodoi J (2002) Overexpression of human thioredoxin in transgenic mice controls oxidative stress and lifespan. *Antioxid. Redox. Sign.* 4: 693–696.

[97] Moller M, Botti H, Batthyany C, Rubbo H, Radi R, Denicola A (2005) Direct measurement of nitric oxide and oxygen partitioning into liposomes and low density lipoprotein. *J. Biol. Chem.* 280: 8850-8854.

[98] Montgomery MK, Hulbert AJ, Buttemer WA (2011) The long life of birds: the rat-pigeon comparison revisited. *Plos One* 6: e24138.

[99] Moosmann B (2011) Respiratory chain cysteine and methionine usage indicate a causal role for thiyl radicals in aging. *Exp. Gerontol.* 46: 164-169.

[100] Moosmann B, Behl C (2008) Mitochondrially encoded cysteine predicts animal lifespan. *Aging Cell* 7: 32-46.

[101] Moskovitz J, Bar-Noy S, Williams WM, Requena JR, Berlett BS, Stadtman ER (2001) Methionine sulfoxide reductase (MsrA) is a regulator of antioxidant defense and lifespan in mammals. *Proc. Natl. Acad. Sci. USA* 98: 12920–12925.

[102] Murphy MP (2009) How mitochondria produce reactive oxygen species. *Biochem. J.* 417: 1-13.

[103] Nabholz B, Glémin S, Galtier N (2008) Strong variations of mitochondrial mutation rate across mammals--the longevity hypothesis. *Mol. Biol. Evol.* 25: 120-130.

[104] Nakamura H, Mitsui A, Yodoi J (2002) Thioredoxin overexpression in transgenic mice. *Methods Enzymol.* 347: 436–440.

[105] Nakamura MT, Nara TY (2004) Structure, function, and dietary regulation of delta6, delta5, and delta9 desaturases. *Annu. Rev. Nutr.* 24: 345-376.

[106] Nelson DL, Cox MM (2005) Lehninger Principles of Biochemistry. WH Freeman and Company, USA.

[107] Ohnishi T, Johnson JE Jr, Yano T, Lobrutto R, Widger WR (2005) Thermodynamic and EPR studies of slowly relaxing ubisemiquinone species in the isolated bovine heart complex I. *FEBS Letts*. 579: 500-506.
[108] Olshansky SJ, Rattan SI (2005) At the heart of aging: is it metabolic rate or stability?. *Biogerontology* 6: 291-295.
[109] Olshansky SJ, Rattan SI (2005) What determines longevity: metabolic rate or stability?. *Discov. Med.* 5: 359-362.
[110] Pamplona R (2008) Membrane phospholipids, lipoxidative damage and molecular integrity: a causal role in aging and longevity. *Biochim. Biophys. Acta* 1777: 1249-1262.
[111] Pamplona R (2011) Advanced lipoxidation end-products. *Chem. Biol. Interact.* 192: 14-20.
[112] Pamplona R, Barja G (2003) Aging rate, free radical production, ans constitutive sensitivity to lipid peroxidation: insights from comparative studies. In: Van Zglinicki T (ed) Biology of aging and its modulation series. Vol. 1. Aging at the molecular level. Kluwer Academic Publisher, New York, pp 47-64.
[113] Pamplona R, Barja G (2006) Mitochondrial oxidative stress, aging and caloric restriction: the protein and methionine connection. *Biochim. Biophys. Acta* 1757: 496-508.
[114] Pamplona R, Barja G (2007) Highly resistant macromolecular components and low rate of generation of endogenous damage: two key traits of longevity. *Ageing Res. Rev.* 6: 189-210.
[115] Pamplona R, Barja G, Portero-Otín M (2002) Membrane fatty acid unsaturation, protection against oxidative stress, and maximum life span: a homeoviscous-longevity adaptation?. *Ann. NY Acad. Sci.* 959: 475-490.
[116] Pamplona R, Portero-Otín M, Requena JR, Thorpe SR, Herrero A, Barja G (1999) A low degree of fatty acid unsaturation leads to lower lipid peroxidation and lipoxidation-derived protein modification in heart mitochondria of the longevous pigeon than in the short-lived rat. *Mech. Ageing Dev.* 106: 283-296.
[117] Pamplona R, Portero-Otín M, Riba D, Ledo F, Gredilla R, Herrero A, Barja G (1999) Heart fatty acid unsaturation and lipid peroxidation, and aging rate, are lower in the canary and the parakeet than in the mouse. *Aging* (Milano) 11: 44-49.
[118] Pamplona R, Portero-Otín M, Riba D, Requena JR, Thorpe SR, López-Torres M, Barja G (2000) Low fatty acid unsaturation: a mechanism for lowered lipoperoxidative modification of tissue proteins in mammalian species with long life spans. *J. Gerontol. A Biol. Sci. Med. Sci.* 55: B286-B291.
[119] Pamplona R, Portero-Otín M, Riba D, Ruiz C, Prat J, Bellmunt MJ, Barja G (1998) Mitochondrial membrane peroxidizability index is inversely related to maximum life span in mammals. *J. Lipid Res.* 39: 1989-1994.
[120] Pamplona R, Portero-Otín M, Ruiz C, Gredilla R, Herrero A, Barja G (2000) Double bond content of phospholipids and lipid peroxidation negatively correlate with maximum longevity in the heart of mammals. *Mech. Ageing Dev.* 112: 169-183.
[121] Pamplona R, Portero-Otín M, Sanz A, Ayala V, Vasileva E, Barja G (2005) Protein and lipid oxidative damage and complex I content are lower in the brain of budgerigards and canaries than in mice. Relation to aging rate. *AGE* 27: 267-280.

[122] Pamplona R, Portero-Otin M, Sanz A, Requena J, Barja G (2004) Modification of the longevity-related degree of fatty acid unsaturation modulates oxidative damage to proteins and mitochondrial DNA in liver and brain. *Exp. Gerontol.* 39: 725-733.

[123] Pamplona R, Prat J, Cadenas S, Rojas C, Pérez-Campo R, López Torres M, Barja G (1996) Low fatty acid unsaturation protects against lipid peroxidation in liver mitochondria from long-lived species: the pigeon and human case. *Mech. Ageing Dev.* 86: 53-66.

[124] Park JW, Choi CH, Kim MS, Chung MH (1996) Oxidative status in senescence accelerated mice. *J. Gerontol. A Biol. Sci. Med. Sci.* 51: B337-345.

[125] Pearl R (1928) The rate of living. London: University of London Press.

[126] Perez-Campo R, López-Torres M, Cadenas S, Rojas C, Barja G (1998) The rate of free radical production as a determinant of the rate of aging: evidence from the comparative approach. *J. Comp. Physiol. B* 168: 149-158.

[127] Portero-Otin M, Bellmunt MJ, Requena JR, Pamplona R (2003) Protein modification by advanced Maillard adducts can be modulated by dietary polyunsaturated fatty acids. *Biochem. Soc. Trans.* 31: 1403-1405.

[128] Portero-Otín M, Bellmunt MJ, Ruiz MC, Barja G, Pamplona R (2001) Correlation of fatty acid unsaturation of the major liver mitochondrial phospholipid classes in mammals to their maximum life span potential. *Lipids* 36: 491-498.

[129] Portero-Otín M, Requena JR, Bellmunt MJ, Ayala V, Pamplona R (2004) Protein nonenzymatic modifications and proteasome activity in skeletal muscle from the short-lived rat and long-lived pigeon. *Exp. Gerontol.* 39: 1527–1535.

[130] Puca AA, Andrew P, Novelli V, Anselmi CV, Somalvico F, Cirillo NA, Chatgilialoglu C, Ferreri C (2008) Fatty acid profile of erythrocyte membranes as possible biomarker of longevity. *Rej. Res.* 11: 63-72.

[131] Richter C, Park JW, Ames BN (1988) Normal oxidative damage to mitochondrial and nuclear DNA is extensive. *Proc. Natl. Acad. Sci. USA* 85: 6465-6467.

[132] Ruan H, Tang XD, Chen ML, Joiner ML, Sun G, Brot N, Weissbach H, Heinemann SH, Iverson L, Wu CF, Hoshi T (2002) High-quality life extension by the enzyme peptide methionine sulfoxide reductase. *Proc. Natl. Acad. Sci. USA* 99: 2748-2753.

[133] Ruiz MC, Ayala V, Portero-Otín M, Requena JR, Barja G, Pamplona R (2005) Protein methionine content and MDA-lysine protein adducts are inversely related to maximum life span in the heart of mammals. *Mech. Ageing Dev.* 126: 1106-1114.

[134] Salway KD, Page MM, Faure PA, Burness G, Stuart JA (2010) Enhanced protein repair and recycling are not correlated with longevity in 15 vertebrate endotherm species. *Age* (Dordr). doi 10.1007/s11357-010-9157-5.

[135] Samuels DC (2005) Life span is related to the free energy of mitochondrial DNA. *Mech. Ageing Dev.* 126: 1123-1129.

[136] Sanz A, Barja G, Pamplona R, Leeuwenburgh C (2009) Free radicals and mammalian aging. In: Jacob C, Winyard PG (eds) Redox Signaling and Regulation in Biology and Medicine. Wiley-VCH Verlag GmbH and Co. KGaA, Germany, pp 433-472.

[137] Sanz A, Fernández-Ayala DJ, Stefanatos RK, Jacobs HT (2010) Mitochondrial ROS production correlates with, but does not directly regulate lifespan in Drosophila. *Aging* (Albany NY) 2: 200-203.

[138] Sanz A, Soikkeli M, Portero-Otín M, Wilson A, Kemppainen E, McIlroy G, Ellilä S, Kemppainen KK, Tuomela T, Lakanmaa M, Kiviranta E, Stefanatos R, Dufour E, Hutz B, Naudí A, Jové M, Zeb A, Vartiainen S, Matsuno-Yagi A, Yagi T, Rustin P, Pamplona R, Jacobs HT (2010) Expression of the yeast NADH dehydrogenase Ndi1 in Drosophila confers increased lifespan independently of dietary restriction. *Proc. Natl. Acad. Sci. USA* 107: 9105-9110.

[139] Scheffler IE (1999) *Mitochondria*. Wiley-Liss, USA.

[140] Sherman PW, Jarvis JUM (2002) Extraordinary life spans of naked mole-rats (Heterocephalus glaber). *J. Zool.* (Lond.) 258: 307–311.

[141] Shmookler Reis RJ, Xu L, Lee H, Chae M, Thaden JJ, Bharill P, Tazearslan C, Siegel E, Alla R, Zimniak P, Ayyadevara S (2011) Modulation of lipid biosynthesis contributes to stress resistance and longevity of C. elegans mutants. *Aging* (Albany) 3: 125-147.

[142] Sohal RS, Svensson I, Brunk UT (1990) Hydrogen peroxide production by liver mitochondria in different species. *Mech. Ageing Dev.* 53: 209-215.

[143] Speakman J (2005) Correlations between physiology and lifespan--two widely ignored problems with comparative studies. *Aging Cell* 4: 167-175.

[144] Stadtman ER, Moskovitz J, Berlett BS, Levine RL (2002) Cyclic oxidation and reduction of protein methionine residues is an important antioxidant mechanism. *Mol. Cell Biochem.* 234-235: 3-9.

[145] Stadtman ER, Moskovitz J, Levine RL (2003) Oxidation of methionine residues in proteins: biological consequences. *Antioxid. Redox. Sign.* 5: 577–582.

[146] Stefanatos R, Sanz A (2011) Mitochondrial complex I: A central regulator of the aging process. *Cell Cycle* 10: 1528-1532.

[147] Stipanuk MH (2004) Sulfur amino acid metabolism: pathways for production and removal of homocysteine and cysteine. *Annu. Rev. Nutr.* 24: 539-577.

[148] St-Pierre J, Buckingham JA, Roebuck SJ, Brand MD (2002) Topology of superoxide production from different sites in the mitochondrial electron transport chain. *J. Biol. Chem.* 277: 44784-44790.

[149] Taylor ER, Hurrell F, Shannon RJ, Lin TK, Hirst J, Murphy MP (2003) Reversible glutathionylation of complex I increases mitochondrial superoxide formation. *J. Biol. Chem.* 278: 19603-19610.

[150] Terman A, Brunk UT (2004) Lipofuscin. *Int. J. Biochem. Cell Biol.* 36: 1400-1404.

[151] Thorpe SR, Baynes JW (2003) Maillard reaction products in tissue proteins: new products and new perspectives. *Amino Acids* 25: 275-281.

[152] Toime LJ, Brand MD (2010) Uncoupling protein-3 lowers reactive oxygen species production in isolated mitochondria. *Free Radic. Biol. Med.* 49: 606-611.

[153] Treberg JR, Brand MD (2011) A model of the proton translocation mechanism of complex I. *J. Biol. Chem.* 286:17579-17584.

[154] Treberg JR, Quinlan CL, Brand MD (2011) Evidence for Two Sites of Superoxide Production by Mitochondrial NADH-Q Oxidoreductase (Complex I). *J. Biol. Chem.* 286: 27103-27110.

[155] Trifunovic A, Wredenberg A, Falkenberg M, Spelbrink JN, Rovio AT, Bruder CE, Bohlooly YM, Gidlof S, Oldfors A, Wiborn R, Tornell J, Jacobs HT, Larsson NG (2004) Premature aging in mice expressing defective mitochondrial DNA polymerase. *Nature* 429: 417-423.

[156] Vance DE, Vance JE (1996) Biochemistry of lipids, lipoproteins and membranes. Elsevier Science BV, Amsterdam, The Netherlands. pp 1-553.

[157] Wakabayashi N, Dinkova-Kostova AT, Holtzclaw WD, Kang MI, Kobayashi A, Yamamoto M, Kensler TW, Talalay P (2004) Protection against electrophile and oxidant stress by induction of the phase 2 response: fate of cysteines of the Keap1 sensor modified by inducers. *Proc. Natl. Acad. Sci. USA* 101: 2040–2045.

[158] Wallis JG, Watts JL, Browse J (2002) Polyunsaturated fatty acid synthesis: what will they think of next?. *Trends Biochem. Sci.* 27: 467-473.

[159] Wang E, Wonq A, Cortopassi G (1997) The rate of mitochondrial mutagenesis is faster in mice than in humans. *Mut. Res.* 377: 157-166.

[160] Winston ML (1987) The biology of the honey bee. Cambridge MA: Harvard University Press.

[161] Yeagle P (1993) The membranes of cells. Academic Press, San Diego, pp 1-349.

Reviewed by Prof. Dr. José Viña, Department of Physiology, Faculty of Medicine, University of Valencia, Av. Blasco Ibañez -15, 46010 València, Spain.

E-mail: jose.vina@uv.es

In: Natural Selection
Editors: J. R. Lynch and D. T. Williamson
ISBN: 978-1-62257-467-4

Chapter 4

EVOLUTIONARY TRANSITIONS IN MATHEMATICAL PRACTICES MODELED BY EVOLUTIONARY SYSTEMICS

Rahman Khatibi*
Consultant Mathematical Modeller, Swindon, UK

ABSTRACT

Mathematical complexity is often regarded as an abstract world and sterile due to its formidable status but this reflects the grip of an ontological mindset. This complexity seems beyond the reach of evolutionary processes, where it sounds counterintuitive to explain it by *evolutionary thinking*. This paper shows that there is nothing unique about mathematics, as it evolves like any other complexity. The modeling approach used is *evolutionary systemics*, which is the integration of *systemic thinking* (the architecture) with evolutionary thinking (the architect). Evolutionary systemics is driven by a set of *axioms*, in which natural selection drives evolutionary transitions through four types of feedback loops from a lower loop to higher ones. The outcome is a modeling capability, which uses historic accounts as evidence. Thus, evidence gathered from historic accounts identifies evolutionary transitions and the approach copes with the roles of *cognitive mathematics of individuals*, *communal mathematical thinking*, and *institutional mathematics*.

The paper offers an insight into mathematical practices and holds that evolutionary transitions in mathematics are in tandem with other human endeavors. It can be highlighted that, during evolutionary transitions, the past knowledge may become obsolete without being unselected. As mathematics is pluralist by its nature and increasing in complexity, feedforward loops are indispensible for proactively reshaping mathematical practices with the evolving culture. The following strategies are identified. (i) Philosopher's aspirations are in asking stimulating questions but they produce mutually exclusive doctrines, which are contentious and devoid of significant practical benefits. (ii) Practitioners' aspirations are fruitful in axiomatization to overhaul existing knowledge riddled with conflicts to self-evident information but these are few and far in-

* Corresponding author: Rahman Khatibi, PhD, MSc, BSc, Consultant Mathematical Modeler, Swindon, UK, rahman_khatibi@yahoo.co.uk

between. (iii) This paper promotes goal-orientation by applying evolutionary systemics through identifying missing links in mathematics, proposing a model to resolve their impacts, outlining the elements of a vision to identify its equivalent cladistics and to manage expectations on mathematics. A new calculus is presented to support the thinking behind goal-orientation in the sequence of natural numbers primed from the building block of "1" by the virtue of new operators.

Keywords: evolutionary-systemics, feedback loops, sub-loops, modeling complexity, internal entropy, external entropy, hierarchy, goal-orientation, vision.

1.

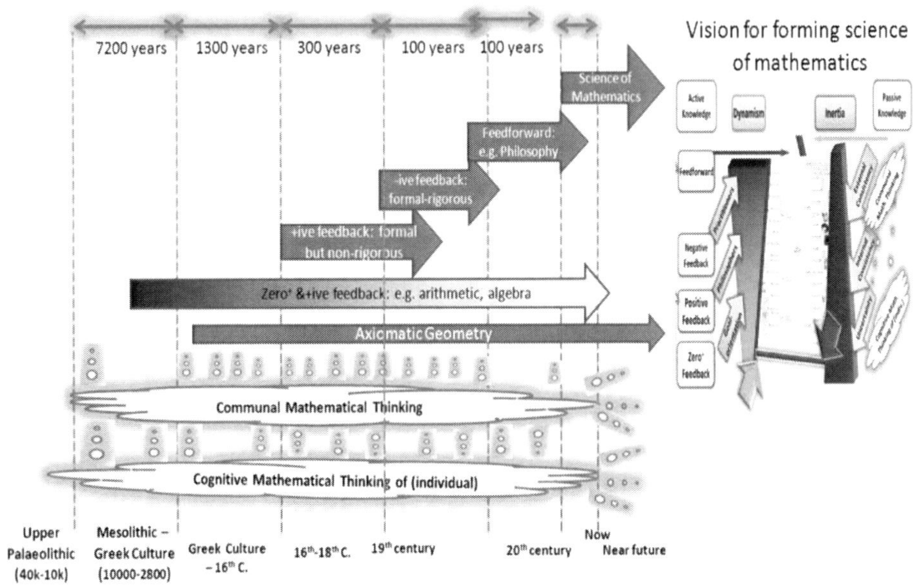

INTRODUCTION

The history of mathematics or its historiography does not provide sufficient insight into changing practices of mathematics, where historiographers interpret changes as they understand and complexity refers to entities without any apparent structure. The doctrine of paradigm shift (Kuhn, 1962) by Thomas Kuhn (1922–1996) is one way of explaining historic accounts of science but Tymoczko (1998, p201) points out that Kuhn holds mathematics to be exempt from revolutions. However, some others disagree, e.g. Grabiner (1974) holds mathematics is an area of human activity, which has at once the least destructive but still the most fundamental revolutions. The evidence provided by Grabiner (1974) is used in this paper but as the modeling data for *evolutionary systemics* presented by Khatibi (2012a and 2011). The focus of this paper is to explain the evolutionary transitions in mathematical complexity in a transparent way to other disciplines and to formulate a vision for the future.

The essence of evolutionary systemics presented by Khatibi (2012a) is the differentiation in variations on the time horizons: (i) *Dynamic timescale*: time starts at an origin; processes take place on continuous or discrete time intervals; and the process ends. (ii) *Long timescale*: this is normally large compared with dynamic timescale, over which adaptation may take place. (iii) *Evolutionary timescale*: changes comprise adding/removing building block to/from the architecture hereditary machinery in many generations. Mathematical complexity takes place at the dynamic timescale but the complexity is formed at the evolutionary timescale beneath dynamic activities. The architect of the forming is natural selection through the architecture of four feedback loops: zero$^+$ feedback, positive feedback, negative feedback and feedforward loops. Each loop has three sub-loops corresponding to each timescale, as summarized in Table 1.

Mathematical complexity appears to be confined to d-loops without any apparent conception of inherent ℓ-loops and e-loops and this explains the reason for inadvertently regarding mathematical practices to be ontological; whereas evolutionary thinking comprises ℓ-loops and e-loops without any conception of feedback loops and without any reference to these two sub-loops. In between d-loops of mathematical complexity on the one hand and ℓ-loops and e-loops of this complexity on the other hand, there is a gap, but to be bridged in this paper by evolutionary systemics, captured in Table 1. This is the integration of natural selection or e-loops, (the architect) with feedback loops (the architecture or d-loops), to be referred as *evolutionary systemics* or *evolutionary systemic modeling* (EvSyM).

Table 1 Variations of Feedback Loops with Respect to Timescale

Loops	Drivers at Dynamic Timescale	Long Timescale	Evolutionary Timescale
Zero$^+$ Feedback	**d-loop**: Randomness	**ℓ-loop**: Obscure	**e-loop**: Spontaneity
	Zero$^+$ Feedback Loop gives rises to simplexes, which grow as below:		
Positive Feedback	**d-loop**: Performance	**ℓ-loop**: Adaptation	**e-loop**: Natural Selection
	Positive Feedback Complexes $=$ ***d_loop*** $+$ ***ℓ_loop*** $+$ ***e_loop***		
Negative Feedback	**d-loop**: Internally-consistent performance	**ℓ-loop**: 'Discretion'	**e-loop**: Natural selection of rules for internal consistency
	Negative Feedback Complexes $=$ ***d_loop*** $+$ ***ℓ_loop*** $+$ ***e_loop***		
Feedforward	**d-loop**: Vision for external consistency	**ℓ-loop**: Mission for external consistency	**e-loop**: Natural selection of rules for external consistency
	Feedforward Complexes $=$ ***d_loop*** $+$ ***ℓ_loop*** $+$ ***e_loop***		
Note 1: Each loop has three integral components of d-loop (real dynamic state of the system), ℓ-loop (manipulation of the phase), or e-loop (changes to hereditary codes)			

Scientific enquiries are focused largely on explaining the complexity of life but this is a narrow line of enquiry for excluding *evolutionary* processes in the *wider* sciences concerning social, cultural and intellectual complexity. For instance, mathematics is an area of complexity in the cultural and intellectual life of human beings, unjustifiably held outside the reach of evolution. The author is not alone to express a concern on overlooking wider evolutionary theories but the following example illustrates the depth of the problem: (i) As argued by Hodgson, (1993, 2004), Darwinism is always required to complete the explanation of populations of varied and replicating entities but despite their necessity, they are never sufficient on their own. This paper provides a solution for the wider application of evolutionary thinking.

Some philosophers of mathematics are implicitly minded with evolutionary thinking, including Grabiner (1974) and Rav (2006), Kitcher (1983) and Wilder (1981). Their works are in the style of historiography or prescriptive philosophy and unable to explain evolutionary transitions in mathematics in a transparent way to other human endeavors. Their works are used as data in this paper to substantiate evolutionary transitions covering from the origins of mathematics until now. Mathematics is inspirational by ensuring its internal consistency and this paper regards this as an expression of feedforward loops for ensuring the external consistency between mathematics and the culture.

The critical view presented in this paper shows that there are a range of missing links in mathematics but it also presents possible solutions as follows: (i) mathematics is seen as the making of ingenious individuals but it is an open enterprise deeply rooted in cultures; (ii) mathematical complexity as a whole and its individual areas are treated ontologically with no explicit sense of time variations in the practice but it is an evolutionary construct; (iii) mathematical complexity seems amorphous for lacking any apparent hierarchy but this paper suggests that the *hierarchical structure* of science can be used as a model; (iv) within each hierarchical level, *interconnectivity* among the building blocks are overlooked but this paper suggests that it is possible to seek for the equivalents of the Periodical Table and cladistics in mathematics.

The seemingly amorphous mathematical complexity is a by-product of opportunism and pluralism and this overshadows its possible hierarchical structure. This can be understood better by using the hierarchical organization of science as a model. Science comprises the hierarchies of physics, chemistry, biology, psychology, sociology and anthropology, qualified as: *physicalistic*, *chemicalistic*, *biologistic*, *psychologistic*, *sociologistic* and *anthropologistic*. If mathematical complexity lends itself to a hierarchical structure with emergent properties from one hierarchy to the next, similar to that of science, overlooking these concepts will be at the expense of losing inherent information but this cannot be justified. The paper is not offering a hierarchical structure to mathematics but the case is highlighted by the sequences of natural numbers. A new calculus is outlined, which makes it possible (i) to prime the sequence of natural numbers from the building block of "1;" (ii) to prime a whole family of sequences of natural numbers with inherent interconnectivity and hierarchies; (iii) to overhaul their conventional calculus to the new analytical framework.

The ability to explain the emergence of mathematics by the wider evolutionary thinking is significant, as mathematics is a prized intellectual world and counterintuitive to regard it as a simple product of evolution. But then a description of mathematics in terms of undefined attributes of infallibility, impeccability and irrefutability signify inherent misconceptions. Practitioners show very little interest in philosophical issues and conversely philosophers are driven by their presuppositions with their discourse often devoid of real issues. The dilemma is that philosophical enquiries are riddled with inconsistencies and controversies but are stimulating by asking fundamental questions, such as the followings. Is mathematics invented or a social product? Does it exist independent of human perceptions? Is it growing cumulatively? This paper aims to provide scientific explanations to philosophical enquiries by evolutionary systemics.

This paper is largely concerned with e-loops and only refers explicitly when necessary. It has a multidisciplinary approach and the reader needs to brace up to concepts and terms adapted from mathematics and interdisciplinary sciences including evolution, systems science (or cybernetics), evolution of science and mythology.

The return is to see *mathematical complexity* as a whole in a new light, capable of explaining of evolutionary transitions in mathematics. These terms are defined in a glossary by Khatibi (2012a).

2. Evolutionary Systemic Modeling of Mathematical Complexity

Modern Pure/rigorous mathematics has not emerged at a stroke but through an evolutionary process. Man's intellectual life is operated by information, which is an emergent property and has a selective advantage of triggering action and being readily understandable. Science produces information by compacting data but mathematics produces information in its own way (as discussed in Section 10.1) without involving much data. This takes the study of mathematics back to prehistory, just like any other evolvable entity.

2.1 Context of Mathematical Complexity

Individuals, communities and institutions make up the context of mathematical complexity, but disaggregating their roles seems intractable. Context is intrinsic to evolutionary systemics and during the zero$^+$ feedback loop, these three contextual contributions are often one and the same.

The origins for the emergence of mathematics are often attributed to numbers, forms and shapes but this seems rather unclear, as there may be other founding building blocks. Figure 1 presents one view of branches of mathematics and Table 1 suggests a similar view of origins. Mac Lane (1981) argues that "Mathematics begins with puzzles and problems dealing with combinatoric and symbolic spects of the general human experience, ... From this starting point, the subject has developed to be a deductive analysis of a large number of very different but interlocking formal structures."

Table 2. Various Origins of Mathematics by Mac Lane (1981)

Origin	Description
Counting	arithmetic and number theory
Measuring	real numbers, calculus, analysis
Shaping	geometry, topology
Forming (as in architecture)	symmetry, group theory
Estimating	probability, measure theory, statistics
Moving	mechanics, calculus, dynamics
Calculating	algebra, numerical analysis
Proving	Logic
Puzzling	combinatorics, number theory
Grouping	set theory, combinatorics

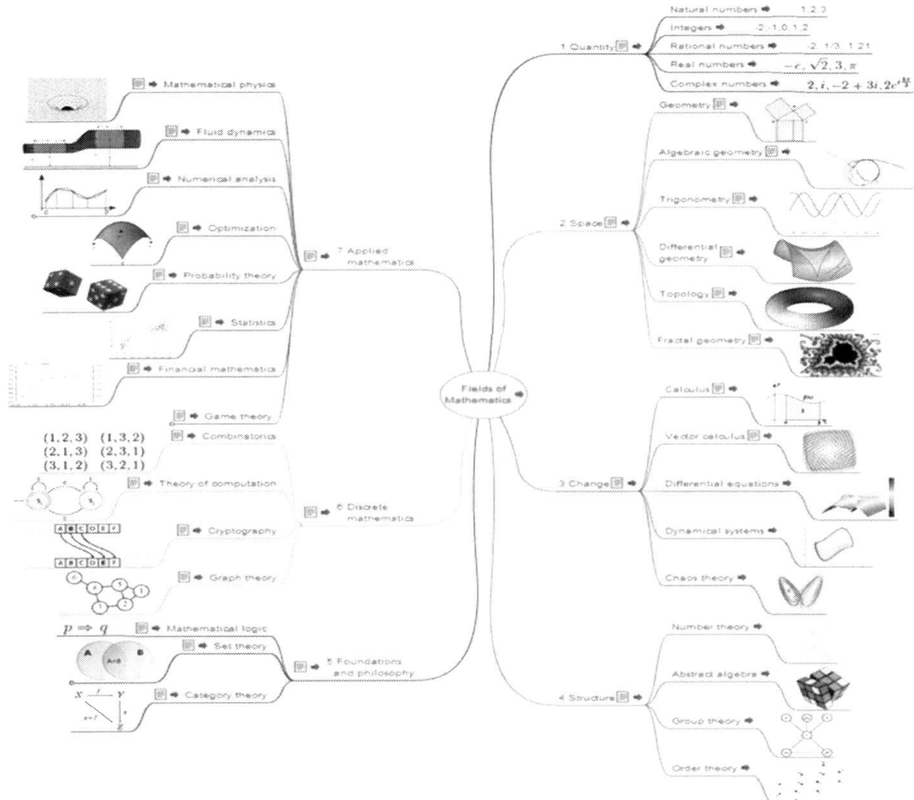

Figure 1. A Representation of Pure Mathematics (http://www.gogeometry.com/education/mathematics_fields_mind_map.html).

The absence of a methodology for the wider evolutionary thinking signifies the grip of ontology over mathematical complexity and most other intellectual endeavors. This paper simply equates any entity to be ontological if its inherent evolution is overlooked. It is argued that ontology is at the expense of evolutionary thinking.

2.2. Evolutionary versus Ontology in Mathematical Complexity

Traditional enquiries on the nature of mathematics are categorized as follows.

2.2.1 Historiography

Historiography of mathematical complexity is an interpretation of changes as understood by historiographers. This approach is rather different than historicism, attaching an overwhelming significance to individual historic events or incidents, or determinism outcomes as inevitable given a set of conditions.

2.2.2 Philosophical enquiries

Philosophical enquiries are reviewed in Section 8.2 of this paper but the overview is that they are largely prescriptive and treat the logic of enquiry in a prescriptive way. Cellucci

(2000) discusses the views of Fregé and Poincaré, opposing and promoting respectively evolutionary thinking. Fregé's ontological mindset emerges from the following quotations (Fregé, 1979): "in mathematics we must always strive after a system that is complete in itself" (p. 279); in this way "chains of inference are formed connecting truths; and the further the science develops, the longer and more numerous become the chains of inference and the greater the diversity of the theorems" (Fregé, 1979, p204). Since systems can develop only by deriving consequences from given axioms, they cannot evolve. Once a system has been set up, it either "must remain, or else the whole system must be dismantled in order that a new one may be constructed" (Fregé, 1979, p242). A system does not admit change: it is an all-or-nothing business and so, if it cannot be all, then it must be nothing. If the system "that has been acknowledged until now proves inadequate, it must be demolished and replaced by a new edifice" (Fregé, 1979, p279).

Even if a philosophical doctrine accounts explicitly for the time dimension, they are purely focused on the d-loop and have no focus on e-loops. This paper is focused on the e-loop of mathematical complexity as a whole, without precluding each of its areas. Thus, each of the disciplines in Table 2 has its individual story of being primed at an origin and of undergoing through a subsequent selection for proliferation within the practitioners of mathematics. Evolutionary systemics is outlined below, abstracting its salient features given by given by Khatibi (2012a), deemed essential to make sense of this paper:

- The context (the phase) of mathematics is furnished by individual, communal and institutional contributions to complexity, each displaying time variations at three timescales and additionally with evolutionary transitions taking place in institutional practices as below.
- *Zero$^+$ feedback e-loops*: this is the process of "priming" the simplicity hereditary machinery of mathematical complexity as a whole or any of its particular areas. In reality, there is no e-loop and d-loop at this stage yet but the potentials for both and the term priming is a key concept signifying the first emergence of such a complex. The primed concept has yet to be selected and reach a critical mass through uncontrolled proliferation before it becomes an *evolvable entity*.
- *Positive feedback d-ℓ-e-loops*: these sub-loops work hand-in-hand making up the bulk of each areas of complexity; d-loops represent the visible activities; e-loops drive selection, diversification and entropy; and ℓ-loops represent adaptation. Entropy continually reduces efficacy and triggers a selective advantage for an evolutionary transition to the next higher loop.
- *Negative feedback d-ℓ-e-loops*: these sub-loops work hand-in-hand in which d-loops drive the processes operated by the addition of a controller subsystem to an areas of complexity to reduce internal entropy for maintaining internal consistency; ℓ-loops comprise adaptations to a concept; and e-loops comprise variations in criteria for ensuring internal consistencies.
- *Feedforward d-ℓ-e-loops*: these sub-loops work hand-in-hand in which d-loops drive goal-orientation to reduce external entropy for maintaining external consistency; ℓ-loops comprise the mission to vary a concept to make a better sense; and e-loops comprise the vision of variations to ensure external consistencies.

- The roles of entropy, pluralism, cognitive and communal mathematical thinking and institutional evolutionary transitions become explicit under evolutionary systemics.
- The e-loops may be seen as a mindset, i.e. as a collection of fundamental codes, assumptions, knowledge or beliefs.
- The evolutionary systemic axioms in ESM1 are essential to take a full benefit from this paper.
- Evolutionary systemics create a bottom-up approach to identify appropriate attributes of mathematical thinking at each stage of its evolutionary transitions. In contrast, historiography of mathematics interprets changes as understood by individual historians; whereas historicists and philosophers of mathematics present prescriptive accounts.

The focus of this paper is to study evolutionary transitions in mathematical complexity as a whole including the problem of origins but one classification of mathematical disciplines is reproduced in Figure 1, as it is eye-opening, although it does not unearth its inherent interconnectivity similar to that of highly-diversified hierarchical taxonomy of biological species: species, genus, family, order, class, phylum and kingdom.

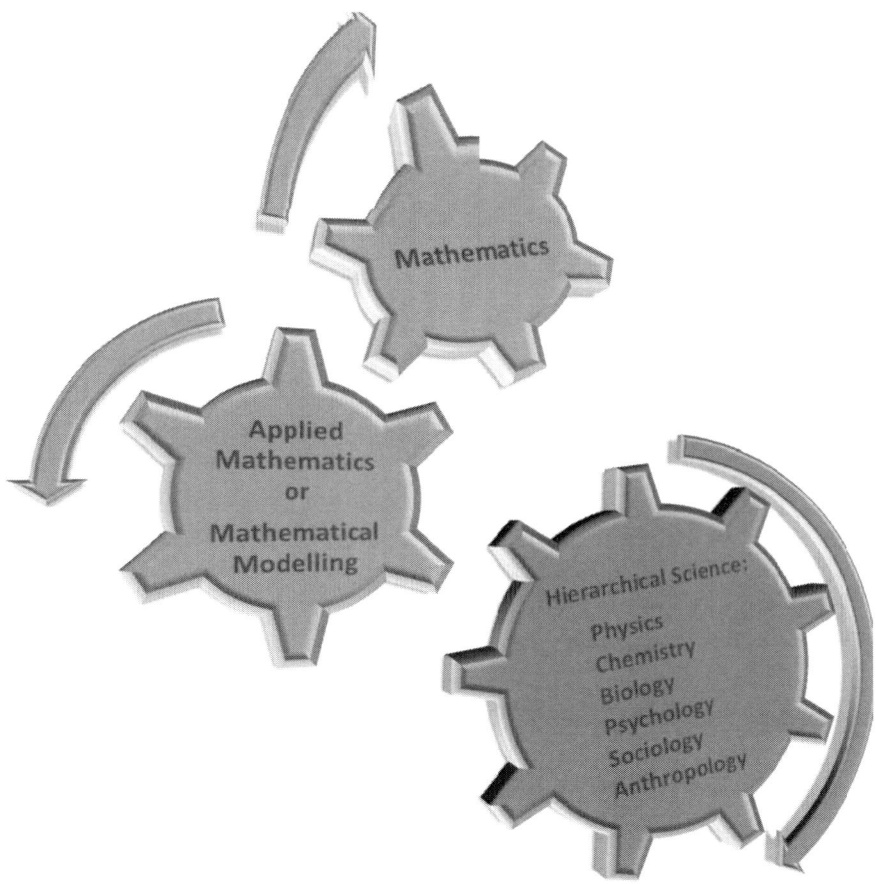

Figure 2. Symbiosis of Science, Mathematical Modeling and Mathematics.

Mathematics is a symbiotic outgrowth of science and applied mathematics, as depicted in Figure 2, where it is seen as the backdrop of applied mathematics and science in the sense that these three endeavors provide impetus to one another but the interaction between science and mathematics is normally indirect and through applied mathematics.

3. Emergence of Cultures and Mathematics

The roots of modern cultures and mathematical complexity are normally traced to the Upper Paleolithic era (50,000–10,000 years ago) in prehistory. During prehistory, the way of life of the human beings was a hunter-gatherer one with only a few ostensible markers (such as language and tool-making) to distinguish them from other species of the animal kingdom. The emergence of a particular intellectual endeavor in a society is not a measure of intelligence but stems from cultural factors characterizing ways of life adapted to local conditions acting as wellsprings for unanticipated contributions emerging spontaneously or through self-organization. This explains why an intellectual endeavor would emerge in some societies but not in others.

The association of human endeavors such as mathematics with intelligence of a particular ethnic people creates undue complications. Many of such unfounded claims in 18th-19th centuries were subsequently challenged and discredited in the 20th century. Eliade (1978) argues that 'The very slow progress in technology does not imply a similar development of intelligence. We know that the extraordinary upsurge in technology during the past two centuries has not found expression in a comparable development of Western man's intelligence. Beside, as has been said, "every innovation brought with it the danger of collective death" (André Varagnac).' The published literature on the history of mathematics is vast but to the best knowledge of the author, they often overlook the role of evolutionary thinking. This paper searches the emergence of mathematical complexity to stem from the culture in the first place before than intelligence.

3.1. The Priming of Numbers and Counting Systems

Emergence of 'One': In search of the roots of *mathematical complexity*, numbers are presented here as an example for their priming process within the past cultural setting, where "priming" signifies extra effort before the selection. Numbers did not emerge in a matter of days but over thousands of years since the start of the Upper Paleolithic Age, during which changes were very slow. Archaeologists and mythologists attribute the emergence of numbers to the way of life. These together with mythological narratives indicate that the notion of the Sky God emerged in Upper Paleolithic times (35,000–10,000 years before the present times) and according to Armstrong (2005) "At some point – we do not know exactly when this happened – people in various far-flung parts of the world began to personify the sky. They started to tell stories about a 'Sky God' who had single-handedly created heaven and earth out of nothing." *One* as a number is thought to be a linguistic imprint of the Sky God.

Emergence of other Numbers: The tendency towards more Gods emerged with diversifying ways of life during the Mesolithic age (10,000-6000 years ago) and their imprints in the form of numbers would have coalesced over millennia, priming a number of unplanned outcomes including a counting system and calendars, where these imprints are without any mathematical thinking yet.

The above explanation seems to be tinted with religion and therefore more rational explanations are likely to capture prehistoric tallies carving notches on objects. The need for numbers was amplified by the emergence of property in the settled way of life during the Mesolithic Age, but there is evidence for the emergence of numbers before the Mesolithic Ages.

Anthropologic studies of contemporary tribal ways of life provide other sources of information. For instance, a primitive society is reckoned functional without numbers but the priming of a counting system is almost a global human experience. Boyer (1991) argues that each culture gave rise to the emergence of a counting system but not necessarily through an orderly process. The base for a counting system is discussed below and based on anthropological study of nonliterate tribes.

3.2. Emergence of Decimal Base for Counting

Although there is no certainty whether numbers emerged in the Paleolithic Age or later, the need for the emergence of a counting system was amplified after the emergence of property in the settled way of life during the Mesolithic Age and this brought human societies nearer to mathematical thinking. The Sumerian culture, going back more than 6,000 years, is the first one to leave behind written records on experimenting with different bases. Probably, one of the first usages of numbers is in the Sumerian calendar divisions, who devised a system by dividing the year to 12 months, 4 seasons of 3 months, and each month to 4 weeks of 7 days. This calendar system has been transmitted into the global cultural life but without transmitting their religious significance.

A decimal counting system was primed in many cultures including the Chinese, Indians, Egyptians and the Greeks but anthropological studies show that this is not universal. According to Boyer (1991), "A study of several hundred tribes among the American Indians showed that almost one third used a decimal base and about another third had adopted a quinary or a quinary-decimal system; fewer than a third had a binary scheme, and those using ternary system constituted less than 1 percent of the group. The vigesimal system, with twenty as a base, occurred in about 10 percent of the tribes." He argues that there was no orderly advance from binary to quinery to decimal system and decimals were essentially the product of the modern age in mathematics, rather than of the ancient period. Also, he states that among nonliterate tribes there seems to have been no need for fractions. Boyer (1991) also discusses the emergence of different bases for counting in the Eurasia in antiquity.

3.3. Priming of other Areas of Mathematical Complexity

Each area of mathematical complexity, shown in Table 2 or Figure 1, was primed in an appropriate cultural setting. Axiomatic geometry is another classic example, which is attributed to Euclid, who flourished 300 B.C. in the ancient Greece by transforming tacit knowledge on geometry inherited since the Sumerians and Egyptians. It comprises building blocks (e.g. points, lines, planes and space); definitions (e.g. focal point, side, angle, distance) and operation rules (e.g. drawing a straight line, circle). These make it feasible to collectively identify a huge number of regularity patterns in the form of observations, posits, conjectures or theorem.

The priming of some of mathematical disciplines and theorems happened a very long time ago and inevitably, historic evidence could suffer from uncertainty. However, the priming of a new can take place any time, e.g. consider the Conway's game of life in 1970 (for detail, refer to Sigmund, 1993). This is known as surreal numbers, composed of a set of dots on a grid obeying some rules. It is played in generations and at each generation, depending on the rules, either the dots replicate or die. The prototype of the game of life needed priming, after which the life-resembling patterns emerged spontaneously as an emergent property. Arguably, dots and grid squares stared at the eyes ever since they emerged but without any avail. After the priming of the game of life, a thriving intellectual activity emerged to aid the study of life-resembling patterns.

4. Co-Evolution of Communal and Institutional Mathematics

4.1. Priming of Mathematical Thinking

As depicted in Figure 3, mathematical complexity operates at three contexts of: individuals, communities and institutions. At an individual level, one's cognitive learning is honed through counting objects, length, area, volume or weight: (i) *communal mathematics* aids the abstraction and the use of same number systems across a diversity of unit systems, (ii) *institutional mathematics* gives rise to mathematical notions to calculate such properties as area and volume.

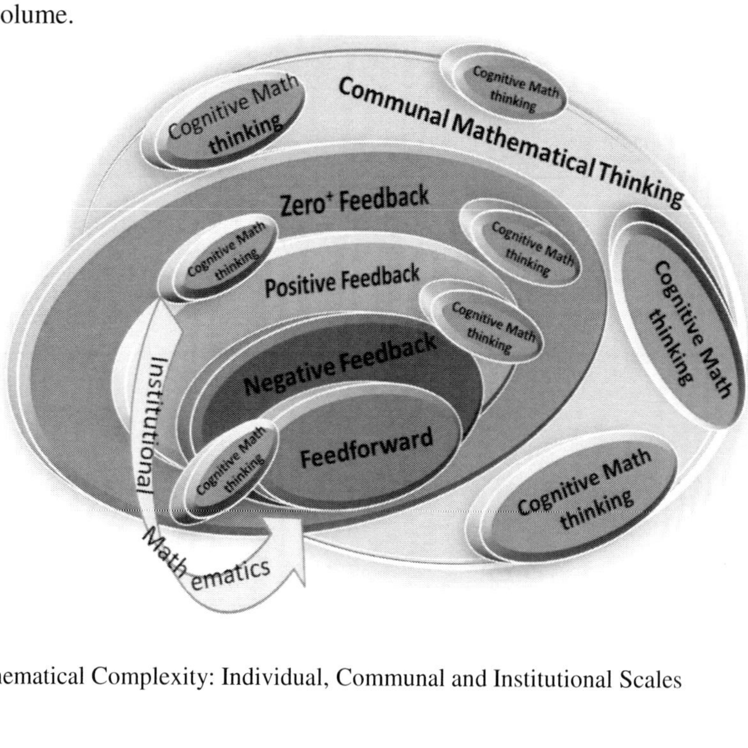

Figure 3. Mathematical Complexity: Individual, Communal and Institutional Scales

The absence of evolutionary thinking in the emergence of mathematical complexity signifies the shadow of an ontological mindset in the mathematical culture. It seems that mathematics has a permanent essence but its variations are superficial accidents. To highlight the role of evolutionary thinking, consider the counting system, as an area of mathematical complexity, which is now almost a global human experience. But this was not so in the past. The Sumerian culture had one number system for counting discrete objects such as animals, tools, and containers but used different systems for counting cheese and grain products, volumes of grain which included fractions, beer ingredients, weights, land areas, time units and calendar units; these systems changed over the years. Nissen et al (1993) express the problem as: 'The Sumerians had a complex assortment of incompatible number systems and each city had their own local way of writing numerals. In the city of Uruk about 3100 BC, there were more than a dozen different numeric systems.' The roadmap from the priming of an area of mathematical complexity to its fruition is outlined below in terms of individuals' cognitive mathematical thinking, communal mathematics thinking and institutional mathematics.

4.2. Priming of Mathematical Thinking

Even if numbers were primed in prehistory during the Upper Paleolithic or Mesolithic Ages and were subsequently selected, the emergence of mathematical thinking was not automatic but was through seeing patterns in numbers and making their use, akin to cause-and-effect. This probably emerged during the Neolithic Age (6000–4200 years ago) but through various events, triggering a realization that numbers have properties beyond the linguistic parts of speech. For instance, Sumerian officials who added and subtracted volumes of grain every day used this arithmetic skill to count other things unrelated to volume measurements.

It was natural to have different bases for counting but the harmonization of the diversity of these systems would have needed some mathematical thinking. The argument in this paper is that the emerging numbers made up some of the building block for mathematics but the mathematical thinking was probably "pump-primed" by the cognitive understanding related to regularity patterns within the numbers (this presumes the stripping any supernatural significance in. the numbers). The priming process itself is a spontaneous process in the sense that there is no guarantee for an outcome to be selected but should there be a selection, there must be a culture teeming with diverse activities. This is why a culture of diversity is very important without exception.

Mathematical thinking is not limited to numbers alone as its building blocks but other ones include shapes, forms and magnitudes and the magnitude may include length, area, volume, angle, a rhythm of time, weight, etc, all expressible by numbers. The first written record of mathematical thinking comes from deciphering a considerable number of tablets baked in clay from the Sumerian culture even showing methods of multiplication and division. These artifact recordings in the Sumerian city of Uruk comprise inventories of crops gathered and other statistical observations and these suggest the emergence of mathematical thinking conforming to social way of life.

4.3. Priming of Institutional Mathematics

Communal mathematical thinking was vulnerable to passing on skills on complex mathematics through generations, which triggered a selective advantage for the priming of *institutional mathematics*, in the ancient Greece or even in Sumer. Until the 19th century, there was no universal education but institutional mathematics and formal education were confined to the elite and some talented individuals. Applied knowledge was not formal but proportional to cognitive or communal knowledge passed on from generation to generation. The status of mathematics was formidable, its range was narrow, e.g. arithmetic, algebra and trigonometry, but this took a long time to penetrate into ordinary life.

4.4. Co-evolution between Communal and Institutional Mathematics

The overview of evolutionary transitions is captured in Figure 4 and the co-evolution of communal mathematical thinking with institutional mathematics is outlined below:

Mesolithic Age: Communal mathematical thinking was probably primed in Mesolithic periods in the form of counting and this was selected almost globally.

Mesolithic–15th century: Mathematical thinking was largely communal and its thrust was on arithmetic and geometry. This was proportional to the way of life then and included awareness of circumference, area and volume. Since the Sumerian times, the beneficiaries of formal education were largely the elite and the subjects were narrow e.g. arithmetic, axiomatic geometry, algebra, and some trigonometry. These were primed by the thriving communal mathematical thinking by elites (e.g. Euclid who lived in antiquity, and Alkhwarazmi, 780-850 AD and many others). The communal/cultural mathematics co-evolved with institutional mathematics but impacts were shallow and the slow developments were by contributions from individuals. There were also considerable obstacles, e.g. those even in simple arithmetic operations. Alkhwarazmi's contributions gave a further impetus to mathematics within the "Islamic" cultures in the 9th century, including arithmetic and algebra, and in particular, his introduction of Arabic-Indian numerals created a conduit for co-evolution between institutional and communal mathematical thinking.

15th to 19th century: Communal mathematical thinking in Europe was the backdrop to primed institutional mathematics, which was largely driven by the elites, as still there was no universal education. The organic growth of pure mathematics emerged from the 17th century onwards first in Europe, triggered by the emergence of reductive science in the 17th century and gave a new lease of life to applied mathematics (or mathematical analysis). Up to the 19th century, the role of mathematics in communal life was limited, as institutional arrangements were very few and far in between.

19th century: The universal education emerged in Europe and rigorous version of mathematical analysis emerged as pure mathematics.

Since 1950: Universal literacy was achieved globally since 1950 creating a more suitable background to the co-evolution of communal and institutional mathematics at a global scale. Since then, mathematics has become pervasive and virtually permeating every quarter of communal/institutional life, similar to, but at a different level of significance than, language.

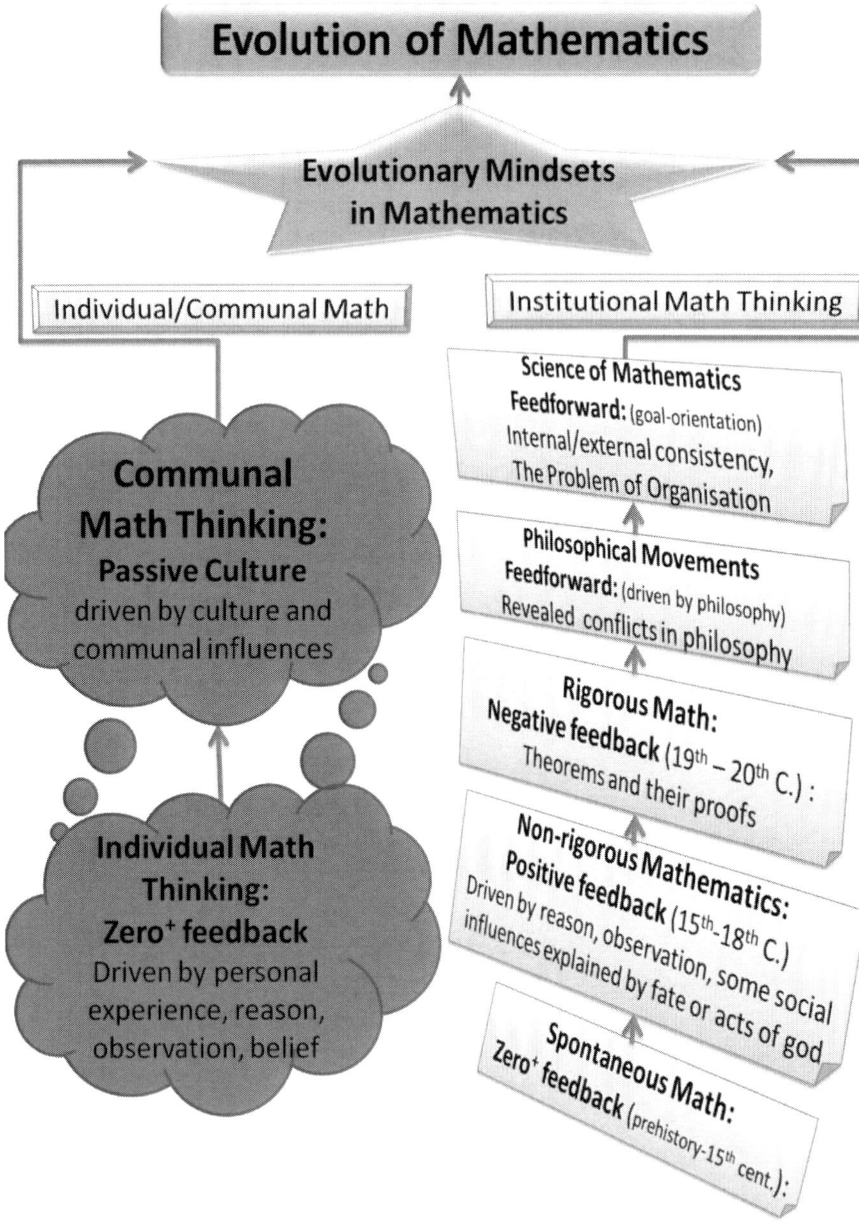

Figure 4. Evolutionary Transitions of Mathematical Complexity.

5. Positive Feedback Driving Non-rigorous Maths

As mathematics remains a pinnacle of human cultures, its association with positive feedback loops seems counterintuitive. At this stage there is no rigor but the goal is to be fit-for-purpose, reminiscent of survival in biology. Tacit knowledge of mathematical regularity patterns rarely involves formal proofing processes but are often made eventually explicit as propositions and concepts logically linked deductively through guessing, intuition or experimentation; analogizing, hypothesizing or abstracting; inductive processes of generalizing, etc.

This section shows that mathematics has not always been rigorous. The d-loop of mathematical complexity at the stage of positive feedback comprises explicit knowledge captured in terms of: observations of some sort, posits, deductions or conjectures; whereas its e-loops (hereditary machinery) can be seen in terms of building blocks, definitions and rules. If these e-loops have not been expressed as a coherent body of knowledge, it exists evidenced by its growth.

The routemap of any area of mathematical complexity at the positive feedback stage comprises: (i) the environment rife for a thriving culture, in which the various building blocks of each area of mathematical complexity would emerge; (ii) motivated mathematicians, driven by a wide range of incentives including the love of knowledge, are engaged in knowledge mining through applying some sort of rules to the building blocks and identifying regularity patterns; (iii) these patterns would be captured as observations, posits or conjectures as an expression of their selective advantage and would enter to the 'chain of reproduction'; (iv) the ℓ-loops would contribute to the subsequent honing of observations, posits and conjectures until they are rigorised; and (v) the captured knowledge would contribute to the growth of complexity and add up to an existing e-loop of the institutional mathematics.

Mathematical thinking underpins the positive feedback mindset (e-loop), which may prevail at individual, communal and institutional levels, as depicted in Figure 4. Arguably, the ability to "think" about a theme (say mathematics, risk, systems, the environment, economic, etc) without requiring professional training is a causative (cause-and-effect) mindset, which builds on (i) individual's cognitive learning going beyond instinctive reflexes (cognitive learning of individuals at the zero$^+$ feedback stage); and (ii) communal mathematical thinking, whereby mathematical thinking is stimulated by the penetration of mathematical constructs.

Positive feedback e-loops signify the rigor of in the building blocks, definitions and rules. Just like the complete absence of any of any meta- or transcendental beings in genotypic machinery of all the living things, no such entities are stipulated in the makeup of the e-loops of mathematical complexity to grant their functionality. Consider the example of the area of a circle. In order to calculate its area, well-formed definitions are needed (radius and the centre and shape of objects, as well as points, lines, curves, locus, areas) and rules (e.g. rotation for creating the locus of the equidistant points from the centre) and mathematical thinking e.g. the number: π. The Sumerians and Egyptians did not wait for rigorous calculations but on the contrary, their mindset driven by positive feedback complexes were the key for the development of rigor a few millennia later. Section 9 gives another example based on number sequences.

Positive feedback also stipulates that once an evolving entity is selected as simplicity, it will grow towards complexity through generations and continual selection of further associated building blocks. This is evident all over mathematics and one example from mathematics relating to π is the Stirling's approximation of $n! = \sqrt{2\pi n}(n/e)^n + \big(1 + O(1/n)\big)$ (see Cameron 1994). The point is that π, *e* and factorials do not individually resemble to one another but through the selection of their individual concepts and the growth of complexity, even the above unexpected formula can arise.

Evolutionary systemics stipulates the routemap of positive feedback sub-loops, as follows: (i) d-loops: selective advantage, proliferation, directions of information flow; (ii) ℓ-loops: adaptation; e-loops: the existence of a hereditary machinery, diversification, and entropy (internal consistency). These are explained in this section, where all of them are important to fully describe the mindset but perhaps, proliferation, entropy and diversification are the key ones.

5.1. Proliferation the Fundamental Character of d-loops

Until the 17th century, developments in mathematical complexity were slow, steady, fragmentary and limited to discrete regularity patterns. Only Euclidean geometry developed well in the antiquity and served as the pinnacle of ingenuity but remained narrow. Algebra had a very significant boost in the 9th century but its development, especially its symbolism, was a gradual process until the modern times. Kline (1980, p153] remarks that: ",By 1800, mathematics was in a highly paradoxical situation. Its successes in representing and predicting physical phenomena were beyond all expectation superlative. On the other hand, as many 18th century men had already pointed out, the massive structure had no logical foundation, and there was no assurance that mathematics was correct."

Kline's account befits the context of evolutionary systemics, according to which regularity patterns are selected by some sort of selective advantage even though they would often lack internal consistency. Many new areas of non-rigorous mathematics were primed between prior to the 16th century but between 16th–18th centuries, mathematical regularity patterns were proliferated at an unprecedented rate albeit susceptible to anomalies. In this period, reductive science, industrialization and mathematics were at a full steam of proliferation in Europe and both reductive science and applied mathematics (and thereby mathematics) were priming new mathematical information and opening up the door for new possibilities with emergent properties. Kline expresses these thriving activities as: "the Hindus, Arabs, and Europeans added floor after floor, complex numbers, more algebra, the calculus, differential equations, differential geometry, and many more subjects" which were without rigor but often fit-for-purpose.

Axiomatic geometry had been inspirational throughout its history but it was unwittingly infested with surprising paradoxes. The positive feedback process of proliferation tended to diversify, as well as throw up anomalies. At the stage, simply there is no foresight and therefore the researchers on a new discipline simply do not know the impact they will be making but are engaged with developments guided by their cognate perceptions of building blocks and the rules of their subject of investigation. The community (or Nature) is the judge of the primed mathematical information, although the professionalism of the researchers guided by their cognate perceptions keeps the momentum. This is what happened in the mathematics before the 19th century and it is what ought to happen if a new mathematical discipline is primed today, a fitting image of evolutionary thinking.

5.2. Attributes of Positive Feedback d-loops

Khatibi (2012a) regards emergent property as the genotypic expression of an evolving entity and selective advantage as its phenotypic expression. In mathematics, emergent property may be regarded as new information related to the emergence of new entities, such as posits, conjectures or theorems. Mathematics has intriguing selective advantages, as outlined below.

5.2.1. Emergent properties of pure mathematics

Consider the simple example of:

$$y = x^2 + 2x + 1 \quad (5.1)$$

Notably, everything about (5.1) can be envisaged in the mind and communicated verbally without needing the aid of the real world, although using a medium, such as paper, aids the communication. A description of (5.1) is possible through two interoperable approaches of conic sections and algebra: (i) The conic sections uses a definition in terms of the locus of equidistance points (the rule) from the directrix and the focus (the building blocks), in which case the *shape and the vertex* may be regarded as the emergent properties of these building blocks and the rule. (ii) Using algebraic techniques, everything about the curve can be determined (emergent property) in terms of the building blocks (dependent and independent variables) and the rule (the definition of function and their operations including differential variations aided by a coordinate system but this is not essential). The emergent properties in this setup are the full description of the y-value for each x-value, the curvature of the shape at any point, the focal point and the directrix.

Emergent properties are routine and taken for granted but it is noted that (i) they do not exist in the individual building blocks and (ii) the building blocks are brought together by rules. Proofs in mathematics comprise: definitions, symbolic operation (rules) and conclusions, normally referred to as theorems when the procedure is rigorous. The paper argues that the conclusions are emergent properties. The arguments put forward above resembles Gödel's Second Incompleteness Theorem Franzén (2005), stating that *for any consistent formal System S within which, a certain amount of elementary arithmetic can be carried out* (analogous to building blocks and rules), *the consistency of S cannot be proved in S itself.* The theorem is much clearer when consistency is regarded as an emergent property of a formal system, whose elements are not necessarily consistent with one another or with System *S*.

5.2.2. Mathematical symbolism as a selective advantage

Although the language and symbolism of mathematics may be intractable for many and act as a barrier, its ideographic nature breaks through national, linguistic and cultural barriers and acts as a universal language, which is one of its selective advantages. Even the symbolism of mathematics is subject to evolutionary processes. There are excellent websites depicting the various symbols with their date and originators (e.g. http://jeff560.tripod.com/). It is often seen here that various notations were suggested for the same functionality, many of them with a potential for selection. An example is shown in Table 3.

5.2.3. Data

A massive amount of coordinate data (x, y) can be produced without actually measuring anything and without producing any error, although human blunders can always occur. Evidently, the world of pure mathematics does not need data but can supplement their creation in abundance as and when necessary. Thus, there is no presumption that data from the real world are needed for the proof of mathematical theorems but its data-generating ability is an emergent property.

5.2.4 Compaction

This is the reverse of the data-generating ability of mathematics, as the masses of (x, y) data can be replaced by an appropriate function such as (5.1). Compaction is one of the ways to use mathematics for the production of information and this is an emergent property.

Table 3. Some of Basic Mathematical Notations in Use Now

+ / - plus / minus	Johannes Widmann (1489) / Luca Pacioli (1494) / Giel Vander Hoecke (1514)
x multiplication	William Oughtred (1631 or 1618)
. (dot) multiplication	Gottfried Leibniz (1698) / Johann Bernoulli (1694)
÷ division	Johann Rahn (1659) / John Pell (1668)
= equals	Robert Recorde (1557) / William Oughtred (1618)
< / > less than / greater than	Thomas Harriot (1631)
x^2 exponent	Nicolas Chuquet (1484) / Pierre Hérigone (1634) / René Descartes (1637)
$\sqrt{\ }$ $\sqrt[3]{\ }$ radical (root)	Christoff Rudolff (1525) / Albert Girard (1629) / René Descartes (1637)
$\frac{a}{b}$ fraction bar	Fibonacci (1202)
. decimal point	Francesco Pellos (1492) / G.A. Magini (1592) / Bartholomaeus Pitiscus (1595) / John Napier (1617)

(–

Source: http://www.storyofmathematics.com/16th.html)

5.2.5. Variability

Functional expressions together with the rules on differentiation and integration provide the ability to gain an insight into variability of mathematical behaviors and to derive stationary points. Variability opens the eye and focuses attention on important features, and as such, it is an emergent property of pure mathematics.

5.2.6. Prediction capability

Implicit in the predictive capability is that the equations transform information from one form to another. For instance, the process of deriving a system of equations includes knowledge of the variables. After the formulation of the system of the equations, the values of the variables are regarded as unknowns but mathematics create the process of "reverse engineering" to uncover (or predict) their original values. Consider the example of:

$$x_1 = 2;\ x_2 = 3 \quad (5.2.a)$$

(2.a) can be used to formulate the following equations:

$$2x_1 + 3x_2 = 13 \quad (5.2.b)$$

$$3x_1 + 4x_2 = 18 \quad (5.2.c)$$

The knowledge of "13" and "18" in (2.b) comes from (2.a) and the mathematical method of solving (5.2.b-c) does not predict the values of x_1, x_2 but uncovers their original values. Sometimes instead of (5.2.a), regularity patterns are used, e.g. the law of gravity with an empirically-determined coefficients. It is very important to realize that mathematical predictions are not a generation of new information or a discovery of new knowledge but the exchange in the form of information. This in itself is a formidable emergent property, where mathematics formalizes the procedure for a gamut of equations potentially usable as predictive tools.

5.3. Directions of information flow concerning both d- and e-loops

Mathematics, as a collection of theorems, conjectures and posits, is a body of transparent and readily intelligible mathematical information capable of invoking appropriate actions but what is the nature of this information? To the trained eye, the function, say, $y = x^2 + 2x + 1$ invokes an immediate image of a parabola with full information on its geometric and algebraic properties. This may be regarded as *one-way* flow of mathematical *information*, from the world of abstract mathematical objects to the world of mathematicians. Literally speaking, man extracts information from the inanimate world of mathematics as if it flows from the world of mathematics to man. This is reminiscent of the rotation of the Earth around the Sun but it seems other way round. However, there are issues with mathematical information as follows.

Consider the trajectory of a projectile following a parabola, where the projectile has no way of knowing its past, present and future positions. Notably, the information on the trajectory is a man's construct combining the information on the time history of the positions of the projectile through the cause-and-effect understanding of the processes. In reality, the trajectory is subject to some uncertainty, which can be ignored in pure mathematics to the extent of assuming a perfect trajectory determined by mathematics. Notwithstanding these physical limitations, the assumption of a perfect curve describing the trajectory breaks down at the subatomic scale but even this can be ignored. The remaining perfect mathematical description of the trajectory only exists in mathematicians' minds or the mathematicians "bend" their minds and regard certain unreal conditions as real, much like some scientists who believe in both evolution and creation at the same time. Therefore, there are two types of information: (i) a *one-way* flow of information from the world of mathematics to the world of mathematicians; (ii) the flow of information among mathematicians to justify the assumptions, which does not interfere with the flow from mathematics to mathematicians – this is often overlooked.

There remains the question that: can information flow from the real world to that of mathematics? When mathematical facts are used as surrogates for objects in the real world (the world of physical and social sciences), the mathematical constructs have to fit to the real world conditions but this is not possible, as the working of the real world is not dictated by the calculator in the pocket. The attempt to fit is called applied mathematics (or mathematical modeling), which necessitates the *flow of information in more than one direction*, including the flow of information from the real world to the world of mathematics. Notwithstanding the flow of information from the real world to mathematical modeling, information can also flow from the real world to mathematics in the proof processes but this is regarded as negative feedback and is discussed in Section 7. Information can also flow from the man's world to the world of mathematics to ensure its external consistency but this is discussed in Section 8.

The next section discusses negative feedback loops having the emergent property of ensuring internal consistency, infallibility, impeccability and irrefutability associated with proving theorems. However, at the stage of positive feedback, mathematical information can be a posit, conjecture or some sort of observed regularity pattern and all expected to be approximately true. At this stage, there is no guarantee for their internal consistency and impeccability. To appreciate internal consistency, consider philosophy, which uses reasoning and logic to arrive at its brand of truth but ends up with discrete doctrines riddled with internal inconsistencies (as philosophy lacks negative feedback loops).

5.4. Consistency of mathematics – negative Feedback Loops

Mathematics prior to the 19th century was not rigorous most of the time, as illustrated by reproducing an example from Grabiner (1974, p203). Now, there are rigorous approaches for the expansions of trigonometric functions, but she presents a procedure by Leonard Euler (1707–1783) arriving at a correct conclusion from a set of arguments that leaves modern mathematicians breathless. Euler's procedure for expanding $\cos v$ is as follows: The following identity holds:

$$(\cos z + i \sin z)^n = \cos nz + i \sin nz \tag{5.3}$$

Expanding the left hand side and taking the real parts, he obtained:

$$\cos nz = (\cos z)^n - \frac{n(n-1)}{2!}(\cos z)^{n-2}(\sin nz)^2 + \frac{n(n-1)(n-2)(n-3)}{4!}(\cos z)^{n-4}(\sin nz)^4 - \ldots \tag{5.4}$$

By assuming z as infinitely small and n infinitely large, the following approximations are obtained:

$$\cos z = 1,\ \sin z = z,\ n(n-1) = n^2, n(n-1)(n-2)(n-3) = n^4 \tag{5.5}$$

Leading to:

$$\cos nz = 1 - \frac{n^2z^2}{2!} + \frac{n^4z^4}{4!} - \ldots \tag{5.6}$$

Euler argues that for z being infinitely small and n being infinitely large, nz must be a finite quantity and equal to v. Hence the familiar expansion of $cos\, vz$ as follows:

$$cos\, vz = cos\, v = 1 - \frac{v^2}{2!} + \frac{v^4}{4!} - \ \ldots \qquad (5.7)$$

This example is very revealing as the quality of the conclusion was sound but that of the arguments was poor and hence the risk of errors and uncertainty.

Grabiner (1974, p204) also gives another example by arguing that the drive for results was more overwhelming than rigor and the emerging popularity of mathematical symbolism was often good enough to develop mathematics. One example is the Fundamental Theorem of Algebra. Its cubic version was known to the ancient Babylonians, Egyptians and Greeks; its first general solution was reportedly found by Niccolò Tartaglia (1500–1557) but published by Gerolamo Cardano (1501–1576) in 1545. The general symbolic notation is now taken for granted now but it was introduced in 1591 by a French mathematician, François Viéte (1540–1603), who expressed the cubic equation as:

$$(x - a)(x - b)(x - c) = x^3 - (a + b + c)x^2 + (ab + ac + bc)x - abc \qquad (5.8)$$

This was then generalized by Albert Girard (1595–1632) for the nth degree equation and its first rigorous treatment was presented by Johann Gauss (1777–1855) in 1799, which also had "holes" in it but they were filled later.

Examples on the evolution of non-rigorous mathematical concepts are many, and one widely known is the calculation of the value of π. This started with the Sumerians and Egyptians and thrived until modern times ensuring rigor. The point is that even approximate regularity patterns have selective advantages and they are selected until the required rigor becomes an issue.

5.5. Diversification

Diversification arises largely due to minor internal differences in reproducing individual entities. Although the reproduced entities are all unique, diversity reflects the health in the underlying processes. Mathematics diversifies in various ways, such as the emergence of different techniques to define the same quantity, e.g. central tendency (mean, median, mode ...). Each discipline in turn would start from simple building blocks and diversify in a variety of ways, such as diversification of the building blocks (e.g. geometry starts with the definition of a point, line, angle, surface, volume etc); diversification of rules and operations; and diversification of theorems. Published research literatures indicate the scale of diversification.

In spite of the obvious diversification in mathematics, there is a remarkable pluralism. Mathematicians do not quarrel with each other on the supremacy of a theorem over another and if there is more than one method of proof for a particular theorem, they all are taken on board. For instance, the Pythagorean Theorem was proven by different geometric methods and later by equally acceptable algebraic or trigonometric methods.

5.6. Entropy

Evolutionary transitions are driven by the spontaneous instigation of entropy but its occurrence in pure mathematics is rather subtle. Entropy is not only a feature of the concept of closed systems in the sense of von Bertalanffy (1969) but is also a feature of any evolving system *at the stage of positive feedback*. This is reviewed in an appropriate level of details by Khatibi (2012a).

Mathematics is a study of logically connected abstract systems and naturally the strength of Euler's argument in the derivation of (5.7) (a finite quantity is obtained by multiplying one infinitely small and another infinitely large quantity) is not logical and a source of entropy: i.e. such loose statements stand the chance of introducing randomness or uncertainty to mathematics. Also, the generalization of (5.8) towards the Fundamental Theorem of Algebra is breathtaking but with the benefit of the hindsight, mathematics escaped unscathed on this occasion. It is important to realize that such a lack of rigor does not undermine the whole foundation of mathematics, as the assemblage of a set of building blocks together creates a system with cognate outcomes, which reduces blunders. However, mathematics being the study of logically connected abstract systems is undermined without rigor and without consistencies.

Entropy may be sought in terms of paradoxes in mathematics but the author is not sufficiently learnt about them, for more information, see Bunch (1997), Clark (2002), Sainsbury (1987) and Sorensen (2003). Entropy may also be sought in terms of concepts undermining the efficacy of mathematics. Although, infinity is a mind-bugling concepts, regarding it in terms of entropy is helpful, signifying a loss of efficacy. Thus, whilst indulgence with thinking about infinity is acceptable for unfettering mathematics, equally one can simply say, "so what", as there is no efficacy in pursuing increasing numbers forever. A logical way to contemplate infinity is to identify a range, beyond which the loss of efficacy becomes significant.

Although Goodman (1998) does not use the term entropy, his description fits well to the context of this section. "In the past decade, however, set theory has been undermined roughly in the sense that geometry was undermined about a hundred years earlier. The independent results, the proliferation of large cardinal axioms, and the construction of increasingly bizarre models for set theory have made mathematicians realize how weak their set-theoretic intuition actually is. In the absence of new insight, the views of set-theorists begin to diverge. Some still follow Cantor in thinking the continuum hypothesis plausible, but others follow Gödel in believing more and more strongly that it must be false. It is becoming truistic that we need a new concept, one more fundamental than that of a set. Unfortunately, no one can imagine where to look for such a concept." Arguably, the above reflect entropy.

Besides the cases outlined above invoking entropy in the world of mathematics, there are further soft aspects that also remind entropy in the world of mathematics and these include: (i) mathematics is intractable for many or difficult to muster; (ii) its symbolism is complex; (iii) its growth seems to be inflationary or a never ending process. All these may be regarded as entropy in mathematics but further important entropy may stem from human's inability to master such diversity of mathematical facts.

5.7. Adaptation concerning ℓ-loops

Adaptation in biology is related to organisms fitting to their environment, and is driven by the diversity in the gene pool. As discussed above, diversity of techniques normally emerge for the same mathematical entity equivalent to gene pools and therefore these would enable adaptation to be feasible in mathematics, e.g. variety of central tendency measures. The whole body of mathematics may be regarded as adaptive by abstracting mathematical analysis as a tool of science. Undoubtedly, mathematics will also adopt more features from the modern mathematical modeling.

6. Evolutionary Transition from Non-Rigorous to Rigorous Mathematics

Kitcher (1983) remarks that the history of mathematics from the 17th century on is underdeveloped, even by comparison with the history of science. Although there are books on the subject, getting a critical view is not easy. The author relies on the critical understanding of some of the scholars who discuss that the mathematics of the 19th century overhauled that of the 18th century infested by inconsistencies. Carl Boyer has called the 18th century as "the period of indecision" in relation to the foundations of mathematics, Boyer (Grabiner 1974, p205). Grabiner (1974, p205) argues that the practice of mathematics did not depend on a perfect understanding of the basic concepts used. There were various imperfections or anomalies associated with inequality-based treatments of limit, convergence and continuity.

Grabiner (1974, pp206-207) recognizes a number of reasons for the transitions in mathematical practices from the 18th to the 19th century but her ideas are recast, consolidated and interpreted below to explain the evolutionary transition in terms of entropy, as follows:

(i) ***The focus on individual contributions***: Although there was not a serious incidence of mathematical errors by then, mathematical rigor was a potential way of reducing errors specially dealing with more complex functions using several variables, which became widespread to the end of the 18th century; there were many plausible conjectures but difficult to establish intuitively their truth; increase in complexity reduced the guiding role of intuition and increased the need for rigor; rigorous mathematics would have acted as a source of information in verifying increasing volume of results.

Using the language of evolutionary systemic axioms, the lack of rigor in the 18th century mathematics was instigating entropy by diffusing intellectual energy. New rigorous techniques offered a selective advantage to get more out of existing knowledge capital then, where the legacy of the Euclidian axiomatic rigor was inspirational.

(ii) ***The focus on generalization***: The analysis of increasing volume of results with a focus on individual sets of results would have created an anecdotal sense and mathematics would have been seen as a heap of numerous locally true conjectures. This would have retarded mathematics and Struik (1987) argues that "At the end of the eighteenth century, several mathematicians thought that the pace of getting new results was decreasing. This feeling had some basis in fact; most of the results obtainable by the routine application of eighteenth-century methods had been obtained. Perhaps, if progress was slowing, it was time to sit back and reflect about what had been done" (cited in Kuhn, 1962). This view is of course better explained by entropy triggering the selective advantage for the emergence of rigor.

(iii) ***A shift in culture***: Grabiner (1974) presents some detailed statistic to assert that there was a shift of culture from the 18th to the 19th century on educating mathematics, as universal education was taking off in the Europe. It is widely known that the 16th–18th centuries' scholars were often elites and largely lived on patronage of royal courts to add glory to their status. She explains this as "A mathematician could understand enough about a concept to use it, and could rely on the insight he had gained through his experience. But this does not work with the freshmen, even in the eighteenth century. Beginners will not accept being told, 'After you have worked with this concept for three years, you'll understand it'."

If the 18th century mathematics was overhauled to the 19th century by rigorizing it, there ought to be some concepts that became obsolete. Kitcher (Kitcher, 1983) illuminates this stating: "If one compares contemporary text in analysis with a classic text of the early part of the 18th century, it is impossible to regard the later work as a simple extension of the earlier." He adds that "we no longer care for the systematic exploration of special functions which our Weirstrassian predecessors loved so well."

Grabiner (1974) argues that the transition from results-oriented 18th century mathematical endeavors to rigor-minded 19th century mathematical practices were facilitated by (i) the right definitions, and (ii) right techniques of proof to derive the known results from the definitions. The 19th century mathematicians largely reinvented the 18th century mathematics. Although Grabiner (1974) does not use evolutionary systemics, she strikes a very harmonious note with this paper in stating that the facts "show that a real change in point of view was required for the rigorization of analysis; it was not an automatic development out of eighteenth-century mathematics." The author argues that entropy triggers a selective advantage for an evolutionary transition through rethinking and setting up negative feedback loops. If intellectual endeavors fail to develop negative feedback loops, the entropy in the particular area of complexity undermines its efficacy but if the appropriate loops are developed, a great deal of the past knowledge capital is reinvented in the new light, as discussed next.

The author distinguishes between unselection, e.g. the unselection of the Ptolemaic geocentric model of planetary motions and becoming obsolete, e.g. the sophisticated epicycles associated with the geocentric model. Past mathematics is not normally unselected but can become obsolete.

7. Negative Feedback Driving Rigorisation

A clear understanding of negative feedback loops was originated in systems science/cybernetics with the effect of underpinning internal consistency. However, the systemic version was unwittingly ontological but this was remedied by Khatibi (2012a) adding an evolutionary dimension to the systemic version of feedback loops, as summarized in Table 1. The d-loop of negative feedback loops are presented in Table 4 and Figure 5 compare their implementations in systems science, reductive science and pure mathematics. The salient features of the figures include:

(iii) Negative feedback loops are subsystems, composed of three sub-loops. The d-loops are *control subsystems* comprising *Transmitters*, *Fact Engines* and *Actuators*;
(iv) The negative feedback d-loops at their point of initial selection overhaul an evolvable entity (say non-rigorous mathematics) by:
  - Making the positive feedback d-e-loops flexible through acquiring a state-aware capability;
  - Inserting *Interfaces* between evolvable entities/control subsystems.

When the flow of information in mathematics is in one direction, the cultural mindset (e-loop) is referred to as *fit-for-purpose* or *non-rigorous* mathematics. This mindset is happily in the currency until anomalies are found. Prior to anomalies, proofs (negative feedback) are welcomed but become necessary after anomalies. The following types of negative feedback d-loops are distinguished.

(i) *Proactive* negative feedback: A proof is an intellectual challenge and its realization means a set of good definitions and operations (e-loops). These are additional flow of information from mathematicians to mathematics, i.e. mathematicians use their experience and devise clear definitions and correct operations to prove rigorously the theorem.
(ii) *Reactive* negative feedback: When paradoxes or anomalies emerge, mathematicians do not defend the old erroneous approaches but refine by going back to the definitions and operations.
(iii) Negative feedback in cybernetics is normally ubiquitous, whereas its implementation in the form of proof can be *reactive* or *proactive*.
(iv) Mathematics also include *intrinsic negative feedback loops*, e.g.:
  (a) The solution of a system of equations often includes a trinity of conditions for solving properly-posed system of equations (e.g. the solution exists, it is unique and it is stable and also it must be shown that the solution is not ill-conditioned), see Khatibi (2001);
  (b) Improperly-posed equations must satisfy identifiability, uniqueness and stability conditions, see Khatibi (2001); and
  (c) There are many recursive methods in mathematics, including the solution of nonlinear systems of equations.
(v) Internal consistency of a problem lacking rigor creates a case for negative feedback: (i) improperly-posed problems in the sense of Hadamard (1952), see Gustafson (1980), (ii) diminishing return problems (e.g. Sider, 1991), (iii) optimization problems, (iv) fractal problems (e.g. Falconer, 1990), (v) chaos problems [e.g. Lorenz, 1963), (vi) catastrophe problems Thom (1972 and 1975), and (vii) a whole branch of mathematics devoted to stochastic and probabilistic problems.

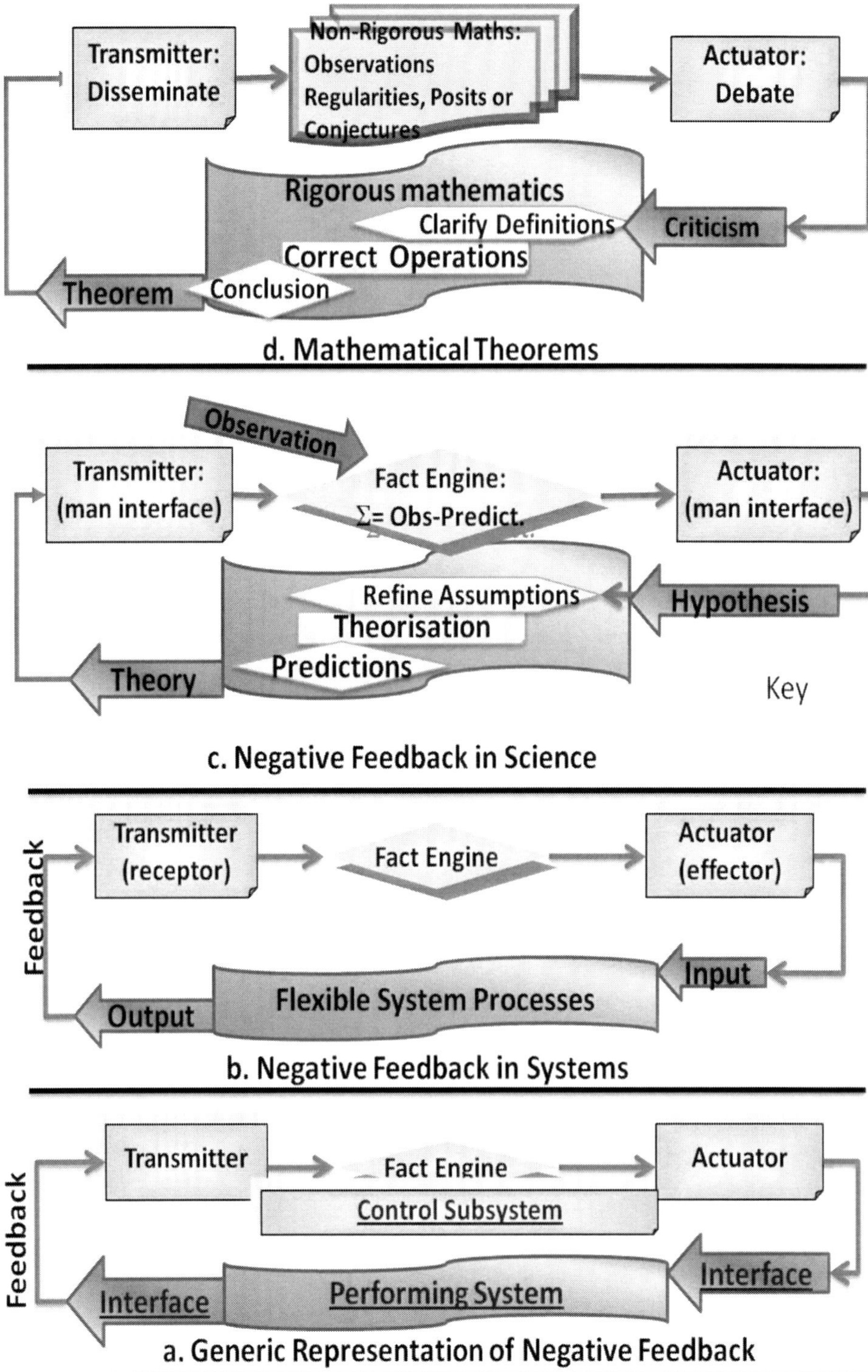

Figure 5. Negative Feedback: (a) Systems Science, (b) Reductive Science (c) Mathematics.

Table 4. Comparing Negative Feedbacks: Systems Science, Reductive Science and Mathematics

Attributes	Systems Science	Reductive Science	Mathematics
Background	• Classic industrial systems in 19th century • Well-defined inputs, processes, output and interfaces • (i) performing subsystem, (ii) control subsystem, (iii) interface	• 1st explicit negative feedback loop primed in the 17th century • Tantamount to: Right to Challenge • Primed within a suppressive culture • (i) performing subsystem, (ii) control subsystem, (iii) interface	• Mathematical proofs reached a pinnacle through Euclidian geometry • Maths proofs are negative feedback but subtly so and not obvious • (i) clear definition, (ii) correct operations (iii) conclusions
Interface	• Interface is automated	• Man acts as the interface	• Man acts as the interface
Data	• Early control machines worked on tacit information • Measurement became essential from 20th century onwards	• Astronomy primed data • Reductive science relied on data • Data proliferated leading to "data rich information poor" situations	• Mathematics is amazing for not requiring much data and perhaps no data • Applied maths require data
Transmitter	• Actual unit in control subsystem • Measures a particular value and passes it to the Fact Engine	• Predictions provide plethora of data/ information, ubiquitously available • Man transfers then to Fact Engine	• Knowledge dissemination invokes debates and challenges and these invoke remedial measures
Fact Engine	• Require target values and computations of state variable to maintain target values • Diversity of techniques, e.g. PID	• Target replaced by observed values • Minimizing (obs-predict.) provide a new setting for Performing System • Minimization can be manual	• Fact Engine is the rethinking, refining, recombining, abandoning or whatever it takes to remove anomaly
Actuator	• Actual unit in Control Subsystem • Physically resets something in the d-loops	• Previous assumptions are refined, making the theory ready for a new prediction	• New theorems or reworked ones are published
Emergent Property	• Control over loss of efficacy • Understanding efficiency	• Objectivity (not depending on individuals) • Internal consistency, transparency • Defensibility	• Pure mathematics, e.g. proven theorems • Emotive references to infallibility, impeccability, irrefutability are misleading!
Examples	• Homeostatic / ecological systems • Many industrial units: flywheel	• Plethora of theories	• Plethora of theorems

(d) Improperly-posed equations must satisfy identifiability, uniqueness and stability conditions, see Khatibi (2001); and

(e) There are many recursive methods in mathematics, including the solution of nonlinear systems of equations.

(vi) Internal consistency of a problem lacking rigor creates a case for negative feedback: (i) improperly-posed problems in the sense of Hadamard (1952), see Gustafson (1980), (ii) diminishing return problems (e.g. Sider, 1991), (iii) optimization problems, (iv) fractal problems (e.g. Falconer, 1990), (v) chaos problems [e.g. Lorenz, 1963), (vi) catastrophe problems Thom (1972 and 1975), and (vii) a whole branch of mathematics devoted to stochastic and probabilistic problems.

Proving a theorem is tantamount to inserting an additional direction of information flow and collectively they are: (i) the flow of information from the world of mathematics to man, as discussed in Section 5.3 and (ii) from man to the world of mathematics to improve its internal consistency. The source of information for the emergence of both proactive and reactive negative feedback is implicitly explained by Kline (1980, Pp328-329]. For proactive negative feedback, his equivalent account includes: the appeal to applications is not as radical as it may seem to mathematical purist. The concepts and axioms came from observation of the physical world. Even the laws of logic are now generally granted to be a product of experience. Problems which led to theorems and even suggestions about methods of proof came from the same source. For the reactive negative feedback Kline's equivalent account includes: A theorem may work in n cases and fail in the $(n+1)^{th}$ case. One disagreement disqualifies the theorem. But modification may lead, and historically has led, to corrections restoring its use.

Evidence for negative feedback in mathematics is implicit in the literature. For instance, Kline (1980, P100) argues that up to the 19th century, the few disagreements caused by mathematics were brushed aside and the equanimity of mathematician in the 19th century was shattered by the various attacks, giving rise to rigorisation. Avigad (2003) argues that "Historians of mathematics usually take the nineteenth century to be the birth of the 'modern' style of mathematical thought that is practiced today." Grabiner (1974) refers to this as rigorisation of mathematics and gives examples from the 19th century such as using the 18th century approximate solutions to devise the condition for existence of their solutions, now known as the Cauchy-Lipschitz method of proving the existence of the solution of a differential equation. Grabiner (1974) provides greater historic details of the rigorisation of calculus. This paper regards rigorisation and proof, as equivalent to the implementation of negative feedback d-loops.

The insight into the internal consistency of mathematics through reactive/ proactive /intrinsic feedback loops signifies that even this seemingly sterile, virtual or platonic world of mathematics on its own is a colorful world and has a good potential to serve as a surrogate to the real world. Even if all the conditions for internal consistency are satisfied, the behavior of a mathematical entity can rapidly depart from one deterministic state to another depending on small changes in initial conditions. This development was a pinnacle of the 20th century, known as chaos theory. This even arises in the simple *logistic equation* of: $x_{n+1} = ax_n(x_n - 1)$, in which depending on values of a and x_0, the behavior of equation is fixed, periodic and chaotic. There are also similarities with catastrophe theory.

8. FEEDFORWARD LOOPS IN MATHEMATICAL COMPLEXITY

Mathematical complexity is a product of human cultures and is continually reshaped by the proliferation of new ideas (positive feedback), seeking better definitions and better rules (negative feedback loops) and systematizing or consolidating it (feedforward). Sections 5-8 focus on positive and negative feedback loops driven by internal entropy towards *internally consistency* mathematics. Likewise, the external consistency of mathematical complexity with the environment is underpinned by feedforward loops driven by "external entropy" stemming from: (i) cumulative growth of each areas of mathematical complexity; (ii) the need for the consistency of different areas of mathematical complexity at their interfaces with one another; (iii) managing the expectation stemming from the illusive attributes such as infallibility, impeccability and irrefutability. However, mathematics is supposed to reside in an insular world, so does it matter whether or not it is consistent with the external world? The responses are obtained from philosophers' and practitioners' viewpoints, as well as on goal-orientation but there is a further coverage of the subject in Section 8.3.

8.1. Philosophers Aspiration – Top-down or Prescriptive Doctrines

Philosophers drive a prescriptive culture, who tend to be democratically exclusionists so that philosophers: (i) prescribe the entire mathematics to follow their presupposition but appreciate the shortfalls; (ii) embark on "reverse engineering" of mathematics (e.g. formalism and logicism) guided by their presuppositions; (iii) preserve instinctively their presupposition when anomalies are observed but retrofit definitions/operations to maintain their presupposition. Thus, philosophical discourses are rich in ideas but unfit for practical purposes.

The mindset in the philosophies of mathematics are by-and-large ontological, in the sense that they overlook evolutionary processes in mathematical complexity. This paper refers to them as a physicalistic model, in which they overlook evolutionary processes in the time dimension. Characteristically, all the philosophies of mathematics are driven by their nonmathematical presuppositions and often using very little evidence.

Philosophical doctrines have emerged in three waves: (i) from antiquity until the modern mathematics emerged in the 16th century; (ii) between 16th-19th centuries; and (iii) doctrines since the late 19th century and particularly, during the first half of the 20th century. Technical details of these doctrines are outside the remit of this paper but these are covered by many good books and papers on the philosophy of pure mathematics, see for example, Russell (1903), Russell (1919), Baum (1973), Courant and Robbins (1978), Kline (1980), Benacerraf and Putnam (1983), Körner (1986), Tymoczko, (1998), Friend (2007). No value is added by reviewing these doctrines but uncovering their thinking is useful.

8.1.1. From antiquity to the emergence of the modern mathematics

Avigad (2007) outlines traditional divide between Plato (424/423-348/347 BC) and Aristotle (384-322 BC) as an early example of tensions between philosophical theories. The division stemmed from presuppositions of the primacy given to experience by Plato, and to abstraction by Aristotle. The discourse was then on mathematical objects and the truth underlining their knowledge, as follows: (i) Plato regarded abstract mathematical objects, like triangles and spheres, as *forms*, to be imperfect reflections in this world but dialectical processes could improve their knowledge; thus, the mysterious abstract nature of mathematical objects found an ontological expression to justify by our physical experiences to mathematical statements. (ii) Aristotle regarded mathematical objects, like triangles and spheres, as *abstractions* from experiences (i.e. spherical objects exist roughly but perfect sphere are deliberate abstraction ignoring certain features like size or weight); this disciplined behavior justifies the conclusions despite their rough experience. The respective discourses made the *received wisdom* until the emergence of the modern mathematics in the 16th century.

8.1.2. Emergence of modern mathematics until the late 19th century

According to Avigad (2007), Gottfried Leibniz (1646–1716) distinguished between *necessary* truths (e.g. mathematical truth) and *contingent* truths (i.e., scientific facts); David Hume (1711–1776) also had similar discourses. Emmanuel Kant (1724–1804) gave a new lease of life to the philosophy of mathematics by distinguishing that: (i) an appropriate *justification* is not needed for assertions of mathematical truth as they cannot be sorted out by definitions, in which case he called them as *a priori* like those in mathematics e.g. the truth about a sphere is independent of definitions; (ii) but he called those needing justification as *a posteriori*. He argued that the justification for 5 + 7 = 12 cannot be found in the definition of 5, or 7, or 12, '+', or '=' and yet no reference is made to experience for its justification. The attack of Bishop Berkeley (1685-1753) on mathematics was on the denial of ontological status of mathematical objects in the first place, but he also regarded them as a science governing the use of *signs*. Arguably, these discourses did not seriously affect mathematics but were invoked by science through the revival of old Greek philosophies of mathematics.

8.1.3. Philosophies of mathematics actively influencing mathematics

Philosophical thinking played an ambitious role in mathematics in the late 19th and early 20th centuries in terms of the *foundational studies* or its denial. Kline (1980, P245) has a vivid narration (but without taking side): "The logicist and intuitionist philosophers, launched during the first decade of" the 20th century "and diametrically opposed in their views on the proper foundations of mathematics, were just the first guns to be fired. A third school of thought, called formalist, was fashioned and led by David Hilbert (1862-1943), and a forth, the set-theoretic school, was initiated by Ernst Zermelo (1871–1953). By the middle of the 20th century, the following doctrines were well established:

(i) Formalism: it was a program of manipulating symbols to show built-in mathematical objectivity and rigor and denied the significance of theorems and their external significance (analogically the relation between theorems and thought can be likened to that between artificial intelligence and human intelligence; its protagonist was Hilbert.

(ii) Logicism: this was a program of "reverse engineering" applied to mathematical truth or theorems, which recognized the existence of mathematical truth as the starting point and transformed them into a long logical expression; its protagonist was Fregé.

(iii) Platonism: the adherents of this philosophy shows that they believe in a complete package that exists but it is "unknowable" and we are better off with their simple construction rather than a detailed attempt;

(iv) Intuitionism**:** it focused on mathematical objects where their collective truth dictated by the particular constructions, (the objects in this doctrine have the sense of individuality as in existentialism). Its protagonist was L. E. J. Brouwer (1881–1966). Goodman (1998) remarks that "Like the formalist, the intuitionist takes the meaning of the theorem to reside in our practice, not in any external reality to which the statement might refer."

Kline (1980, P197) paints the picture as: after centuries of wandering, by 1900 mathematicians seemingly imparted to their ideal structure delineated by Euclid in his *Elements*. They recognized the need for undefined terms, definitions were purged of vague or objectionable terms; the several branches were founded on rigorous axiomatic bases; and valid, rigorous, deductive proofs replaced intuitively or empirically based conclusions. These were even extended to logics. At this time, their equanimity was disturbed in the first half of the 20th century and it is argued in this paper that the efficacy of the philosophical doctrines was lost by these mutually exclusive doctrines. The above philosophical doctrines may be likened to the following equation:

$$w + x + y + z \tag{8.1}$$

with four independent variables of w, x, y and z. This has no solution unless three other equations are provided. If a mathematician elects to solve (8.1) by fixing three of the variables, the solutions will not be unique but the artifact of the fixed values. Philosophers are like such a supposed mathematician, for fixing a host of issues by their presuppositions. Each and all of the above doctrines have considerable grains of truth in them but they are not stated explicitly. The outcome is discrete, contentious and mutually exclusive doctrines. Nonetheless, each of the above doctrines is interesting and stimulating and indeed parts of their tenets are agreeable to a pluralist. The fundamental problem of the philosophy of mathematics is its tendency to ignore inherent evolution, reminiscent of the situation prior to 1859 before Darwin's publication the law of natural selection.

At least, three philosophical doctrines touch on changes in the time dimension and they are: existentialism, dialectics and an application of dialectics to the philosophy of science by Thomas Kuhn (1962) under the doctrine of paradigm shifts, which modifies Marxian revolution with Kantian constructivism.

The prime presuppositions (i.e. an explanation based on some prior knowledge without ever revisiting them) of existentialism is a view of matter (inner world) and mind (outer world) such that *matter is indeterminate* but *the outer world determinable* but through only *the revered ones* who are *revealed with an understanding of matter*, apparently to save the mankind from a *nihilist* prospect, which is the driver of change. To Heidegger (1962), nihilism is a historical movement, and not just any view or doctrine advocated by someone or other. Nihilism is thought in essence. It is the world-historical movement of the people of the Earth which has been drawn into the power realm of the modern age. Heidegger (1962, p19) on his elaboration of *Being* and difficulties in its understanding argues "Our provisional aim is the Interpretation of time as the possible horizon for any understanding whatsoever of Being." The author regards nihilism as a mystic entity but analogous to entropy in a systemic philosophy.

A conception of change is inherent in the Marxian dialectics and is often referred to as historicism. It is based on the antagonism of thesis and antithesis driven by the concept of historic inevitability and holds that antitheses increase in quantity and is transformed violently into theses leading to a new cycle of antagonism. It is obvious that the time dimension is intrinsic to the Marxian dialectics. Marxian philosophy has an immediate problem in explaining transitions in science, mathematics and philosophy for viewing violence as a means to end, which is not entertained in these disciplines. Generally, the philosophical doctrines including those on mathematics are in contention with one another but display some sort of cultural pluralism. It is fortunate that the contentions among philosophers are just verbal. This means that philosophers are often pluralists but their doctrines are not and this is some food for thought.

Thomas Kuhn (1922-96) mended this problem in 1962 by combining the Marxian/Hegelian dialectics with Kantian constructivism and proposed paradigm shifts in science, as a doctrine of change. He differentiated between permanent contributions of an older science to the vantage of the present and the *historical integrity* of that science to a group of scientists. This was profound, as it seemed that Marxian *historical inevitability* had undergone a welcome transformation into *historical integrity*. It invoked an evolutionary routemap dotted with a series of paradigm shifts. The influence of Kuhnian doctrines on mathematics materialized recently, e.g. (Gillies, 1995 and Quinn, 2012) but these are in the tradition of historiography and unable to explain evolutionary transitions.

Both existentialistic and Marxian forms of explaining changes in history and possibly foreseeing the future are stimulating but speculations. Both doctrines can be mapped to the evolutionary theory that the author is advocating but this is the subject of another paper. The author's past approach to the philosophy of science (Khatibi, 2003) used systemic thinking but evolutionary thinking was presented by revisiting Kuhn's paradigm shifts but shifting through feedback loops. However, the author has gone through a considerable rethinking and now avoids philosophical mindsets by formulating evolutionary systemics (see Khatibi 2011 and 2012a, 2012b and Khatibi et al., 2012) and this paper is another application. It is noted that Kitcher's ideas are very close to that of the author's but he does not use the language of systemic thinking and evolutionary thinking. He aims to explain the history of mathematics without reflecting how this is related to the wider human intellectual, social and cultural endeavors.

8.2 Practitioners' aspirations – Axiomatisation of Mathematics

Characteristically, practitioners' culture is based on pluralism, as they do not reject different methods bearing information; they are often blinkered and suffer from inertia to change; they churn out mathematics without foresight. Nonetheless, this culture has given rise to axiomatization, as a bottom-up approach to systematize areas of mathematics, e.g. the axiomatic geometry.

The intension here is not to provide a technical review of axiomatization, as for more details refer to published information, e.g. Easwaran (2008) and Cellucci (2000) or textbooks. Whilst there is clarity in Euclidian geometry being the first source of axiomatization, there seems to be a lack of clarity in the drivers for priming it. For instance, Baron (1969, p25) seems to seek the origins of the Euclidian axiomatization (flourished circa 300 B.C.) in the influence by the philosophy of Plato (424/423-348/347 BC) and the logic of Aristotle (384-322 BC), as follows: "any attempt to assess the influence of the philosophy of Plato and the logic of Aristotle on the work of systematization undertaken by Euclid and culminating in the thirteen books of the *Elements* is fraught with difficulty."

Whilst Plato's philosophy and Aristotle's logic would make the culture rife to axiomatisation, arguably it was driven by inconsistency (entropy) in the body of geometry at the time. This is captured by Royster (2008) remarking that "In the beginning geometry was a collection of rules for computing lengths, areas, and volumes. Many were crude approximations derived by trial and error. This body of knowledge, developed and used in construction, navigation, and surveying by the Babylonians and Egyptians, was passed to the Greeks. The Greek historian Herodotus (5th century BC) credits the Egyptians with having originated the subject, but there is much evidence that the Babylonians, the Hindu civilization, and the Chinese knew much of what was passed along to the Egyptians."

A complete picture for the Euclidian axiomatization emerges when it is viewed as a strategy (i) to address the unacceptable inconsistencies accumulating over thousands of years and (ii) responding to expectation and foresight set by Plato's philosophy and Aristotle's logic. Axiomatization emerged as a bottom-up approach to resolve conceptual inconsistencies. Villiers (1986) summarizes the drivers as follows: (i) mostly the theorems and powerful techniques are already in existence and applied to many problems, long before their eventual reorganization into an axiomatic deductive system; (ii) the discovery of new results during such periods are mostly made through the inductive processes of generalizing, abstracting, analogizing, guessing, hypothesizing, etc. and not by formally using logical deduction; (iii) many topics are initially abstracted from practical situations, leading to eventual logically linked propositions and concepts.

Axiomatization has been taken up further in modern times and they are outlined by Easwaran (2008)], as follows: Fregé axiomatized first-order logic in 1879 (Frege, 1979), Peano axiomatized the natural numbers in 1889, Zermelo axiomatized set theory in 1908, Russell and Whitehead axiomatized foundations in 1910, and Zermelo–Fraenkel set theory with the axiom of choice (nearly simultaneously axiomatized by Skolem and Fraenkel) in 1922. The author agrees with Villiers (1986) that axiomatization does not normally make new knowledge but arguably, it consolidates existing knowledge and only a few areas of mathematics has been axiomatized.

8.3 Goal-orientation as feedforward in Mathematical Complexity

Based on systems science/cybernetics, feedforward loops are mechanisms for anticipation. These can be used for creating consistency with the environment by anticipating adverse impacts and controlling system performance. Their implementations require extra flexibility, so that system performances can meet multiple objectives. A similar case can be made in mathematics, as follows.

Evolution of mathematical knowledge: Evolutionary variations in the body of mathematical complexity signify that external consistency is already implicit within it. Fauvel and Gray (1987) present extracts on the choice of decimal base in counting from Aristotle in the 4th century BC to some of the 20th century views. They remark that "It is noticeable how the accounts reflect the period in which they were written, both in attitude towards early societies and their life conditions and in the kind of historical and mathematical analysis exemplified." Thus, one is minded that the rigor of mathematics is related to the cultural values of the time, which drives the requirement for the *external consistency* of mathematical information. Arguably, the emergence of new doctrines of the philosophy of mathematics by the middle of the 20th century signify the absence of goal-orientation, which emerged by systems science since the middle of 20th century. Still there is no direct movement now to apply it to mathematical complexity.

Cumulative increase of mathematics: Cumulative increases in mathematical knowledge are topical, see Davis and Hersch (1980, Pp20-23) who introduce Ulam's dilemma related to the number of theorems published annually as follows:

$$\begin{aligned}&\textit{Number of theorems per annum}\\&\quad = \textit{number of journals} \times \textit{number of yearly issues}\\&\quad \times \textit{number of papers per issue}\\&\quad \times \textit{average number of theorems per paper}\end{aligned}$$

This led to a staggering number of 200,000 without necessarily reflecting new ideas. This is a naive linear extrapolation, not accounting for: (i) diminution due to irrelevance or obsolescence; and (ii) the possibility of internal saturation. They ask the question that "If the number of theorems is larger than one can possibly survey, who can be trusted to judge what is 'important'?" They argue that "It would appear from the record that mankind can go on and on generating mathematics;" and foresee no end to all these mathematical production.

An underlying concern is that the growing mathematics may collapse under its own weight in the sense that it will be too big for many to grasp the totality of any part of it in a sensible manner. Some sort of consolidation is needed, e.g. axiomatization as discussed above. Goal-orientation is another approach, implying a flexible approach of reforming mathematics towards the goals without rejecting the knowledge that are not used presently but keeping them for a rainy-day. Section 9 shows one possible way of goal-orientation in an area of mathematics.

Gaps between communal and institutional mathematical capabilities: Arguably, the gaps between communal and institutional capabilities are increasing, as a great deal of mathematics is being developed in institutions that are intractable at a communal level. Too much gap is unsustainable and therefore a certain amount has to trickle back to the community to enrich it. Evolutionary systemics offers a methodology to enhance communal mathematical thinking.

Goal-orientation requires flexibility in mathematical complexity, so that knowledge is assembled and used as per need, without prejudice against the knowledge not being used presently. In this way, knowledge maintains its external consistency with the goals of the day without overturning the principle of diversity. For instance, the author experiences an undue difficulty in publishing his ideas both on evolutionary systemics and the new calculus (outlined in Section 9) and attributes this to an unhealthy culture of filtering out any out-of-kilter scientific and mathematically rigorous contributions. With goal-orientation, out-of-kilter contributions have also to be in the mainstream.

In one sense, feedforward in mathematics is already in action in terms of the movement for axiomatization. These movements are home-grown within the mathematical community to hone the practice and resolve controversy. The philosophy of mathematics also strived to reform mathematics but instigated more controversy, although its pragmatic findings remain valid. Thus, the pragmatic findings of the developments driven by the philosophical of mathematics may be overarched by pluralism, i.e. some problems can have structures and order, some can be formalized, and some can be susceptible to disorder.

To the author's best knowledge, goal-orientation is yet to prevail in mathematics, similar to science, where the principle of sustainability and risk-based decision-making serve as the tools of goal-orientation. Whilst the world of mathematics is inherently pluralist due to the pragmatism of mathematical communities, without goal-orientation there is a risk of inconsistency among the various areas of mathematical complexity. This paper takes the preliminary steps to show that evolutionary systemics may be used to develop goal-orientation capabilities in mathematics. Section 9 overhauls the conventional calculus of sequences of natural numbers into a new calculus using a building block approach directly structured by evolutionary systemics. Section 10 formulates the basis of a vision for mathematical complexity as a whole.

9. A New Calculus to Overhaul that of Sequences of natural Numbers

Conventional calculus of the sequences of natural numbers exemplifies sophisticated physicalistic developments overlooking hierarchical structures. This area of mathematical complexity can be overhauled by a new mathematics developed by the author, to serve as the language of evolutionary systemics with immense potentials to cope with the particulate phase of complexity. This development has not been fully published yet but the author developed the idea in 1995 and published one paper by applying this mathematics to analytically determining sample sizes (Khatibi, 2001). Since 2000, the author has been unable to publish them in mathematical journals despite his repeated attempts. A glimpse of the new mathematics is outlined by Khatibi (2012a) accounting for interconnectivity among building blocks and by Khatibi et al (2012). This section outlines mathematizing hierarchies in overhauling physicalistic calculus of number sequences. Prerequisite definitions and operators are outlined in Electronic Supplementary Material (ESM).

9.1. Conventional Physicalistic Treatment of Number Sequences

Number sequences are preoccupied with the basic operation of finding the sum of *i*-number of consecutive terms, where *i* is an integer number ranging from 0 or 1 to any positive number. Such a category of mathematical problems have captured the imagination for many millennia and the methods dealing with the problems are often classic but complex. A classical sequence of natural number is: $\{1^m \quad 2^m \quad 3^m \quad i^m \quad ... \quad n^m\}$, which is treated by (i) the *theorem of binomial enumeration* often referred to as the calculus of finite differences or the umbral calculus; Bucchianico *el al* (2004), present an interesting historical review, tracing it to the 17th century and outline the developments since then but the umbral calculus operates in the algebraic space and manipulation is difficult; (ii) *Stirling numbers* of second kind is detailed in many textbooks, e.g. Brualdi (1992), which are not reviewed.

This paper refers to the above treatments as physicalistic for the simple reason that the subsequent techniques attempt to solve the mathematical entity as visible by the eye, very much similar to treating each species or each chemical compound on its own without understanding their inherent interconnectivity or hierarchies. This is a classical ontological mindset but this mindset retards mathematics.

9.2. Hierarchical Structure

Science explains the complexity of life, which has acquired a hierarchical organization comprising physics, chemistry, biology, psychology, sociology and anthropology. As mathematics is normally driven by science and provides a successful surrogacy for it, the absence of its hierarchical structure is surprising. The new calculus unravels the existence of such a structure for a family of number sequences based on natural numbers and unearths a world paralleling the world of science with the following hierarchical structure, as outlined in Table 5.

Table 5. Comparison of Hierarchical Structure in Mathematical and Life Complexity

Hierarchy	The Complexity of Life	Mathematical Complexity
Quantumistic level	In subatomic world: in principle, all chemical elements can be formed from: • The basic hydrogen atoms • Under very special conditions	A whole family of sequences can be generated from: • A kernel (equivalent to atoms) • A set of rules conditions)
Physicalistic/ chemicalistic levels	Elements formed out of the quantum world enter chemical reactions & produce compound, where compounds/ elements have: • Physical properties • Chemical interconnectivity/ properties	Sequences formed above can form compounds in a variety of ways (e.g. bondage, regeneration) with: • Physicalistic emergent properties • Chemicalistic interconnectivity
Biologistic/ psychologistic levels	The working of compounds together gives rise to higher emergent properties: • *Biology*: reproducibility, spontaneity, internal consistency cladistics • *Psychology*: the ability to cope with external world	Sequences working together can trigger higher emergent properties: • *Biologistic* level: reproducibility, internal consistency, cladistics, • *Psychologistic* level: coping with external world

9.3. New Calculus of Sequences of Natural Numbers

The analytical calculus for of sequences of natural numbers factorizes a sequence into a *sequence of invariants* (the architecture) and a *sequence of counters* (the rules or architect) by developing new operators, which are capable of unraveling the inner structures of number sequences.

The sequences of invariants (and thereby their *kernels*) provide an understanding of the uniqueness of each individual sequence, similar to understanding the subatomic structure of each chemical element or chemical compounds. Within a particular family of sequences of natural numbers, their *sequences of invariants* are interoperable, i.e. one sequence of invariants can be *transformed* into the other.

The *sequences of counters* can be defined in countless ways. The easiest form of the *sequence of counters* is the simple binomial expansion terms using combinatoric operators.

This approach makes it possible to distinguish the bondage between the building blocks, their layer-by-layer physical collections of lower layers in higher ones by creating of hierarchies. There are many ways to present the new calculus but the aim here is to outline a framework for hierarchy. However, the new calculus has new terms, definitions, operators and rules, which are presented in ESM as a prerequisite for this section. The terms defined in the ESM include: kernels (building blocks), sequence of invariants, sequence of counters (rules); regenerating products, convoluting products, direct (nominal) products, algebraic products and recursive products. These are all italicized, as a reminder that their definitions are given in ESM.

9.3.1. Quantumistic Level

The *kernel* in the world of the sequences of natural numbers is equivalent to quarks or gluons. This can be any number or a set (as opposed to sequence) of numbers. Example 1 generates the sequence of natural numbers, as follows:

Example 1: The sequence of natural numbers is *primed* from its *kernel*, $\langle 1 \rangle$, at the quantumistic level in three steps, as follows.

Step 1 – Generation 0: The operations consist of multiplying the *kernel*, $\langle 1 \rangle$, by *sequence of counter*, represented by an over-bar $\{\bar{1}\}$, using one of the multiplication rules:

$$\langle 1 \rangle \oplus \{\bar{1}\} = \langle 1 \rangle \odot \{\bar{1}\} = \langle 1 \rangle \otimes \{\bar{1}\} = \{1\} = \underbrace{\begin{Bmatrix} \underbrace{N}_{g=0} \end{Bmatrix}^{d=0}}_{\varphi=0} \qquad (9.1.a)$$

Kernel: $\langle 1 \rangle$ is not a sequence but $\{\bar{1}\}$ is a *sequence of counter*, where its over-bar signifies that it is not any sequence but the *sequence of counters*. Their product, $\{1\}$, is a sequence specified as: *Generation*=0, *Order*=0, *Degree*=0.

Step 2 – Generation 1: The product, $\{1\}$ in (9.1.a) is *regenerated*, as follows:

$$\{1\} \oplus \{\overline{1 \quad 1 \quad 1 \quad 1 \quad \ldots}\} = \{1 \quad 1 \quad 1 \quad 1 \quad \ldots\} = \underbrace{\begin{Bmatrix} \underbrace{N}_{g=1} \end{Bmatrix}^{d=0}}_{\varphi=0} \qquad (9.1.b)$$

(9.1.a) expresses the sequence of unitary numbers, reduced as:

$$\{\Re|\{1 \quad 1 \quad 1 \quad 1 \quad \ldots\}\} = \{1\} \odot \left\{\binom{n}{n}\right\} = \{1\} \odot \overbrace{\left\{\binom{n}{n}\right\}}^{Red} \qquad (9.1.c)$$

$\{\Re|\{\ \}\}$ denotes the reducemental of a sequence, e.g.: $\{1 \quad 1 \quad 1 \quad 1 \quad \ldots\}$. This term is proposed by the author to denote a mathematical expression that generates each and all of the terms in a sequence and in this example the reducemental is: $\{1\} \odot \left\{\binom{n}{n}\right\}$ or $\{1\} \odot \overbrace{\left\{\binom{n}{n}\right\}}^{Red}$. Although the sequence of counter: $\{\overline{1 \quad 1 \quad 1 \quad \ldots \quad 1}\}$ is numerically equal to the product: $\{1 \quad 1 \quad 1 \quad 1 \quad \ldots\}$, they are not the same. The elements in the sequence of counters are by choice and can be any sequence. Note that the *order* of Generation 1 is zero

(Order=Generation-1). The Order is created when the act of summing is carried out. Generation 1 involves no summation.

Step 3 – Generation3: When the first *i* terms of Generation 1 are summed, Generation 2 is created, which is Order 1 and this regenerates the sequence of natural numbers, as follows:

$$\{1 \quad 1 \quad 1 \quad 1 \quad ...\} \oplus \{1 \quad 1 \quad 1 \quad 1 \quad ...\} = \{1 \quad 1+1 \quad 1+1+1 \quad 1+1+1+1 \quad ...\} =$$

$$\{1 \quad 2 \quad 3 \quad 4 \quad ...\} = \underbrace{\left\{\underbrace{N}_{g=2}\right\}^{d=0}}_{\varphi=1} \qquad (9.2.a)$$

In an evolutionary systemic context, (9.1) generates a sequence from the kernel: $\langle 1 \rangle$ by a rule and converts it to the sequence of natural numbers in three steps. Although these numbers are used in everyday discourse, irrespective of the newborn evolutionary systemic algebra, this algebra formalizes something is in the everyday discourse. Khatibi (2012a) shows that (9.2.a) is an evolvable entity.

9.3.2. Emergence of Hierarchies

As soon as an evolvable entity emerges, it is intrinsically associated with a potential hierarchical structure in a self-organized way. The actual hierarchy is an evolutionary outcome and it is not possible to predict its details for an area of complexity. However, there is a generic layout, within which the hierarchy of any evolvable entity can be accommodated and its evolutionary systemic algebra is outlined below. This is presented in two steps through Example 2 and 3.

Example 2 regenerates a lower order sequence to higher ones through direct operations but Example 3 is the reverse process of unfolding the structure through the rules given in Box 1 and thereby unfolding their hierarchical structure.

Box 1.

The rules for working out the inner structure of each term is rather simple but laborious and one really needs a software to do it. The rules are:

(i) Each term is identified by its rank (position) in the sequence;
(ii) There are as many sub-terms (ST) for each term as the rank of the term;
(iii) Each sub-term is composed of sub-sub-terms (SST) and there are as many SSTs as its rank within the ST;
(iv) Rule (iii) relates to the process of nesting within each hierarchy but each higher hierarchy also nests within itself the lower hierarchy terms and the process is carried on until reaching the very building blocks.

Example 2: **Regenerating the sequence of natural numbers**

The sequence of natural numbers can be regenerated (9.2.a) expresses the sequence of natural numbers, which is reducible as:

$$\left\{\Re\left|\left\{1 \quad 2 \quad 3 \quad ... \quad \binom{n}{n-1}\right\}\right.\right\} = \{1\} \odot \left\{\binom{n}{n-1}\right\} = \{1\} \odot \overbrace{\left\{\binom{n}{n-1}\right\}}^{Red} \qquad (9.2.b)$$

(9.2.a) can be regenerated as:

$$\{1\ \ 2\ \ 3\ \ 4\ \ ...\} \oplus \{1\ \ 1\ \ 1\ \ 1\ \ ...\} = \{1\ \ 3\ \ 6\ \ 10\ \ ...\} = \underbrace{\left\{\underbrace{N}_{g=3}\right\}^{d=0}}_{\varphi=2} \quad (9.3.a)$$

(9.3.a) expresses *Order* 2 Sequence of Triangular numbers and reducible as:

$$\{\Re|\{1\ \ 3\ \ 6\ \ 10\ \ ...\}\} = \{1\ \ 1\} \odot \left\{\binom{n}{n-1}\ \ \binom{n}{n-2}\right\} = \{1\} \odot \overbrace{\left\{\binom{n+1}{n-1}\right\}}^{Red} \quad (9.3.b)$$

(9.3.a) can be regenerated as:

$$\{1\ \ 3\ \ 6\ \ 10\ \ ...\} \oplus \left\{1\ \ 1\ \ 1\ \ ...\ \ \binom{n-1}{n-1}\right\} = \{1\ \ 4\ \ 10\ \ 20\ \ ...\} = \underbrace{\left\{\underbrace{N}_{g=4}\right\}^{d=0}}_{\varphi=3} \quad (9.4.a)$$

(9.4.a) expresses *Order* 3 Sequence of tetrahedron numbers and reducible as:

$$\{\Re|\{1\ \ 4\ \ 10\ \ 20\ \ ...\}\} = \{1\ \ 2\ \ 1\} \odot \left\{\binom{n}{n-1}\ \ \binom{n}{n-2}\ \ \binom{n}{n-3}\right\} = \{1\} \odot \overbrace{\left\{\binom{n+2}{n-1}\right\}}^{Red} \quad (9.4.b)$$

(9.3.a) can be regenerated as:

$$\{1\ \ 4\ \ 10\ \ 20\ \ ...\} \oplus \{1\ \ 1\ \ 1\ \ 1\ \ ...\} = \{1\ \ 5\ \ 15\ \ 35\ \ ...\} = \underbrace{\left\{\underbrace{N}_{g=5}\right\}^{d=0}}_{\varphi=4} \quad (9.5.a)$$

(9.5.a) expresses *Order* 4 Sequence of pentatope numbers and reducible as:

$$\{\Re|\{1\ \ 5\ \ 15\ \ 35\ \ ...\}\} = \{1\ \ 3\ \ 3\ \ 1\} \odot \left\{\binom{n}{n-1}\ \ \binom{n}{n-2}\ \ \binom{n}{n-3}\ \ \binom{n}{n-4}\right\} = \{1\} \odot \overbrace{\left\{\binom{n+3}{n-1}\right\}}^{Red} \quad (9.5.b)$$

(9.1)-(9.5) are widely known except for the unique notation proposed by the author. The return for the notations/operators includes the following subtle findings:

1. The regenerated sequences of (9.1.b), (9.2.a), (9.3.a), (9.4.a) and (9.5.a) are all reducible to the kernel, $\langle 1 \rangle$, i.e. the regenerating products do not add anything to their complexity.
2. Example 3 below unearths their hidden hierarchical structure hidden.
3. There are countless *kernels* (at the quantumistic level) and countless *sequences of counters* and therefore there are countless potential sequences to be generated similar to the examples in (9.1)-(9.5).

Example 3: Layer-by-layer Physical Growth

Consider (9.1)-(9.5). This example unfolds their inner structure only to find that they are hierarchically organized. The inner structure of (9.1.b) is self evident but that of (9.2.a) - (9.5.a) rather complex. These hierarchies are further illustrated in Figure 6.

The inner structure and thereby hierarchy of (9.1.b) is unfolded as follows:

$$\underbrace{\left\{\underbrace{N}_{g=1}\right\}}_{\varphi=0}=\left\{1\quad 1\quad 1\quad 1\quad 1\quad \ldots\quad \binom{n}{n}\right\}=\left\{\overbrace{1}^{T1}\quad \overbrace{1}^{T2}\quad \overbrace{1}^{T3}\quad \overbrace{1}^{T4}\quad \overbrace{1}^{T5}\quad \ldots.\quad \overbrace{1}^{T-n^{th}}\right\} \qquad (9.6.a)$$

(9.6.a) illustrates Rule (i) in Box 1. The inner structure and thereby hierarchy of (9.2.a) is unfolded as follows:

$$\underbrace{\left\{\underbrace{N}_{g=2}\right\}}_{\varphi=1}=\{1\quad 2\quad 3\quad 4\quad 5\quad 6\quad \}=\left\{\overbrace{1}^{T1}+\overbrace{\begin{matrix}\overbrace{1}^{S1T1}+\\ \overbrace{1}^{s2t1}+\end{matrix}}^{T2}\quad \overbrace{\begin{matrix}\overbrace{1}^{S1T1}+\\ \overbrace{1}^{s2t1}+\\ \overbrace{1}^{s3t1}+\end{matrix}}^{T3}\quad \overbrace{\begin{matrix}\overbrace{1}^{S1T1}+\\ \overbrace{1}^{s2t1}+\\ \overbrace{1}^{s3t1}+\\ \overbrace{1}^{s4t1}+\end{matrix}}^{T4}\quad \overbrace{\begin{matrix}\overbrace{1}^{s1t1}+\\ \overbrace{1}^{s2t1}+\\ \overbrace{1}^{s3t1}+\\ \overbrace{1}^{s4t1}+\\ \overbrace{1}^{s5t1}\end{matrix}}^{T5}\quad \ldots\right\} \qquad (9.6.b)$$

The expansion in (9.6.b) follows Rule (ii) in Box 1. The inner structure and thereby hierarchy of (9.3.a) is unfolded as follows:

$$\underbrace{\left\{\underbrace{N}_{g=1}\right\}}_{\varphi=2}=\{1\quad 3\quad 6\quad 10\quad 15\quad 21\quad \ldots\}=$$

$$\left\{\overbrace{1}^{T1}+\overbrace{\begin{matrix}\overbrace{1}^{s1t1}+\\ \overbrace{1}^{s2t1}+\overbrace{1}^{S2T2}\end{matrix}}^{Term\ 2}+\overbrace{\begin{matrix}\overbrace{1}^{s1t1}+\\ \overbrace{1}^{s2t1}+\overbrace{1}^{s2t2}+\\ \overbrace{1}^{s3t1}+\overbrace{1}^{S3T2}+\overbrace{1}^{S3T3}\end{matrix}}^{Term\ 3}+\overbrace{\begin{matrix}\overbrace{1}^{s1t1}+\\ \overbrace{1}^{s2t1}+\overbrace{1}^{s2t2}\\ \overbrace{1}^{s3t1}+\overbrace{1}^{s3t2}+\overbrace{1}^{s3t3}\\ \overbrace{1}^{s4t1}+\overbrace{1}^{s4t2}+\overbrace{1}^{s4t3}+\overbrace{1}^{s4t4}\end{matrix}}^{Term\ 4}+\overbrace{\begin{matrix}\overbrace{1}^{s5t1}\\ \overbrace{1}^{s2t1}+\overbrace{1}^{s2t2}\\ \overbrace{1}^{s3t1}+\overbrace{1}^{s3t2}+\overbrace{1}^{s3t3}\\ \overbrace{1}^{s4t1}+\overbrace{1}^{s4t2}+\overbrace{1}^{s4t3}+\overbrace{1}^{s4t4}\\ \overbrace{1}^{s5t1}+\overbrace{1}^{s5t2}+\overbrace{1}^{s5t3}+\overbrace{1}^{s5t4}+\overbrace{1}^{s5t5}\end{matrix}}^{Term\ 5}\ \ldots\right\} \qquad (9.6.c)$$

The expansion in (9.6.c) follows Rule (iii) in Box 1.

$$\underbrace{\left\{\underbrace{N}_{g=1}\right\}}_{\varphi=3}=\left\{\left\{1\quad 4\quad 10\quad 20\quad 35\quad \ldots\quad \binom{n+2}{n-1}\right\}\right\}=$$

$$\overbrace{1}^{T1}+\overbrace{\begin{matrix}\overbrace{1}^{s1t1}+\\ \overbrace{1}^{s2t1}+\overbrace{\overbrace{1}^{s2t1}+\overbrace{1}^{s2t2}}^{s2t2}\end{matrix}}^{T2}+\overbrace{\begin{matrix}\overbrace{1}^{s1t1}+\\ \overbrace{1}^{s2t1}+\overbrace{\overbrace{1}^{s2t1}\ \overbrace{1}^{s2t2}}^{s2t2}+\\ \overbrace{1}^{s3t1}+\overbrace{\overbrace{1}^{s3t1}\ \overbrace{1}^{s3t2}}^{s3t2}+\overbrace{\overbrace{1}^{s3t1}\ \overbrace{1}^{s3t2}\ \overbrace{1}^{s3t3}}^{s3t3}\end{matrix}}^{T3}\ \overbrace{\begin{matrix}\overbrace{1}^{S1T1}+\\ \overbrace{1}^{s2t1}+\overbrace{\overbrace{1}^{s2t1}\ \overbrace{1}^{s2t2}}^{s2t2}\\ \overbrace{1}^{s3t1}+\overbrace{\overbrace{1}^{s3t1}\ \overbrace{1}^{s3t2}}^{s3t2}+\overbrace{\overbrace{1}^{s3t1}\ \overbrace{1}^{s3t2}\ \overbrace{1}^{s3t3}}^{s3t3}\\ \overbrace{1}^{s4t1}+\overbrace{\overbrace{1}^{s4t1}\ \overbrace{1}^{s4t2}}^{s4t2}+\overbrace{\overbrace{1}^{s4t1}\ \overbrace{1}^{s4t2}\ \overbrace{1}^{s4t3}}^{s4t}+\overbrace{\overbrace{1}^{s4t1}+1+1+\overbrace{1}^{s4t4}}^{s4t4}\end{matrix}}^{T4} \qquad (9.6.d)$$

The expansion in (9.6.d) follows Rule (iv) in Box 1.

The above expansions of the inner structures of number sequences of natural numbers unearth a hierarchical structure based on the *kernel* $\langle 1\rangle$ using the simple rules given in Box 1. The hierarchical nature of the sequences is further illustrated in Figure 6. Mathematically-astute readers can easily spot the inherent logic and this logic is general. This hierarchical structure is the outcome of the regeneration product, the elements of which are all '1.'

The above sequences are simple outcomes of regeneration of the kernel of ⟨1⟩ by using the *regeneration products*. In principle, it is possible to work out all the elements, its sub-terms within each hierarchy based on the kernel ⟨1⟩. In practice, the base-10 natural numbers are used, which hide the base-1 (i.e. the direct product from the kernel ⟨1⟩) and this also hides the inherent hierarchical structure. Some of the implications of this structure are as follows: (i) it preserves the hierarchical structure as a mould for a unitary kernel; and (ii) this model just shows the physicalistic arrangements with no bondage between the elements; (iii) for bondage between the elements, the *kernel* can be given some sort of complexity, as illustrated in Example 5.

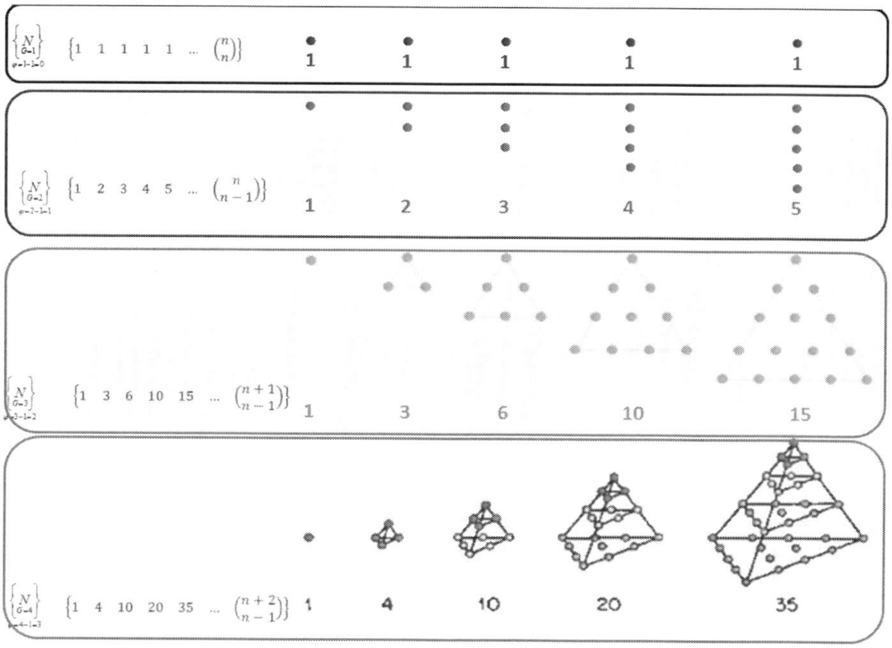

Figure 6. Illustration of Hierarchical Structure in Sequences of Natural Number.

9.3.3. Chemicalistic/Physicalistic Level

Any number of different sequences based on natural numbers can be operated upon to produce new compounds using operators given in ESM. The outcome can be handled by the new calculus in a chemicalistic level. The best way is to illustrate the problem through examples. Three examples are presented:

Example 4: Consider the following complex sequences defined by:

$$\underbrace{\left\{\underbrace{N}_{g=1}\right\}^{d=3}}_{\varphi=1} \setminus \underbrace{\left\{\underbrace{N}_{g=3}\right\}^{d=2}}_{\varphi=2} \setminus \underbrace{\left\{\underbrace{N}_{g=4}\right\}^{d=1}}_{\varphi=3} \setminus \underbrace{\left\{\underbrace{N}_{g=5}\right\}^{d=1}}_{\varphi=4} \qquad (9.7.a)$$

This leads to the following sequence:

$$\begin{Bmatrix} 1 \quad 1440 \quad 145800 \quad 4480000 \quad 68906250 \quad 672126336 \quad 4743607680 \\ 26276659200 \quad 120570676875 \quad 475832500000 \quad 1659839277096 \quad 5223564702720\ldots \end{Bmatrix} \quad (9.7.b)$$

The *reducemental* of this synthetic sequence is the direct product of:

$$\{1 \quad 1425 \quad 124305 \quad 2443745 \quad 16361415 \quad 44556183 \quad 53166575 \quad 28028895 \quad 6172680 \quad 486760 \quad 9216\} \odot$$
$$\left\{\binom{n+10}{n-1} \binom{n+9}{n-2} \binom{n+8}{n-3} \binom{n+7}{n-4} \binom{n+6}{n-5} \binom{n+5}{n-6} \binom{n+4}{n-7} \binom{n+3}{n-8} \binom{n+2}{n-9} \binom{n+1}{n-10} \binom{n}{n-11}\right\} \quad (9.7.c)$$

Notably, (9.7.a-9.7.c) are quite intractable by the conventional technique but the analytical new calculus identifies its building blocks and the rules of the synthetic product.

Example 5: A simpler example is the power sequence of natural numbers, i.e.:

$$\{1^m \quad 2^m \quad 3^m \quad 4^m \quad \ldots \quad n^m\} \quad (9.8)$$

Table 6. Kernels of Power Sequences

	Individual terms in the Sequence of Kernels	Complexity	Gain
	$\{1 \quad 2 \quad 3 \quad \ldots \quad n\} = \{1\} \odot \overbrace{\left\{\binom{n}{n-1}\right\}}^{\text{Red}}$ $1! = \{1\}$	1	0
	$\{1^2 \quad 2^2 \quad 3^2 \quad \ldots \quad n^2\} = \{1 \quad 1\} \odot \overbrace{\left\{\binom{n+1}{n-1} \binom{n}{n-2}\right\}}^{\text{Red}}$ $2! = \{1\} \widehat{\oplus} \{1 \quad 1\}$	1	1
	$\{1^3 \quad 2^3 \quad 3^3 \quad \ldots \quad n^3\} = \{1 \quad 4 \quad 1\} \odot \overbrace{\left\{\binom{n+2}{n-1} \binom{n+1}{n-2} \binom{n}{n-3}\right\}}^{\text{Red}}$ $3! = \{1 \quad 1\} \widehat{\oplus} \left\{1 \quad \begin{matrix} 2 & 3 \\ 3 & 2 \end{matrix} \quad 1\right\} = \{1\times1 \quad 1\times2+2\times1 \quad 1\times1\} = \{1 \quad 4 \quad 1\}$	1	2
	$4! = \{1 \quad 4 \quad 1\} \widehat{\oplus} \left\{1 \quad \begin{matrix} 2 & 3 \\ 3 & 2 \end{matrix} \quad 1\right\} = \{1\times1 \quad 1\times3+4\times2 \quad 4\times2+1\times3 \quad 1\times1\} = \{1 \quad 11 \quad 11 \quad 1\}$	1	3
	$5! = \{1 \quad 11 \quad 11 \quad 1\} \widehat{\oplus} \left\{\left\{1 \quad \begin{matrix} 2 & 3 & 4 \\ 4 & 3 & 2 \end{matrix} \quad 1\right\}\right\} = \{1 \quad 26 \quad 66 \quad 26 \quad 1\}$	1	4
	$6! = \{1 \quad 26 \quad 66 \quad 26 \quad 1\} \widehat{\oplus} \left\{\left\{1 \quad \begin{matrix} 2 & 3 & 4 & 5 \\ 5 & 4 & 3 & 2 \end{matrix} \quad 1\right\}\right\} = \{1 \quad 57 \quad 302 \quad 302 \quad 57 \quad 1\}$	1	5
	$7! = \{1 \quad 57 \quad 302 \quad 302 \quad 57 \quad 1\} \widehat{\oplus} \left\{\left\{1 \quad \begin{matrix} 2 & 3 & 4 & 5 & 6 \\ 6 & 5 & 4 & 3 & 2 \end{matrix} \quad 1\right\}\right\}$ $= \{1 \quad 120 \quad 1191 \quad 2416 \quad 1191 \quad 120 \quad 1\}$	1	6
	$8! = \{1 \quad 120 \quad 1191 \quad 2416 \quad 1191 \quad 120 \quad 1\} \widehat{\oplus} \left\{\left\{1 \quad \begin{matrix} 2 & 3 & 4 & 5 & 6 & 7 \\ 7 & 6 & 5 & 4 & 3 & 2 \end{matrix} \quad 1\right\}\right\}$ $= \{1 \quad 247 \quad 4293 \quad 15619 \quad 15619 \quad 4293 \quad 247 \quad 1\}$	1	7
	The sequence of invariants: $9! = \{1 \quad 502 \quad 14606 \quad 88234 \quad 156190 \quad 88234 \quad 14606 \quad 502 \quad 1\}$ The sequence of counters: $\left\{\binom{n+8}{n-1} \binom{n+7}{n-2} \binom{n+6}{n-3} \binom{n+5}{n-4} \binom{n+4}{n-5} \binom{n+3}{n-6} \binom{n+2}{n-7} \binom{n+1}{n-8} \binom{n}{n-9}\right\}$	1	8

This synthetic sequence can be factorized into the products of two sequences, just like any other sequences: (i) the sequence of invariants, as given in Table 6; and (ii) the sequence of invariants, as given in Table 7. There are countless ways of representing both the sequences of invariants and counters but they are outside the scope of this paper, although their systematic formats are referred to as alleles and discussed below.

Table 6 shows the results for the sequences up to *m*=9 are presented. It involves three different approaches and therefore care is needed to understand it. The bottom line of these sequences is the kernels, which are not a sequence but bare building blocks. The sequence of invariants (and literally, the kernels) of one sequence can be transformed into that of higher or lower ones through transformers.

One approach for the sequences of counters is given in Table 7 is the most *parsimonious* form and it is simply the sequence of *natural numbers* and its various *generations*. Another sequence of counters is given by the binomial expansion terms.

Table 7. *Sequence of Invariants* for Power Sequences (up to *m*=4)

m=1	*m*=2		*m*=3			*m*=4			
$\binom{n}{n-1}$	$\binom{n+1}{n-1}$	$\binom{n}{n-2}$	$\binom{n+2}{n-1}$	$\binom{n+1}{n-2}$	$\binom{n}{n-3}$	$\binom{n+3}{n-1}$	$\binom{n+2}{n-2}$	$\binom{n+1}{n-3}$	$\binom{n}{n-4}$
1	1	0	1	0	0	1	0	0	0
2	3	1	4	1	0	5	1	0	0
3	6	3	10	4	1	15	5	1	0
4	10	6	20	10	4	35	15	5	1
5	15	10	35	20	10	70	35	15	5
6	21	15	56	35	20	126	70	35	15
7	28	21	84	56	35	210	126	70	35
8	36	28	120	84	56	330	210	126	70
9	45	36	165	120	84	495	330	210	126

Example 6: Bondage within the Building Blocks

Consider Example 5 of the sequence:

$$\{1^3 \quad 2^3 \quad 3^3 \quad 4^3 \quad \ldots \quad n^3\} \tag{9.9.a}$$

This sequence and thereby each term can be factorized into the *direct products* of two sequences: (i) the sequence of building blocks (kernel): $\{1 \quad 4 \quad 1\}$, which are primed from the kernel of $\langle 1 \quad 4 \quad 1\rangle$; (ii) the sequence of counters as defined in Table 7. However, it may be assumed that the elements in the *kernel* enter into bondage towards some sort of emergent property and in this example, the emergent property is the power 3 sequence. In practice, countless sequences may be devised but those with a selective advantage are likely to be selected. The following example illustrates the idea of hierarchy and the bondage within each hierarchy.

The internal structure within this hierarchy for the 1st, 2nd and 3rd terms are as follows:

$$\overbrace{1}^{T1} + \overbrace{\overbrace{1}^{s1t1} + \overbrace{1}^{s2t1} + \overbrace{\overbrace{1}^{s2t1} + \overbrace{1+4}^{s2t2}}^{s2t2}}^{Term\ 2=2^3} + \overbrace{\overbrace{1}^{s1t1} + \overbrace{1}^{s2t1} + \overbrace{\overbrace{1}^{s2t1} \ \overbrace{1+4}^{s2t2}}^{s2t2} + \overbrace{1}^{s3t1} + \overbrace{\overbrace{1}^{s3t1} \ \overbrace{1+4}^{s3t2}}^{s3t2} + \overbrace{\overbrace{1}^{s3t1} + \overbrace{1+4}^{s3t2} + \overbrace{1+4+\underbrace{1}_{TBB}}^{s3t3}}^{s3t3}}^{Term\ 3=3^3} \quad (9.9.b)$$

where TBB is the Trailing Building Block and in this way the two values of 1 are differentiated. The 4th term (4^3) is expanded as follows:

$$\overbrace{\begin{array}{ll} Sub-Term\ 1 & \overbrace{1}^{s1t1}+ \\ Sub-Term\ 2 & \overbrace{1}^{s2t1} + \overbrace{\overbrace{1}^{s2t1} \ \overbrace{1+4}^{s2t2}}^{s2t2} \\ Sub-Term\ 3 & \overbrace{1}^{s3t1} + \overbrace{\overbrace{1}^{s3t1} \ \overbrace{1+4}^{s3t2}}^{s3t2} + \overbrace{\overbrace{1}^{s3t1} \ \overbrace{1+4}^{s3t2} \ \overbrace{1+4+\underbrace{1}_{TBB}}^{s3t3}}^{s3t3} \\ Sub-Term\ 4 & \overbrace{1}^{s4t1} + \overbrace{\overbrace{1}^{s4t1} \ \overbrace{1+4}^{s4t2}}^{s4t2} + \overbrace{\overbrace{1}^{s4t1} \ \overbrace{1+4}^{s4t2} \ \overbrace{1+4+\underbrace{1}_{TBB}}^{s4t3}}^{s4t3} + \overbrace{\overbrace{1}^{s4t1} + \overbrace{1+4}^{s4t2} + \overbrace{1+4+\underbrace{1}_{TBB}}^{s4t3} + \overbrace{1+4+\underbrace{1}_{TBB}}^{s4t4}}^{s4t4} \end{array}}^{Term\ 4=\ 4^3} \quad (9.9.c)$$

The *reducemental* of each term is as follows:

$$1^{st}\ \text{Term}\ 2^3 = \binom{1+2}{1-1} \times \mathbf{1} \quad (9.9.d)$$

$$2^{nd}\ \text{Term:}\ 2^3 = \binom{2+2}{2-1} \times \mathbf{1} + \binom{2+1}{2-2} \times \mathbf{4} = 4 \times \mathbf{1} + 1 \times \mathbf{4} = 8 \quad (9.9.e)$$

$$3^{rd}\text{Term:}\ 3^3 = 1 \times \binom{3+2}{3-1} + 4 \times \binom{3+1}{3-2} + 1 \times \binom{3}{3-3} = 10 \times \mathbf{1} + 4 \times \mathbf{4} + 1 \times \underbrace{\mathbf{1}}_{\mathbf{TBB}} = 27 \quad (9.9.f)$$

$$4^{th}\text{Term:}\ 4^3 = \binom{4+2}{4-1} \times \mathbf{1} + \binom{4+1}{4-2} \times \mathbf{4} + \binom{4}{4-3} \times \underbrace{\mathbf{1}}_{\mathbf{TBB}} = 20 \times \mathbf{1} + 10 \times \mathbf{4} + 4 \times \underbrace{\mathbf{1}}_{\mathbf{TBB}} = \mathbf{64} \quad (9.9.g)$$

The above individual *reducementals* are generalized as follows:

$$\left\{n^3 = \binom{n+2}{n-1} \times \mathbf{1} + \binom{n+1}{n-2} \times \mathbf{4} + \binom{n}{n-3} \times \mathbf{1} = \right\}\{1 \quad 4 \quad 1\} \odot \left\{\binom{n+2}{n-1} \quad \binom{n+1}{n-2} \quad \binom{n}{n-3}\right\} \quad (9.9.h)$$

Complexity gain: In comparing the inner structure (9.1-9.5) and (9.9.a-h) reveals something fundamental that:

- The complexity of any of the sequences in (9.1-9.5) is unity as they are obtained by *regeneration products*.
- The complexity of the sequence in (9.9.a) is 3 and Table 6 presents the complexity and the complexity gain for the power sequences from power 1-9.

- The first complexity element of "1" in (9.1)-(9.5) shares that with (9.9) or any of the sequences in Table 6, whereas the gains are purely by the virtue of the *direct product.*
- The inner structure of all the terms from 4^3 to above lack new building blocks and all the terms and sub-terms are in terms {**1 4 1**} through nesting. The higher the terms, the greater is the nesting. This reveals that the term at which the complexity is matured (for (9.9.a) this is 3^3), the higher terms have nothing original anymore but they keep nesting inside and reiterating those of the term: 3^3.

Example 7: Alleles

If the power 3 is considered as the emergent properties of the direct products of the two sequences of

$\{1 \quad 4 \quad 1\} \odot \left\{\binom{n+2}{n-1} \quad \binom{n+1}{n-2} \quad \binom{n}{n-3}\right\}$, the power 3 sequence has also alleles, which show different arrangements of the building blocks but producing the same sequence. These are as follows:

$$\{\mathcal{A}_1 | 1^3 \quad 2^3 \quad 3^3 \quad \ldots \quad n^3\} = \{1 \quad 6 \quad 6\} \odot \left\{\binom{n}{n-1} \quad \binom{n}{n-2} \quad \binom{n}{n-3}\right\} \qquad (9.10.b)$$

$$\{\mathcal{A}_2 | 1^3 \quad 2^3 \quad 3^3 \quad \ldots \quad n^3\} = \{1 \quad 5 \quad 1\} \odot \left\{\binom{n+1}{n-1} \quad \binom{n+1}{n-2} \quad \binom{n}{n-3}\right\} \qquad (9.10.a)$$

Overview: In the world of science, each hierarchy is made up of many building blocks, making compounds by their bondage or by making physical collection of the same elements. Chemistry identifies how these compounds are composed of the basic elements, the interconnectivity of which is given by the Periodic Table. Likewise, the inner structure of the sequences of natural numbers displays a parallel world, where equivalents of the Periodic Tables can be derived for this area of complexity to show their inherent interconnectivity but this is outside the scope of this introductory outline.

9.3.4. Biologistic Level

Conventional calculus of sequences of natural numbers based on the *theorem of binomial enumeration* or on *Stirling numbers* is nothing better than formalized observations using *ad hoc* mathematical formulations. The author was surprised to learn that these conventional approaches form the *orthodoxy* in this area of mathematics, to the extent that his preliminary papers were not even reviewed by mathematical journals and in one case, it was reviewed but the manuscript was rejected on the ground of not being novel. The orthodoxy has no foundations; whereas the new calculus unearths the hierarchical structure in this area of mathematics displaying a parallel with hierarchically-organized science.

The new calculus is extremely simple and well within the capability of students of secondary schools. Without any trouble, power *m* of the sequence of natural numbers can be treated by the above transformations. If each term of a sequence can be explained by its inherent building blocks and the sequence of counters making up its chemicalistic hierarchy, the various sequences put side-by-side display unique and similar features, which may be used to derive their cladistics, as an expression of their biologistic hierarchy (the author's unpublished paper). However, this is rather an involved job and only a few examples are presented for a simplistic cladistics, depicted in Figure 7, just to illustrate the idea.

The following examples suffice to illustrate the tree in Figure 11:

$$\left\{\Re\middle|\left\{1^2 \quad 3^2 \quad 6^2 \quad \dots \quad \binom{n+1}{n-1}^2\right\}\right\} = \{1 \quad 4 \quad 1\} \odot \left\{\binom{n+3}{n-1} \quad \binom{n+2}{n-2} \quad \binom{n+1}{n-3}\right\} \qquad (9.11.a)$$

$$\{\Re|\{1 \quad 3 \quad 6 \quad 10 \quad \dots\} \bullet \{1 \quad 4 \quad 10 \quad 20 \quad \dots\}\} = \{1 \quad 6 \quad 3\} \odot \left\{\binom{n+4}{n-1} \quad \binom{n+3}{n-2} \quad \binom{n+2}{n-3}\right\} \qquad (9.11.b)$$

$$\{\Re|\{1^1 \quad 3^1 \quad 5^1 \quad \dots \quad (2n-1)^1\}\} = \{1 \quad 1\} \odot \left\{\binom{n}{n-1} \quad \binom{n-1}{n-2}\right\} \qquad (9.11.c)$$

$$\{\Re|\{1^2 \quad 3^2 \quad 5^2 \quad \dots \quad (2n-1)^2\}\} = \{1 \quad 6 \quad 1\} \odot \left\{\binom{n+1}{n-1} \quad \binom{n}{n-2} \quad \binom{n-1}{n-3}\right\} \qquad (9.11.d)$$

$$\{\Re|\{1^3 \quad 3^3 \quad 5^3 \quad \dots \quad (2n-1)^3\}\} =$$
$$\{1 \quad 23 \quad 23 \quad 1\} \odot \left\{\binom{n+2}{n-1} \quad \binom{n+1}{n-2} \quad \binom{n}{n-3} \quad \binom{n-1}{n-4}\right\} \qquad (9.11.e)$$

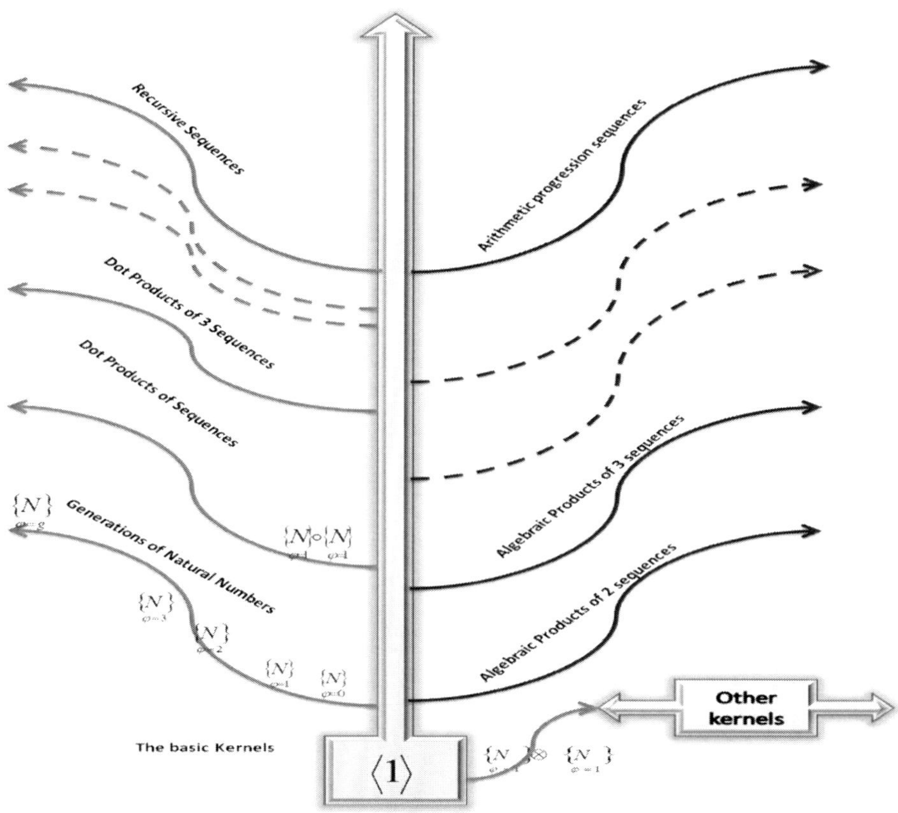

Figure 7. Primitive Cladistics of Sequence of Natural Numbers.

These sequences are highly interoperable. For instance, for any value of m, the sequences of invariants of (9.12.a) can be described by the transformer (9.12.b), as follows.

$$\{\mathcal{T}|\{1^m \quad 3^m \quad 5^m \quad \dots \quad (2n-1)^m\}\} = \tag{9.12.a}$$

$$\left\{1 \quad \frac{3}{2m-1} \quad \frac{5}{2m-2} \quad \frac{7}{2m-3} \quad \frac{\cdots}{\cdots} \quad \frac{2m-3}{7} \quad \frac{2m-2}{5} \quad \frac{2m-1}{3} \quad 1\right\} \tag{9.12.b}$$

The pattern for the transformers emerges, as evident in (9.13.a)-(9.13.b) or (9.14.a)-(9.14.b).

$$\{\mathcal{T}|\{1^m \quad 4^m \quad 7^m \quad \dots \quad (3n-2)^m\}\} = \tag{9.13.a}$$

$$\left\{1 \quad \frac{4}{3m-2} \quad \frac{7}{3m-5} \quad \frac{10}{3m-8} \quad \frac{\cdots}{\cdots} \quad \frac{3m-8}{10} \quad \frac{3m-5}{7} \quad \frac{3m-2}{4} \quad 1\right\} \tag{9.13.b}$$

$$\{\mathcal{T}|\{1^m \quad 5^m \quad 9^m \quad \dots \quad (4n-3)^m\}\} = \tag{9.14.a}$$

$$\left\{1 \quad \frac{5}{4m-3} \quad \frac{9}{4m-7} \quad \frac{13}{4m-11} \quad \frac{\cdots}{\cdots} \quad \frac{4m-11}{13} \quad \frac{4m-7}{9} \quad \frac{4m-3}{5} \quad 1\right\} \tag{9.14.b}$$

It is noted that the base sequences re-emerge in the transformer sequences in two directions and at the same time. The above is sufficient to show that there is a systematic way for the number sequences that their interconnectivity can be defined as the basis for some sort of cladistics.

9.3.5. Psychologistic and Higher Levels

It seems rather surreal to seek psychologistic traits and higher emergent properties of number sequences but when they are put side-by-side, patterns emerges to identify their external features. For instance, the individuality of the power sequences may be expressed by the sum of the terms of the kernels, and as shown in Table 6 these sums equate to factorial numbers. Similarly, the kernels of (9.12.a), (9.13.a) and (9.14.a) are analyzed in Table 8, as follows.

Table 8.a. Analysis of Patterns in the Kernels of (9.12.a)

Power	Sequence of Invariants	Sum	Analysis
m=0	{1}	1	2^0x0!
m=1	{1 1}	2	2^1x1!
m=2	{1 6 1}	8	2^2x2!
m=3	{1 23 23 1}	48	2^3x3!
m=4	{1 76 230 23 1}	384	2^4x4!
m=5	{1 237 1682 1682 237 1}	3480	2^5x5!

Table 8.b. Analysis of Patterns in the Kernels of (9.13.a)

Power	Sequence of Invariants	Sum	Analysis
m=0	{1}	1	3^0x0!
m=1	{1 2}	3	3^1x1!
m=2	{1 13 4}	18	3^2x2!
m=3	{1 60 93 8}	162	3^3x3!
m=4	{1 251 1131 545 16}	1944	3^4x4!
m=5	{1 1018 10678 14498 2933 32}	29160	3^5x5!

Table 8.c. Analysis of Patterns in the Kernels of (14.a)

Power	Sequence of Invariants	Sum	Analysis
m=0	{1}	1	4^0x0!
m=1	{1 3}	4	4^1x1!
m=2	{1 22 9}	32	4^2x2!
m=3	{1 121 235 27}	384	4^3x3!
m=4	{1 620 3446 1996 81}	6144	4^4x4!
m=5	{1 237 1682 1682 237 1}	122880	4^5x5!

Interestingly, the kernels of Power 2 of the sequences (3.a)-(3.e) render the Catalan Number sequences, known as {1 1 2 10 14 42 132 429 1430 ...}, as given in Table 9.

Table 9. Relating Catalan Numbers to the Kernels of the Second Power of Increasing Orders of Sequences of Natural Numbers

Catalan No.	Kernel	The kernel terms (T)	Sum (T)	Factor (F)	Ratio (T/F)
1	$\underbrace{\langle N\rangle}_{\varphi=0}^{D=1}$	{1}	1	1	1
1	$\underbrace{\langle N\rangle}_{\varphi=1}^{D=2}$	{1 1}	2	2	1
2	$\underbrace{\langle N\rangle}_{\varphi=2}^{D=2}$	{1 4 1}	6	3	2
5	$\underbrace{\langle N\rangle}_{\varphi=3}^{D=2}$	{1 9 9 1}	20	4	5
14	$\underbrace{\langle N\rangle}_{\varphi=4}^{D=2}$	{1 16 36 16 1}	70	5	14
42	$\underbrace{\langle N\rangle}_{\varphi=5}^{D=2}$	{1 25 100 100 25 1}	252	6	42
132	$\underbrace{\langle N\rangle}_{\varphi=6}^{D=2}$	{1 36 225 400 225 36 1}	924	7	132
429	$\underbrace{\langle N\rangle}_{\varphi=7}^{D=2}$	{1 49 441 1225 1225 49 1}	3432	8	429

9.4. Overview

The above is a reflection of hierarchical emergent properties, the exploration of which is outside the remit of this paper but it is enough to show that an analytical building block approach to number sequences is feasible and conforms to evolutionary systemic axioms. The author's current research is focused on identifying on cataloguing the properties of the various sequences in a systematic way and indeed this is feasible. When equivalents of the Periodic Table and cladistics emerge, as well as the features identifying traits of individual sequences, it is likely that patterns will emerge that one can see how different sequences would suit different requirements, cases or goals. The author feels that goal-orientation in mathematics is feasible and the results in hand indicate that this is not a mirage.

10. Formulating a Vision for Mathematics

10.1. Mathematical Truth

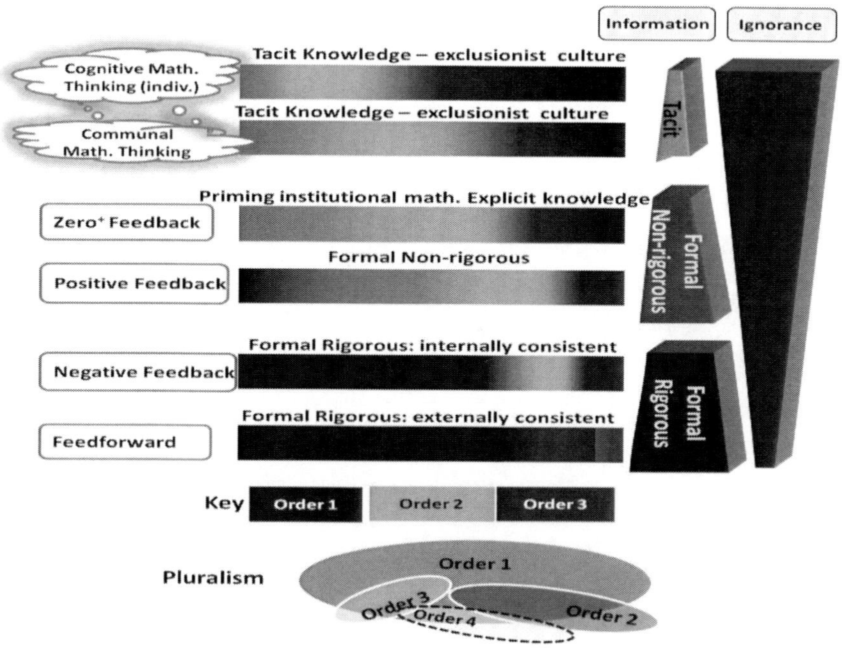

Figure 8. Indicative Evolution in Information Associated with Evolutionary Transitions

Mathematical truth may be confusing for being implicated with philosophical connotations and for semantic twists using such attributes as infallibility, impeccability and irrefutability. The author regards truth as information for decision-making and in this sense, truth, selective advantage and emergent property are different connotations of the same concept. In science, information is obtained by condensing data but in mathematics information is obtained by using a set of definitions and clearly laid down operations leading to a net and condensed statement in the form of regularity patterns, posits, conjectures, theorems or axioms. The attributes of infallibility, impeccability or irrefutability fails to account for the variability in mathematical truth but the variability is captured by evolutionary systemics depending on feedback loops, which are summarized in Table 10 with respect to each feedback loop and Figure 8 transforms its associated tacit knowledge to explicit knowledge depending on feedback loops.

As shown in Figure 8, mathematical truth used in judging or decision-making may be considered as the composition of a number of components. For instance, after the emergence of negative feedback and feedforward mathematical truth may be broken down into four components defined in Box 2.

Box 2

Total Information = **Order 1** (information such as observations, posits, conjectures) +
Order 2 (information on internal consistency – theorems) +
Order 3 (information on internal/external variability – axioms) +

Table 10. Attributes of Evolutionary Systemic Truth

Mindset	Truth Content of Mindsets – General	Truth Content of Mindsets – Mathematics
Cognitive / communal	• Tacit knowledge: "I know the truth when I see it."	• Tacit knowledge: $1+1=2$; $3^2+4^2=5^2$;
Zero$^+$ feedback	• There is no truth, other than mutual agreement but without coercion	• No truth except those primed but not selected yet. • Primed information is just a potential
Positive feedback	• Pluralism is the only truth • Pluralism is not anarchy or relativism • Information with selective advantage can be selected and unselected • Yet there is no *right to challenge*	• Formal but non-rigorous knowledge • Regularity patterns • Observations • Posits • Conjectures
Negative feedback	• Truth underpinned by internal consistency subject to agreed external criteria • Internally consistent selected information not expected to be unselected	• Reactive, proactive and intrinsic loops • Formal-rigorous & internally consistent • Definition, operations and conclusions • Internally consistent selected information is not expected to be unselected
Feedforward + Negative feedback	• Truth is underpinned by feedforward for maintaining external consistency driven by goal-orientation • Feedforward depends on negative feedback	• Formal-rigorous and internally/externally consistent information • Axiomatic mathematics
"Principle of flexibility"	• Internal consistency is by virtue of flexibility to refine and reconfigure the building blocks to comply with the conditions set by negative feedback and feedforward loops • Without negative feedback, adaptation is slow	• Internal consistency of mathematics depends more on ingenuity than flexibility • Flexible definitions and operation rules also help
Principle of Pluralism	• All the loops act pluralistically with each other • Each feedback loop and their transitions affect the nature of information	

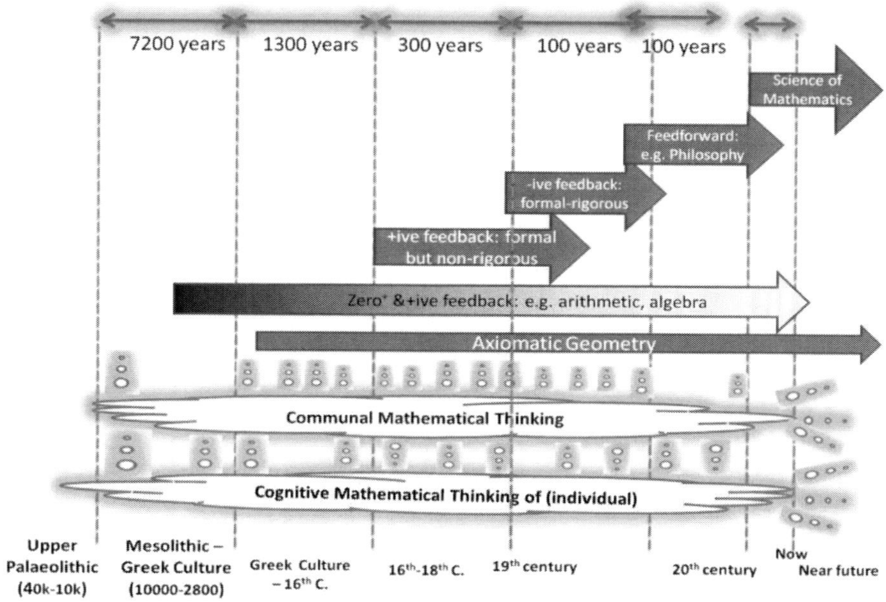

Figure 9. Timeline of Major Transitions in Mathematical Thinking.

Figure 8 indicates the orders of information for each loop, which may be explained as follows: (i) there is no knowledge at e-loops both knowledge and information emerge out of building blocks at the e-loops levels and become evolvable entities; (ii) at d-loops, knowledge/information is not inherited but gained by various means including cognition, education or by creation; (iii) knowledge is defined as a compact information and arguably, information in mathematics is an emergent property commonly known as observed regularity patterns, posits, conjectures, theorems and axioms (iii) knowledge/information may be tacit, formal but non-rigorous and formal-rigorous; and (iv) Box 2 shows that information is an evolvable entity, there are various orders of information.

10.2. Time Dimension in Evolutionary Transitions

Evolutionary transitions and the length of time for each stage after transitions are of a considerable significance and these are shown in Figure 9. The figure captures the evolution of mathematics but the followings are highlighted: (i) the zero$^+$ feedback loop for classic mathematics may be over but there is no reason to assume that new topics are not being primed today or will not do so in the future (e.g. evolutionary systemic algebra); (ii) this also applies to positive feedback (15th-18th century) and negative feedback (19th century); (iii) axiomatic geometry is rather unique, as it is a discipline of mathematics which was primed in antiquity, selected then and has remained selected and inspirational until now. The paper regards axiomatisation as a form of feedforward loops. Although evolutionary systemics can explain the rare emergence of feedforward before that of negative feedback loops, Section 8.2 explains that axiomatic geometry is the actual outcome of thousands of years of inconsistency in antiquity.

10.3. Understanding Mathematical Complexity

Mathematical complexity seems amorphous and in an inflationary state, which is a by-product of opportunism and pluralism. These complexity by-products are driven by the selective advantage of knowledge over ignorance but increasing knowledge is not a simple process of reducing ignorance as increased knowledge is associated with certain overlooked by-products to be discussed below. Ignorance encourages inertia to change, but their reduction accelerates change. For instance, before the global use of "0", the mathematical world was very small and was only grasped by the elite. The selective advantage of *zero* gave rise to a parsimonious structure within the world of numbers, as it becomes possible to write very big numbers with a very few cognate mathematical notations. The global use of zero triggered co-evolution in the communal mindset and resulted in a considerable dynamism for mathematical operations.

Mathematical complexity encompasses both communal mathematical thinking and institutional mathematics, where the communal mindset is passive but co-evolves with the institutional mindsets. Their major difference is that the institutional mindset undergoes evolutionary transitions, as shown in Figure 10. This is the basis to transform the author's tacit knowledge on mathematical complexity into explicit knowledge, as follows: (i) At the d-loop level, there is an obvious reciprocity between dichotomous properties of ignorance-knowledge, parsimony-complexity and inertia-dynamism and their significance depends on feedback loops or the co-evolution of communal mathematical thinking and institutional mathematics. (ii) At the d-loop level, each feedback loop is a mindset and capable of acting as attractors/repellents in the sense of chaos/ catastrophe theories, instigating dynamic behaviors through a sudden loss of temporal correlations among the variables. These would make the culture unstable and the paper refers to them as further aspects of *external entropy*. (iii) The e-loops of for mathematical complexity as a whole is a tacit knowledge yet to be articulated and this is one of the greatest problems of mathematics.

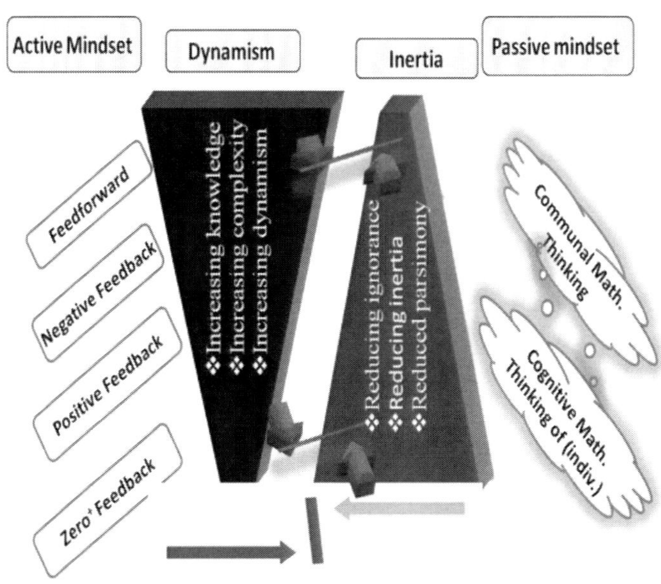

Figure 10. Capturing Changes Associated with Mathematical Thinking.

For a single entity, entropy measures internal disorder and its natural tendency to increase is seen as a natural tendency to instigate a loss of efficacy to the input of external energy and similar effects are in evidence in information systems. Mindsets associated with zero$^+$ and positive feedback loops (i.e. mathematical observation, regularity patterns, posits and conjectures) would inherently contain spontaneity and entropy in the form of internal inconsistencies within the conceptualization process. Pure mathematics is platonic as internal entropy is technically removed through the proof processes (negative feedback) or the massive process of rigorisation. The process of proof has the effects of eradicating attractor/repellents after rigorisation. However, an entire cleanup is impossible as some inconsistency may unwittingly be left in.

When a diversity of internally-consistent entities coalesce together to form complexity, the entities do not necessarily automatically coalesce consistently together. The subsequent inconsistencies are given a euphemistic term of paradox. As mathematics lacks the equivalence of cladistics, there is no true picture of such external consistencies and therefore despite the invincibility of pure mathematics, it is vulnerable.

The most subtle risk may stem from possible attractors/repellents within the culture of mathematics. In the first place, the growing research contribution to mathematics creates a dynamical world, where keeping the abreast of knowledge is a problem similar to other disciplines. This may create "islands of knowledge" within each complexity with poor transfer of knowledge in between. Thus, the islands of knowledge can instigate repellents. The institutional culture is very complex and often inflated by a bias towards new findings but without any significant tendency towards consolidation. In this culture, any paradox attracts great attentions, which may be a reflection of possible attractors/repellents between communal mathematical thinking and institutional mathematics. These are depicted in Figure 10.

10.4. Elements of a Vision for Mathematics

This paper shows that even mathematics, regarded with aurora and branded as insular or sterile, can be explainable by evolutionary systemics. The study identifies the elements for formulating a vision for mathematics as a whole, as follows.

Mathematics is naturally a pluralist discipline and this provides grounds for proliferating areas of complexity without necessitating its consolidation. Evolutionary transitions of mathematics include: proliferating mathematics driven by opportunities created through articulating tacit knowledge into explicit knowledge and then dealing with internal and external inconsistencies. Although internal entropy in mathematics is treated well, external entropy poses problems and therefore managing expectations is an important issue.

The practitioners in their current mindset have the potential to be refocused from the opportunism towards the culture of goal-orientation. The problems can be addressed by a vision through forming the "science of mathematics," the main features of which are depicted in Figure 11. Some of the elements of this vision includes (i) characterizing the amorphous-looking mathematical complexity growing under pluralism; (ii) critically examining the *increasing complexity* of mathematics and mapping it out to identify an equivalence of cladistics in mathematics; (iii) identifying the role and manner of negative feedback loops in creating internal consistency in mathematics and also of feedforward loops in implementing external consistency; (iv) managing expectation by showing the evolutionary nature of mathematics and highlighting the variability of mathematical truth; and (v) identifying the e-loops in mathematical complexity.

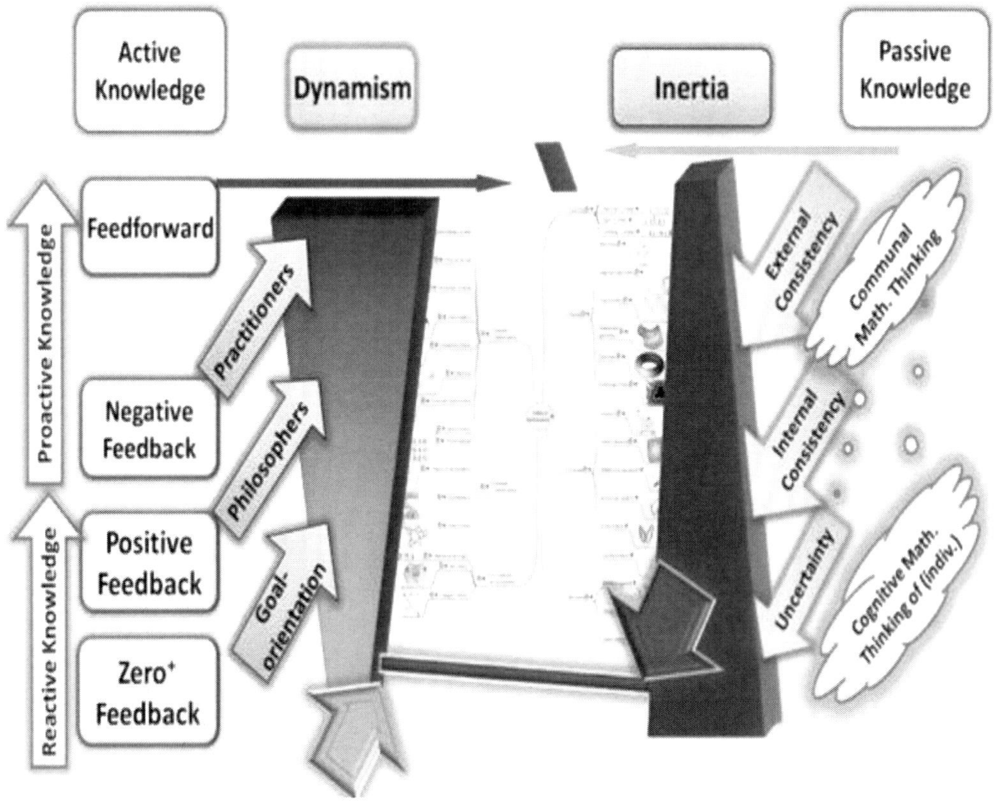

Figure 15. The Elements for Formulating a Vision for Mathematical Complexity.

11. DISCUSSIONS

The comparison of the implications of negative feedback loops in Table 4 in reductive science and systems science with mathematics highlights the strength of evolutionary systemics to explain these three areas of complexity, as a whole. Thus, thanks to evolutionary thinking, it is no longer necessary to seek exactly the same way of evolution in all areas of complexity by moving from one discipline to another but evolutionary thinking provides the flexibility to accommodate variations as per each particular area of complexity.

Rigorised mathematics is referred to certain, infallible, impeccable and irrefutable mathematics but such superlatives convey erroneous messages. In contrast, negative feedback is a mindset underpinning internal consistency or rigorisation and this is exactly the role it played in reductive science and many technological systems, such as the flywheel mechanism regulating engine speeds.

Attention is drawn to the risks of exclusionist cultures. One example is given by Kline (1980, p308) corresponding to the dawn of axiomatic geometry. He describes the Greek conception of irrational numbers such as $\sqrt[2]{2}$. According to him, the Greeks recognized only the ordinary whole numbers, and not accepted irrational numbers such as: $\sqrt[2]{2}$. They resolved their dilemma by ostracising these irrational numbers and opted to their geometric representations. This exclusionist culture can explain the lack of any serious exploration of the Greeks on arithmetic and algebra. In exclusionist cultures, there is a serious risk of overlooking diversity and creating barriers against the selection of newly primed concepts. Unlike exclusionism, diversity enriches the culture. One example of diversity is from the Middle Ages, during which the Indian culture was better developed in arithmetic, Chinese in algebra, the Greek in geometry. The Islamic cultures then integrated and developed these three disciplines with a considerable innovation.

One problem of rigorised mathematics is its exponential growth. The following passage is assembled from the information collated by Rollet and Nabonnand (2003) highlighting specific problems at a certain time. The first scientific journals appeared during 17th century, which did not exclusively contain mathematical publications, e.g. *Journal des scavants* (1665) or the *Philosophical Transactions of the Royal Society of London* (1665). Whilst in 1700, only 15 journals contained mathematical articles; in 1900, probably more than 600 journals published mathematical works and this triggered the need for setting up various bibliographic catalogues. Rollet and Nabonnand (2003) provide an account of seeking to resolve the incapacity of mathematicians regarding the increase of mathematical publications through a project in 1885 by the *Société Mathématique de France* to catalogue a bibliographical mathematics within the international mathematical community. This project was active between 1894 and 1912. They note that 113 of 157 codes used for classifying mathematics referred to bibliographical items that were even active before 1840 but 11 codes emerged exclusively to mathematical works published after 1860; also, 8 codes emerged in the later parts of 19th century. The growth in mathematics has been much greater since then and almost explosive now.

The paper compares the implications of the negative feedback loops in reductive and systems science with mathematics, highlighting the strength of evolutionary systemics to explain these diverse areas of complexity. Philosophers of mathematics have posed deep philosophical enquiries but proliferated top-down and discrete doctrines undermining their efficacy. The author argues that philosophical enquiries are stimulating but holds that evolutionary systemics can respond to all of the philosophical enquiries in a scientific way and this potentially can bring the demise of the top-down philosophy of mathematics. The author's response to philosophical enquiries is outlined in Appendix I.

With the benefits of evolutionary systemics, the paper unraveled that: (i) perfection in mathematics arises by man's preparedness to "bend" his mind and not reference mathematical truth to real world conditions; (ii) mathematics breaks down in the real world, as that and real world are not interchangeable for differing in the nature of their entropies; (iii) if mathematics is to emulate the entropy of the real world, it becomes applied mathematics (mathematical modeling). These key features of mathematics are normally overlooked but this sets an undue expectation and the retardation of science. For instance, mathematics identified with impeccable truth inspired philosophers with similar expectations as they aimed to replicate similar standards of truth in philosophy but truth without referencing to the driving feedback loop is poor. If philosophers were inspired by the pluralism of mathematics, the world today could have been different.

Conclusion

Mathematics may be seen as a sterile discipline and renowned for infallibility, impeccability and irrefutability but this paper substantiates that it is an outcome of man's way of life and evolves through a symbiotic relationship with science and applied mathematics. *Mathematical complexity* overarches the diverse disciplines of mathematics, which seems amorphous, in an inflationary state and its growth is driven by opportunism and pluralism.

Mathematics treated as a discipline outside the reach of evolutionary processes only perpetuates the grip of the ontological mindset over it. This paper substantiates the applicability of evolutionary systemics to mathematical complexity as a whole. It shows evidently that there is nothing unique about mathematical complexity, as it evolves just like any other complexity. Without a theory of evolution for every areas of complexity, it is very likely to be wrong-footed. In contrast, Morton (2001) remarks: "philosophy used to compare itself with mathematics and hence aimed for certainty, proof and an aristocratic oversight above the rest of knowledge." This paper argues that only evolutionary thinking is capable of oversights, as the status quo in both philosophy and mathematics is overshadowed by an ontological mindset. This paper unravels that they are simply outcomes of evolution. Fregé is the furthest from evolutionary thinking in mathematics and with the ambition of reverse engineering logic but Kline (1980, P197) quotes him stating that "just as the building was completed, the foundation collapsed." Notably, the collapsing foundation was ontological in the first place.

This paper uses *evolutionary systemics*, where systemic thinking is integrated with evolutionary thinking. The evolutionary component explains the source of mathematics to be originated from communal mathematical thinking, which emerged in prehistory with very elementary building blocks of counting, primed and selected over thousands of years. The systemic component explains that institutional mathematics has undergone evolutionary transitions comprising: (i) mathematics has been primed since the Sumerians but the priming and selection of the axiomatic geometry by the Greek civilization gave a new impetus to institutional mathematics; (ii) mathematical topics have been proliferated over the time up to the 19th century (particularly the great creativity during the 16th-18th centuries), during which time they were largely non-rigorous (positive feedback); (iii) rigorisation of mathematics was carried out in the 19th century to ensure its internal consistency (negative feedback); and (iv) various initiatives emerged since the middle of the 19th century to strengthen the external consistency of mathematics, e.g. axiomatisation (feedforward loops).

With the benefits of evolutionary systemics, the paper unraveled that: (i) perfection in mathematics arises by man's preparedness to "bend" his mind and not reference mathematical truth to real world conditions; (ii) mathematics breaks down in the real world, as that and real world are not interchangeable for differing in the nature of their entropies; (iii) if mathematics is to emulate the entropy of the real world, it becomes applied mathematics (mathematical modeling). These key features of mathematics are normally overlooked but this sets an undue expectation and the retardation of science. For instance, mathematics identified with impeccable truth inspired philosophers with similar expectations as they aimed to replicate similar standards of truth in philosophy but truth without referencing to the driving feedback loop is poor. If philosophers were inspired by the pluralism of mathematics, the world today could have been different.

The paper is based on the premise that mathematical complexity encompasses contributions of individuals, communal mathematical thinking and institutional mathematics. Accordingly, the institutional mindset undergoes evolutionary transitions but normally passive communal mathematical thinking does not. External inconsistencies may stem from (i) cumulative growth triggering the need for consolidation; (ii) mismatches of different areas of mathematical complexity at their interfaces; (iii) undue expectations from the seemingly sterile mathematics. The most subtle source of inconsistency may stem from possible attractors/ repellents within the culture of mathematics, in a background where the growing research contribution to mathematics creates a dynamical world difficult to keep abreast of knowledge and hence the creation of "islands of knowledge." In this culture, any paradox attracts great attentions, as a reflection of possible attractors/ repellents between communal maths thinking and institutional mathematics.

The external consistency of mathematical complexity with the contextual culture is an expression of the need for feedforward loops. The paper presents three strategies to this end: (i) philosophical enquiries, which stimulate debate but their mutually exclusive doctrines retard their efficacy; (ii) axiomatisation of some areas of mathematics through a process akin to "reverse engineering" in the sense that the body of a certain area of knowledge in mathematics is expressed in terms of consensual, clear and readily acceptable statements, but they are few and far in between; (iii) goal-orientation, advocated by this paper, is yet to prevail in mathematics, as a proactive measure against external inconsistencies among the various areas of mathematical complexity.

This paper makes case for goal-orientation in mathematical complexity for reasons including: (i) its areas of complexity on their own and as a whole are evolving; (ii) although its growth is believed to be cumulative, there are consolidations and evolutionary transitions; and (iii) the gaps between communal and institutional mathematical capabilities needs to be recognized and reduced. This requires flexibility and a positive culture not prejudiced against the knowledge not used presently. It also takes a preliminary step to show that evolutionary systemics may be used to develop goal-orientation capabilities in mathematics by overhauling the conventional calculus of sequences of natural numbers into a new calculus. It also presents the elements of a vision towards goal-orientation in mathematical complexity.

ACKNOWLEDGMENT

The author is grateful to Mr. John Little of New College, Swindon, UK, for critically reviewing and commenting on an early version of the paper. The thanks are also due to Dr. Mark Everard of the Environment Agency, and Prof. Oleg Makarynskyy, of Asia-Pacific Applied Science for reading and commenting the earlier version. The development is a contribution to open scientific and mathematical culture. No moral right is given to any usage of the emerging calculus for any development, software or otherwise, that uses the new calculus and closes their outcome to the public.

Appendix I: Evolutionary Systemic Response to Doctrines of Philosophy of Mathematics

Avigad (2007) is minded that mathematical practices have evolved over time with some developments even evoking controversy and debate but he reflects that "Mathematical objects like numbers and sets are archetypical examples of abstracta, since we treat such objects in our discourse as though they are independent of time and space; finding a place for such objects in a broader framework of thought is a central task of *ontology*, or *metaphysics*." The paper argues that ontology and metaphysics are minefields of problems and evolutionary systemics offers a methodology to discard ontology or metaphysics and even study the evolution of abstract concepts. This section does not touch on any of philosophical doctrines but only responds to their enquiries through the alternative modeling capability of evolutionary systemic axioms, as follows:

Is mathematics invented or a social product? Mathematics is a product of evolutionary systemic processes and nothing more. This may be explained using (5.8), referring to the Fundamental Theorem of Algebra. This is an equation of pure mathematics and the emergent properties of configuring particular building blocks subject to rules. The building blocks include: (i) representing numbers symbols, (ii) devising rules to carry out the operation of addition and multiplication on the symbolically-represented-numbers to obtain compound symbolic expression; (ii) discarding any knowledge of the operations by expression the compound expression as an equation, and (iv) reverse engineering the process in terms of solving the equation. The point is that as soon as these building blocks are invented, (5.8) is a spontaneous emergent property but as a potential, which needs to be conceived, expressed and refined. Section 5 outlines the long routemap of (5.8) substantiating that this equation was indeed conceived, expressed and refined. This paper gave the evolutionary systemic interpretation of this long history, where its refinements mean the process of developing negative feedback loops to hammer out entropy in the concept. The outcome is an elegant and parsimonious equation (5.8). There is nothing unique about it even if this is regarded as a perfect product of human intellect. Even if this equation stands the test of the time, Nature has given rise to species that has been standing the test of the time, e.g. the crocodile species has seen many calamities and yet has survived without much change in 700 million years. The fact is that both are products of assemblage of building blocks subject to rules but the attachment of human values alters nothing.

Does mathematics exist independent of human perceptions? Mathematics does not exist independent of human perceptions but as a potential this may be true. Mathematics is a product of natural selection just like biological species. There are so much potential developments in mathematics but they have not been primed and selected yet and they may never emerge. Thus, natural/social worlds perceived within the perimeter of human perceptions are out there and man tries to understand them including any aspect of the physical world or man's activities create the potential for new building blocks and thereby the creation of new complexity, e.g. mathematics. It is important not to be trapped by metaphysical and ontological mindsets and escaping its grips by regarding any area of complexity as evolvable entity.

Does mathematics has a foundation? There is only one foundation: the assemblage of building blocks with some rules subject to natural selection leading to a possible selective advantage, which cannot be known before hand but susceptible to a continuous evolution. Any doctrine proclaiming foundation will only be credible for consideration by having *zero assumptions*, *no presumptions* and *no predispositions*.

Is mathematics growing cumulatively? It is certainly growing but not linearly and not cumulatively. The paper provides evidence that in the enterprise of mathematics, the process of selection and unselection is a fact and at work. Its growth is subject to evolutionary transitions, including consolidation, systematization or reverse engineering.

Is mathematics perfect or susceptible to anomalies? Mathematics is an abstract enterprise and if it is claimed to be perfect, it just shows the failure on communicating the ongoing simplifications.

What is the status of mathematical truth? Mathematical truth is non-real and fails to be reproduced in the real world. But then so is language, which is an assemblage of vowels and consonants and subject to rules. Both have utilities acting as their selective advantages and they have been selected proportional to their selective advantages. Both are evolving depending on social needs. Mathematics has evolved as a backdrop to applied mathematics and driven by science, technology and cultural way of life.

What is the nature of mathematical objects? They are evolving entities just like any other evolving entities with own peculiarities without possessing any magical quality.

What gives mathematical statements their aurora of infallibility? The author argues that ignorance is responsible to attaching transcendental qualities to the various entities. Consider the parts of speech in a sentence. Some nine or ten parts of speech fulfill mutually exclusive functions. By removing each of them in turn, the meaning will lose proportionately in it quality. These parts of speech are rightly regarded as ordinary entities and without any transcendental. Mathematics based on numbers started its journey as linguistic artifacts and found wider uses. Attachment of aurora to mathematics has no real basis and noting other than cultural values. Human cultures have given rise to language, politics, economics, technology, sports, architecture, art, literature and so on; similar to mathematics or language. Elements of these products of natural selection are also visible in other species.

REFERENCES

Armstrong, K. (2005), *A short history of mythology*, Pp 63-65, Publ. By Canongate, Edinburgh

Avigad, J., (2003) Number theory and elementary arithmetic. Philosophia Mathematica, (3) Vol. 11 (2003), pp. 257-284

Avigad, J. (2007), "Philosophy of Mathematics", *The Edinburgh Companion to the 20th Century Philosophies*, Ed. Boundas, Constantin, Edinburgh University Press.

Baron, M.E. (1969), *The origins of infinitesimal calculus*, Dover Phoenix Editions, originally published by Pergamon Press

Baum, R. (1973), *Philosophy and Mathematics.* San Francisco: Freeman, Cooper.

Benacerraf, P., and Putnam, H. (1983). *Philosophy of Mathematics, Selected Readings.* Cambridge: Cambridge University Press, 1983

Boyer, C. (1959), The History of the Calculus and Its Conceptual Development, Dover Publications, Chap VI)

Boyer, C.B. (1991), *A history of mathematics*, 2nd edition, John Wiley and Son Inc.

Brualdi, R. A. (1992), *Introductory Combinatorics*, published by Pearson.

Bucchianico, A. Di., Loeb, D. (200), "A Selected Survey of Umbral Calculus," the Electronic Journal of Combinatorics, http://www.combinatorics.org/index.html.

Bunch, B.H., (1997), *Mathematical Fallacies and Paradoxes*, Dover Publications

Clark, M. (2002), *Paradoxes from A to Z*, Routledge

Cellucci, C., (2000)."The Growth of Mathematical Knowledge: An Open World View," in Emily Grosholz & Herbert Breger (eds.), *The Growth of Mathematical Knowledge*. Kluwer (http://w3.uniroma1.it/cellucci/documents/Growth.pdf).

Courant R., and Robbins, H., (1978), *What is Mathematics?* Oxford: Oxford University Press.

Davis PJ and Hersh R. (1980), *The mathematical experience, Penguin Books*

Easwaran, K. (2008), "The role of axioms in mathematics," Erkenntnis (1975)

Vol. 68, No. 3, May, 2008, Towards a New Epistemology of Mathematics, Pub.Springer

Eliade, M., (1989), *A history of Religious Ideas*, Vol 1, The University of Chicago Press

Falconer, K. (1990), *Fractal geometry*, Wiley, New York, NY

Fauvel, J. and Gray J. (1987), *The history of mathematics: a reader*, Open University, Macmillan Press Ltd,

Franzén, T. (2005), *Gödel's theorem, an incomplete guide to its use and abuse*, A.K. peters, Ltd, Wellesley

Fregé, G. (1979), Posthumous Writings. Oxford: Blackwell. (cited in Easwaran) (2008).

Friend, M. (2007), *Introducing philosophy of mathematics*, Acumen Publishing Limited

Gillies, G. (1995), Revolutions in Mathematics, Oxford University Press

Goodman, N.D., *Mathematics as objective science*, reprinted in Tymoczko, T. (1998), *New Directions in the Philosophy of Mathematics*, 1986. Revised edition, 1998, Princeton University Press.

Grabiner, J.V. (1974), *Is mathematical truth time-dependent?*, American mathematical Monthly, Vol 81, No4, April 1974

Gustafson, K.E. (1980), *Introduction to partial differential equations and Hilbert space methods*, John Wiley and Sons, New York,

Hadamard, J., (1952), *Lectures on Cauchy's problem in linear partial differential equations*, Dover, New York, 1952. MR 14:474f

Heidegger, M., (1962) Being and Time, translated by John Macquarrie and Edward Robinson, Harper and Row Publishers, Inc. (http://www.amazon.com/Being-Time-Martin-Heidegger/dp/0060638508#reader_0060638508), accessed on 9 Oct 2010

Hodgson, G. M., (1993), *Economics and Evolution: Bringing Life Back into Economics*. Cambridge, UK and Ann Arbor, MI: Polity Press and University of Michigan Press.

Hodgson, G. M., (2004), *The Evolution of Institutional Economics: Agency, Structure and Darwinism in American Institutionalism*. London and New York: Routledge.

Kicher, P. (1983), *Nature of mathematical knowledge*, Oxford University Press, Inc.

Khatibi, R.H., Wormleaton, P.R., and Williams, J.J.R., (2001), "Parameter quality conditions in open channel inverse problems", Journal of Hydraulic Research, Vol. 38, No. 1, pp1-12 (http://www.iahr.org/publications/assets/jhr38-6/RAHMAN.pdf.

Khatibi, R.H. (2001), "Sample Size Determination in Open Channel Inverse problems", ASCE J. of Hyd. Eng., Vol. 127, No.8, Pp 678-688

Khatibi, R. (2003) Khatibi, R.H., (2003), "Systemic knowledge management in hydraulic systems: I. A postulate on paradigm shifts as a methodological tool", Journal of Hydroinformatics; Vol. 5.2, PP 127-140; (http://www.iwaponline.com/jh/005/0127/0050127.pdf).

Khatibi, R., (2011), "Evolutionary Systemic Modelling for Flood Risk Management Practices," Journal of Hydrology Vol. 401 Issue 1-2, Pp 36–52 (http://dx.doi.org/10.1016/j.jhydrol.2011.02.006)

Khatibi R., (2012a), "Learning from Natural Selection in Biology: Reinventing Existing Science to Generalise Theory of Evolution – *Evolutionary Systemics*," in press, Chapter 1: *Natural Selection: Biological Processes, Theory and Role in Evolution,* Ed. Lynch, J.R and Derek T. Williamson, D.T. to be published by Nova Publishers this summer https://www.novapublishers.com/catalog/product_info.php?products_id=32051

Khatibi, R., (2012b), "Evolutionary Transitions in Mathematical Modelling Complexity by using Evolutionary Systemic Modelling – Formulating a Vision," in press, Chapter 5: *Natural Selection: Biological Processes, Theory and Role in Evolution,* Ed. Lynch, J.R and Derek T. Williamson, D.T. to be published by Nova Publishers this summer

https://www.novapublishers.com/catalog/product_info.php?products_id=32051

Khatibi, R., Surendran, S., Everard, M., Ware, J.A., and Wilcox, S. (2012), "Vision for Intelligent Agent Capabilities Based on Evolutionary Systemic Rational," in press, Chapter 2: *Natural Selection: Biological Processes, Theory and Role in Evolution,* Ed. Lynch, J.R and Derek T. Williamson, D.T. to be published by Nova Publishers this summer

https://www.novapublishers.com/catalog/product_info.php?products_id=32051

Kline, M, (1980), *Mathematics: the loss of certainty*, PP 520-525, pp 163-169 reprinted in *The World Treasury of Physics, Astronomy and mathematics*, edited by Timothy Ferris, Little Brown and Company, 1991.

Körner, S., (1986), The Philosophy of Mathematics, an Introductory Essay. New York: Dover.

Kuhn, T.S., (1962), *The structure of scientific revolution*, 3rd edition, The University of Chicago Press.

Lorenz, E. N. (1963) Deterministic non-periodic flow. *J. Atmos. Sci.* 20,130-141.

Mac Lane, S. (1981), "Mathematical Models: A Sketch for the Philosophy of Mathematics" *The American Mathematical Monthly*, Vol. 88, No. 7., pp. 462-472.

http://home.dei.polimi.it/schiaffo/TFIS/philofmaths.pdf

Morton, A. (2001), "Philosophy as Engineering" Chap. 3 in Two Roads to Wisdom?: Chinese and Analytic Philosophical *Tradition*, Ed. Bo Mou, Open Court Publishing Company.

Nissen, H. J., Damerow, P., Englund, R. K. (1993) *Archaic Bookkeeping*, Chapter 3: Archaic Numerical Sign Systems, The University of Chicago Press, pages 25–29

Quinn, F. (2012), A Revolution in Mathematics? What Really Happened a Century Ago and Why It Matters Today, http://dx.doi.org/10.1090/noti787

Rav, (2006), Philosophical Problems of Mathematics in the Light of Evolutionary Epistemology, Chapter 6 in *18 Unconventional Essays on the Nature of Mathematics*, ed. By Reuben Hersh

Royster, D. C. (2008), Non-Euclidean Geometry, Chap 1: The origins of geometry, http://education.uncc.edu/droyster/courses/spring02/classnotes/Lecture01.pdf; accessed on 25 December 2010.

Rollet, L. and Nabonnand, P., (2003), "An Answer to the Growth of Mathematical Knowledge? The Répertoire Bibliographique des Sciences Mathématiques", *Publié dans l'*European Mathematical Society Newsletter *47 (mars 2003), pp. 9-14.* (http://poincare.univ-nancy2.fr/digitalAssets/12592_Answer_Growth_Mathematical_Knowledge.pdf)

Russell, B. (1903), *The principles of mathematics*, vol. I., II and II, Cambridge: University press, 1903 (http://quod.lib.umich.edu/cgi/t/text/text-idx?c=umhistmath;idno=AAT3201.0001.001, accessed 15 November 2010).

Russell, B (1919) *Introduction to Mathematical Philosophy* (London: George Allen and Unwin, New York: The Macmillan Company (http://people.umass.edu/klement/russell-imp.html accessed 15 November 2010).

Sainsbury, R.M. (1987), *Paradoxes*, Cambridge University Press.

Sider, T. (1991) "Might Theory X Be a Theory of Diminishing Marginal Value?" *Analysis,* 51, Pp: 265-271

Sigmund, K., (1993), *Games of life, explorations in ecology, evolution and behaviour*, Oxford University press, Oxford

Sorensen, R. (2003), *A Brief History of the Paradox: Philosophy and the Labyrinths of the Mind*, Oxford University Press, http://www.amazon.com/Brief-History-Paradox-Philosophy-Labyrinths/dp/0195179862/ref=pd_sim_b_5 accessed on 9 Oct 2010

Struik, D.J. (1987), *A Concise History of Mathematics*, Dover Publications, New York

Thom, R. (1972) Stabilite Structurelle Et Morphogenese, Benjamin, New York.

Thom, R. (1975) Structural Stability and Morphogenesis: An Outline of a General Theory of Models; Fowler H, translator. Reading; Benjamin: 1975.

Tymoczko, T. (1998),*New Directions in the Philosophy of Mathematics*, 1986. Revised edition, 1998, Princeton University Press.

Villiers, M. De, (1986), "The Role of Axiomatisation in Mathematics and Mathematics Teaching", Originally published by Research Unit for Mathematics Education (RUMEUS), University of Stellenbosch, South Africa (http://mzone.mweb.co.za/residents/profmd/axiom.pdf)

Von Bertalanffy, L. (1969) General System Theory: Foundations, Development; George Braziller, N.Y.

Wilder, R. (1981), *Mathematics as a cultural system*, Pergamon Press, Oxford.

In: Natural Selection
Editors: J. R. Lynch and D. T. Williamson

ISBN: 978-1-62257-467-4

Chapter 5

EVOLUTIONARY TRANSITIONS IN MATHEMATICAL MODELING COMPLEXITY BY EVOLUTIONARY SYSTEMICS

Rahman Khatibi
Consultant Mathematical Modeller, Swindon, UK

ABSTRACT

Since the emergence of computational mathematical modeling in the 1950s, its developments have been suffering from fragmentary visions, if any. Still there is no insight into, or foresight of, modeling complexity as a whole. This paper formulates a vision for modeling complexity using evolutionary systemics, which integrates evolutionary and systemic thinking. This paper recognizes cognitive model thinking of the individuals assisted by their instincts; communal model thinking due to cohesion of individual mindsets; and institutional modeling practices capable of evolutionary transitions. The following transitions are identified: (i) primitive modeling, prevailed until the 17th century; (ii) prototype modeling until 1950, which started afresh by unselecting its past primitive modeling practices; and (iii) computational modeling since 1950, which largely reinvented past capabilities under the computational environments. Over the years since 1950, modeling complexity has seemingly been growing organically but whilst this signifies healthy growth, it underlines potential problems. The practice-data is used to analyze this complexity by applying evolutionary systemics. This helps referencing a vision of modeling complexity to the full evolutionary contexts and identifying relevant issues. In this way, the paper shows that modeling complexity is infested with increasing complexity and increasing dynamism. The emerging complex systems suffer from a new kind of entropy retarding growth. The paper also studies the structure of complexity exploring the feasibility of deriving a cladistics for modeling complexity. This is a radical departure from conventional thinking, as new investments on model developments can avoid bottlenecks. The gained insight invokes the need for rethinking and making a research case for the formation of the science of modeling .

Keywords: modeling complexity, evolutionary-systemic, evolutionary dynamism, new kind of entropy, hierarchy, vision

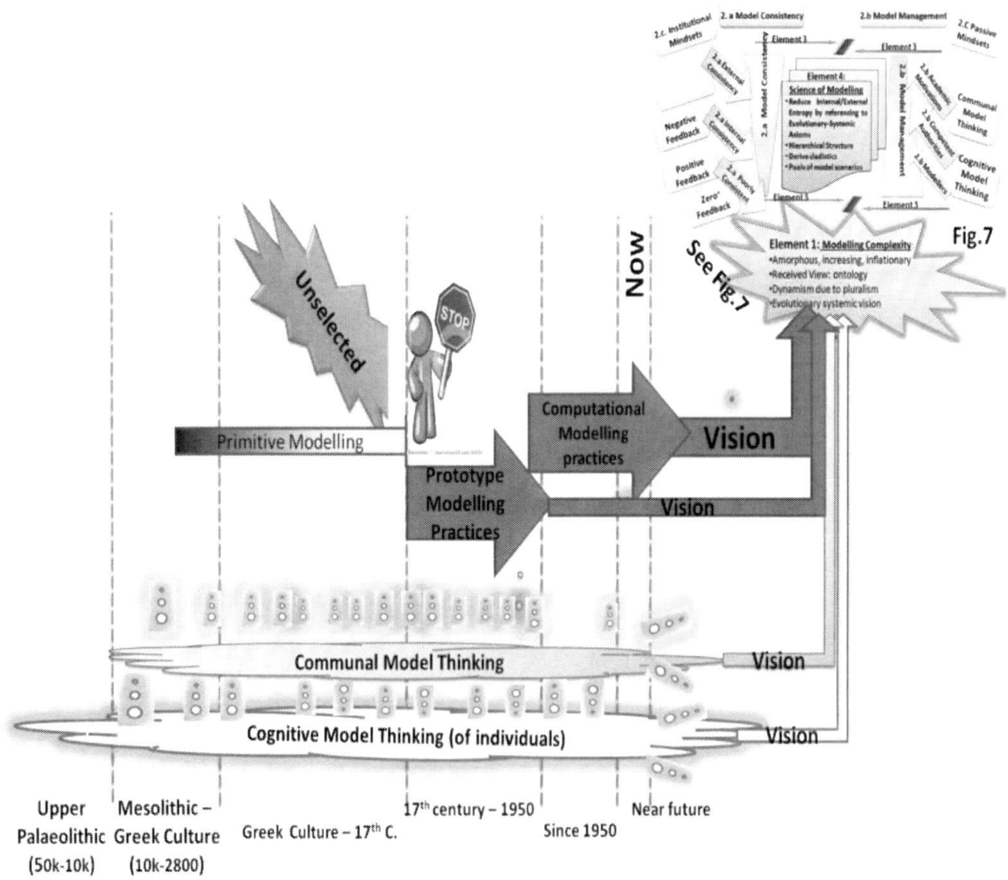

1. Introduction

Modern sophisticated modeling practices use evidence, data, space, time, objects and/or abstract concepts as a surrogate to make a better sense of the environment, where the emergence of this capability is an outcome of a never-ending process of a continual change. Models may be seen of a recent origin (say, since 1950) but are being built on the knowledge capital accumulated since the 17th century by scientific enquiries and evolving over a timeline that goes back to prehistory with many *evolutionary transitions*. These evolutionary transitions are overlooked as a norm and hence the prevailing mindset is ontological, where ontology refers to permanently fixed complexity in a given time horizon and a "mindset" refers to a body of fundamental assumptions, knowledge, beliefs and values. Ontology is the old Aristotelian mindset but survives in modern thinking and retards science by encouraging blinkered and fragmentary visions. Ontology signifies the absence of an evolutionary thinking. The paper aims to break the grip of ontology on modeling complexity by applying evolutionary systemic modeling as given by Khatibi (2011) and Khatibi (2012a).

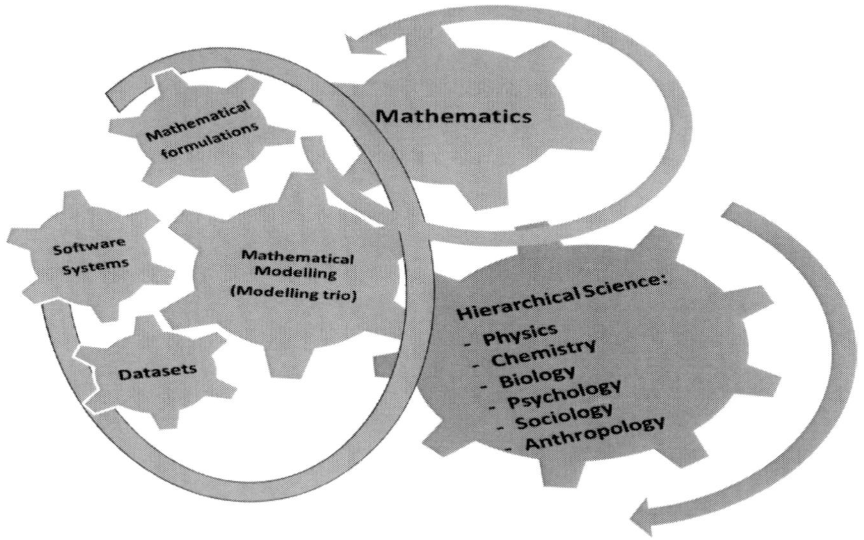

Figure 1. Symbiotic Interactions of Science, Mathematical Modeling and Mathematics.

Khatibi (2012a) differentiates variations on the following time horizons: (i) *Dynamic timescale*: time starts at an origin, processes are triggered on continuous or discrete time intervals of fractions of a second to hours, days and so on, the process ends. (ii) *Long timescale*: this is normally large compared with dynamic timescale, over which adaptation may take place. (iii) *Evolutionary timescale*: changes comprise adding/removing building block to/from the architecture hereditary machinery) in many generations. Mathematical modeling practices are carried out at the dynamic timescale but the practice is honed beneath dynamic activities at the evolutionary timescale. The architect of honing the practice is natural selection through the architecture of four feedback loops: zero$^+$ feedback, positive feedback, negative feedback and feedforward loops. Each loop has three manifestations corresponding to each timescale, as summarized in Table 1.

Table 1. Variations of Feedback Loops with Respect to Timescale

	Drivers at Dynamic Timescale	Long Timescale	Evolutionary Timescale
Zero$^+$ Feedback	**d-loop**: Randomness	**ℓ-loop**: Obscure	**e-loop**: Spontaneity
	Zero$^+$ Feedback Loop gives rises to simplexes, which grow as below:		
Positive Feedback	**d-loop**: Performance	**ℓ-loop**: Adaptation	**e-loop**: Natural Selection
	Positive Feedback Complexes = d_loop + ℓ_loop + e_loop		
Negative Feedback	**d-loop**: Internally-consistent performance	**ℓ-loop**: 'Discretion'	**e-loop**: Natural selection of rules for internal consistency
	Negative Feedback Complexes = d_loop + ℓ_loop + e_loop		
Feedforward	**d-loop**: Vision for external consistency	**ℓ-loop**: Mission for external consistency	**e-loop**: Natural selection of rules for external consistency
	Feedforward Complexes = d_loop + ℓ_loop + e_loop		
Note 1: Positive feedback has three integral components of d-loop (real dynamic state of the system), ℓ-loop (manipulation of the phase), or e-loop (changes to hereditary codes)			

Modeling practices are confined to d-loops without any conception of their inherent ℓ-loops and e-loops and this is how modeling practices are inadvertently ontological; whereas ℓ-loops and e-loops are all that define the horizon of theory of evolution with no conception of feedback loops. In between them, there is a gap, but to be bridged in this paper according to Table 1. This is the integration of natural selection e-loop, (the architect) with feedback loops (the architecture), to be referred as *evolutionary systemics* or *evolutionary systemic modeling* (EvSyM).

Formulating a vision for mathematical modeling by understanding its evolutionary context is the focus of this paper. The paper unearths that the mindset over modeling complexity is ontological for contending to its amorphous growth and thereby overlooking its possible hierarchical-organization and interconnectivity at each hierarchy. Outside ontology, there is more to modeling complexity than meets the eye. This paper has two lines of enquiry: (i) to understand evolutionary dynamics through the evolutionary systemics to identify real bottom-up issues, see Section 9; (ii) to understand the structure of complexity by suggesting that the hierarchical organization of science can serve as a model of its hierarchy, (these comprise physics, chemistry, biology, psychology, sociology and anthropology), see Section 10. To avoid misunderstanding, these terms are qualified as: *physicalistic*, *chemicalistic*, *biologistic*, *psychologistic*, *sociologistic* and *anthropologistic*. The combination of both of these lines of enquiry leads to a vision, as the focus of this paper. In this way, it can become feasible to understand these emergent properties at each hierarchy. The return for this is significant. For instances, if there is a cladistics (tree of life) of modeling complexity at the biologistic level, the interconnectivity of the various models can be mapped out.

Attention is also drawn to the remit of the paper, which touches on: science, evolutionary thinking, systems science including cybernetics, cognitive development, mathematics /mathematical modeling, philosophy/philosophy of science/philosophy of mathematics, philosophy of mathematical modeling, prehistory and mythology. The paper is inclusive of all types of modeling practices including mathematical ones without specifically referring to a particular field of expertise but few examples are given from hydrology and hydraulics.

Before presenting a summary of the paper, the definition of models is outlined. One commonsense definition of models is given by Schmuller (2004): "when you create a model you're using something that you know a great deal about to help you understand something you know very little about." There are many other definitions, see for example Frigg (2008) and McCarty (2004). This paper gives an evolutionary systemic definition of models, in which models are seen as a world acting as a surrogate for any other world, such as physical or social systems and as such models emulate in themselves the behavior in the other worlds to aid understanding and decision-making. The salient features of the evolutionary features of mathematical modeling are captured in Figure 2 and described as follows:

- Section 3 describes Cognitive Model Thinking of Individuals
- Section 4 describes Communal Model Thinking
- Section 5 introduces institutionalization, a prerequisite to modeling practices
- Sections 6, 7 and 8 present major evolutionary transitions in modeling and modeling practices
- Section 9 *outlines possible cladistic structures in modeling complexity*
- Section 10 makes a research case for forming modeling science

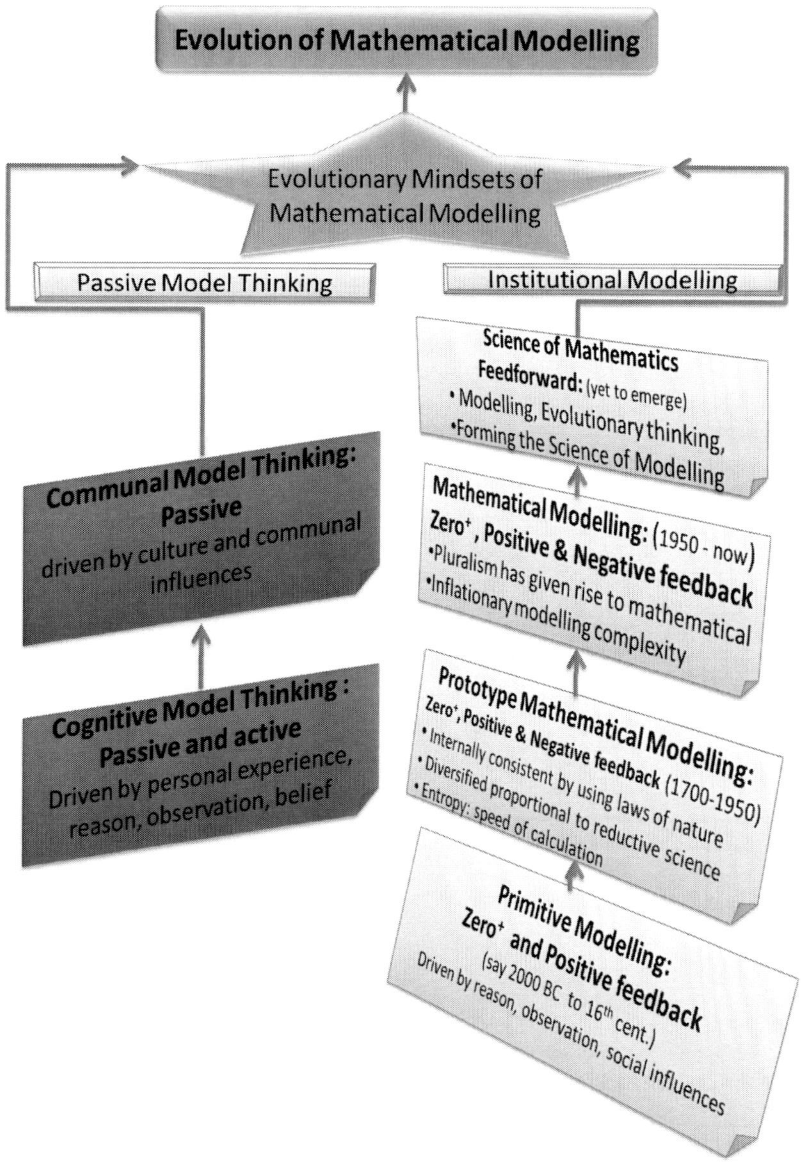

Figure 2. Evolutionary Systemic Model of Mathematical Modelling.

2. Philosophical Views on Mathematical Modeling

To the best knowledge of the author, the complexity of mathematical modeling, as a whole, is not the subject of any research as the developments are normally focused on best practice guidelines, e.g. EPA(2005), Aumann (2008), Beven (2002), Collier et al (2005), Todini el at (2005). The review below outlines studies that tend to take a bird's eye view of mathematical modeling to identify issues.

2.1. Philosophical Doctrines on Mathematical Modeling

Morton (1993) argues that "Mathematical modeling is one of the striking and characteristic intellectual developments of our age" and in order to press its importance, he issues a call to philosophers to take interest in a topic in which real philosophy of science really matters to questions of public policy. Similarly, there are a number of philosophical expositions on mathematical modeling but they are yet to translate into a thriving philosophy. The review here includes: (i) extrapolating Popper's views towards mathematical modeling, e.g. see Jarvie et al. (2002); (ii) the collection of responses to Nancy Cartwright (1983) making references to the use of mathematical models in science, as compiled in Hartmann et al. (2008); (iii) the realist position presented by Beven (2002); (iv) existentialist position expounded by Abbott (1991) and (v) other isolated papers. These are summarized below.

Popper's Verification and falsification in Models: Discussion on theoretical models goes back to Pierre Duhem (1906) referred to the fruitfulness of models for discovery, Rivadulla (2002). Whilst challenging assumptions was explicit in Galileo's thinking, the status quo in the philosophy of science was based on the verification doctrine and virtually "all philosophical systems in the Western tradition, from the time of Descartes until the twentieth century, have been justificationist ones" Diller (2002). The problem with justification is that once a theorem has been verified, the urge to justify it again is removed and thereby the theorem becomes an ontological (a timeless) entity. However, as science is an evolving entity, the doctrine of falsifiability introduced by Karl Popper (1902–1994) in 1963 is in reality the need for a perpetual challenge (negative feedback).

Falsifiability is based on the principle that a theorem can be falsified by evidence and it is true only if it survives falsification. Arguably, this is a welcomed development but it encourages the dogma of the universal truth and becomes an incentive for philosophers to search for the truth withstanding all falsifications. The question is then: has Popper's legacy of falsification and the search for truth been transferred to mathematical modeling? Popper has not developed a detailed account of theoretical models but Rivadulla (2002, Pp87-88) gives extract of some relevant passages from Popper as follows: (i), models consist of certain elements related to each other by some universal laws of interaction; (ii) models may be called theories, but not all theories are models; (iii) models cannot be completely true, since they are necessarily rough, schematic, and vast oversimplifications and approximation; (iv) models omit much, they overemphasize much, and they do not represent facts truly; (v) two competing models can be tested to decide which is better in the sense of better approximation to truth.

Nancy Cartwright's Empiricism and Modeling: In a series of essays, Nancy Cartwright (1983) argues that she holds a kind of anti-realism. Typically, it is an antirealism, which accepts the phenomenological and rejects the theoretical. She states that "But it is not theory versus observation that I reject. Rather it is the theoretical as opposed to the phenomenological." Hoeffer (2008) summarizes Cartwright's philosophy of science to be empiricism in the style of Otto Neurath (1882-1945) and John Stuart Mill (1806–1873) and concerned with the way actual science achieves metaphysical and epistemological understanding. She considers categories of laws as: fundamental physical laws, less-fundamental equations, high-level phenomenological laws, and causal laws. She grant the status of as literally true to only causal laws, but regards some high-level phenomenological laws in physics to be true only when conditions are right.

Bailer-Jones (2002) expresses that Cartwright's "writings on models are a bit of a patchwork in that they come to the topic from different angles and with different approaches, all ultimately belonging together, while it is not always easy to envisage the entire design in the process." Again, Bailer-Jones states that Cartwright's "view on models underwent some development over the years." Indeed, the 15 essays collected by Hartmann et al (2008) comprising 15 different critiques and Cartwright's responses are a testimony for the lack of a concrete discourse in the philosophy of mathematical modeling, so it is not easy to summarize Cartwright's works or these essays.

Beven's Realism in Modeling: Beven (2002) advocates realism in mathematical modeling without being keen in philosophy for the sake of it. He has a vision on a new philosophical approach to environmental modeling but this does not stem from any deep notions of environmentalism "nor from any higher level holistic principles of nature, nor directly from philosophical arguments about the concept(s) of realism itself, but from experience of practical applications of modeling in various areas of environmental science." He argues that "The philosophical subtleties are not really necessary to the practicing environmental modeler, who only needs to know that achieving realism is still difficult in the practical prediction of complex environmental systems. ...The approach combines elements of instrumentalism, relativism, Bayesianism and pragmatism, while allowing the realist stance that underlies much of the practice of environmental modelling as a fundamental aim." Some of the tenets of his vision include: (i) a formal environmental model can only be approximation to the perceptual model; (ii) places are unique and their uniqueness is unknowable; (iii) There is always the possibility of equifinality in modeling a system. The above vision seems pragmatic enough but they have significantly hard philosophy by using such terms as "unknowable", "perfect model" and recommending the use of simple models.

Abbott's Existentialism: Abbott (1991) captures the waves of changes in numerical hydraulics in terms of five generations of computer programming by then, to describe the historic developments. He coined the term hydroinformatics to mark the arrival of:

$$Fithth\ Generation\ Modeling = (Computational\ Hydraulic) \cup (Artificial\ Intelligence)$$

His view on modeling may be summarized: (i) the course of water (arteries and vein of the biosphere) is *coming-to-presence* (analogous to emergent properties) of man's innermost but collective thoughts as material objects in the outer world; (ii) there is a recurrent stress that "numbers have begun to function in another way," presumably meaning that numbers have triggered an evolutionary transition in data.; and (iii) numerical models as number allegories applied to water and ecology involving the social dimension in the first and third world countries. His vision (Abbott, 1991) includes: (i) a concern with the transition of mythos into logos and later into the doom of "dismal" science since the 17th century, minimizing the mythical, romantic, illogical, and arbitrary dimensions in the world (it seems that Abbott mourns the arrival of science!); (ii) a provision of escaping from nihilism by restoring myth and resurrecting or reifying the number myth; (iii) a restoration of dogmatic categories stating that "The problem really then comes down to this: that one cannot even begin to talk about the informational revolution and all that is entrained with it, without resorting to dogmatic categories, even though this approach is highly uncongenial to the usual ways of thinking of modern science;" (iv) an open ticket to religion to put science in order, where proponents of pure logic (people like Frege and Carnap) receive harsh treatments in quotation from Barth (1960, cited in Abbott).

Paradigm Shifts: Khatibi (2005) discusses modeling trio in terms of (i) models referring to mathematical equations and modelling practices, (ii) software tools and (iii) data management. He argues that each of the *modeling trio* is an evolving entity, driven by the four types of feedback loops, articulated by using the metaphor of paradigm shift going through the stages of forming (equivalent to zero$^+$ feedback), storming (positive feedback), norming (negative feedback) and performing (feedforward). The concept behind paradigm shifts in this paper is driven the feedback loops of systems science and as such, these are not the same of those given by Kuhn in terms of the Marxian revolution and Kantian constructivism. This metaphor also used to discuss evolutionary transitions in software development and data management, Khatibi (2003) and Khatibi et al (2003). The argument in these papers is that during modeling activities from the 1950s to 1990s, software developments were based on closed architecture, creating commercially-friendly cultures but islands of models and datasets.

2.2. Evolutionary Systemic Overview of Modeling Complexity

Evolutionary systemics (Khatibi, 2012a) can be applied to any complexity and this paper builds on the literature, ostensibly suggesting *model thinking* to deeply penetrate the timeline of humanity. Sebeok and Danesi (2000) argue that "One of the traits that distinguishes human beings from other species is an instinctive ability to make sophisticated, ingenious, resourceful *models*. Model-making typifies all aspects of human intellectual and social." This paper uses evolutionary systemics or Evolutionary Systemic Modeling (EvSyM) to understand the routemap of mathematical models throughout history. The preliminary form of EvSyM is presented by Khatibi (2011) and further elaborated by Khatibi (2012a). This is an analytical (building block) approach and its core idea is that evolution by natural selection is the architect of natural, social and cultural complexity and systemic feedback loops are their architecture, where their integration are by a set of preliminary *evolutionary systemic axioms*. This builds on the pluralism of individuals, community and institutions as contexts of evolution. This paper is built on Khatibi (2012a) and a full appreciation of its Sections 1-3 is required for this paper. The architecture for evolutionary transitions in institutions or systems is summarized in Table 1 and explained by the four types of feedback loops:

- The context (the phase) of modeling is defined by individual, communal and institutional approaches to complexity, where institutional practices undergo evolutionary transitions as follows.
- Zero$^+$ feedback loops: the process of "priming" of modeling as a whole or its particular brand, where the term priming is a key concept signifying the emergence.
- Positive feedback loops: describe the routemap of an evolving entity after being primed and selected, giving rise to selection by generation after generation, diversification, adaptation and entropy, where entropy continually reduces efficacy and triggers a selective advantage for an evolutionary transition to the next loop.

- Negative feedback loops: describe the routemap of an evolving entity by adding a controller subsystem to modeling complexity, after which inherent entropy can be reduced to maintain internal consistency, e.g. model calibration.
- Feedforward loops: describe a routemap for goal-orientation in modeling complexity, providing additional flexibility to reconfigure the system towards external goals.
- Modeling *practices* are outcomes of three components: (i) *cognitive model thinking* of individuals, (ii) *communal model thinking* providing impetus to thriving communal and cultural activities; (iii) *institutional modeling practices* for developing and using models as tools.
- The roles of entropy, pluralism of the loops and cognitive model thinking, communal model thinking and institutional evolutionary transitions cannot be overemphasized. Equally, each of the above loops refers to the three components of d-loops, ℓ-loops and e-loops.

2.3. Historiography Relevant to Modeling

Historiography of the complexity of mathematical modeling is not reviewed in this paper but the related data are used extensively as evidence to substantiate the application of evolutionary systemic axioms to modeling complexity. It is therefore important to have an overview of historic stages in the development of man's culture. The cultures of Homo sapiens since the Upper Paleolithic Age are undergoing evolutionary transitions. Section 3 reviews the literature on instinctive and cognitive model thinking, according to which the reviewed literature is suggestive of the instinctive nature of model thinking. The origin of language is ostensibly attributable to instinctive model thinking during the Middle Paleolithic Age (circa 200,000 to 50,000 years ago). Languages with simple grammar are believed to have emerged in the Upper Paleolithic Age, which provided impetus for cultures and further model thinking among the community, the salient features of which are shown in Figure 3, as follows:

1. Individuals develop their cognitive model thinking with respect to their communal cultures, which is a passive mindset and is not expected to undergo any evolutionary transitions but co-evolution takes place, as discussed below.
2. Communal mindsets may prime institutionalized mathematical modeling but this is an active process and goes through evolutionary transitions of zero$^+$, positive and negative feedback and feedforward loops, where each e-loop acts as a mindset behind the d-loop dynamic activities.
3. Evolutionary systemic modeling provides a scientific explanation for evolutionary transitions in mathematical modeling, according to which (i) communal mindset was primed in prehistory and coevolves with institutional mindsets since history began; (ii) the first institutional mindset was *primitive modeling* primed during the early civilizations and lasted until the 17th century; (iii) primitive modeling was largely unselected by being replaced with *prototype mathematical modeling* from 17th century until 1950; (iv) the advent of computers gave rise to computational modeling, as a new lease of life for modeling.

4. As the focus of this paper is to present a vision on mathematical modeling as a whole, only the computational modeling practices are detailed in this paper.

2.4. Summary of Philosophical Treatments

It is possible to be immersed into the philosophical doctrines of mathematical modeling and end up with advocating or developing a doctrine. As soon as one develops a critical view of these doctrines, as the author has done, their most striking feature is the contentions among philosophers, stemming from the presuppositions within the philosophical doctrines, where the contentions have normally nothing to do with the subject of enquiry. The most striking shortfall of philosophical discourses on mathematical modeling is the poverty of discourses with no trace of practical issues such as the role of feedback loops, defensibility and pluralism practiced in modeling. Philosophers are often motivated by the search for a universal truth, instigating their doctrines: to be thoroughly removed from relevance, to retard science, or to be trapped in circular arguments. For instance, constructivism seems reasonably sound but the methodology for pluralism is often left vague.

Whilst the author is critical of philosophical enquiries, but is also stimulated by their enquiries and debates and as such the author also regards philosophy, as a way of setting goals. Indeed, the 20th century is the battleground of the application of philosophical doctrines to science, as a whole, and to its various disciplines. By its nature, philosophy in this outlook is prescriptive or top-down. If goals of modeling practices are set by philosophers, modeling will suffer badly by responding to many conflicting goals, e.g. the disastrous objectivism of Ayn Rand (1905–1982). It is not surprising that modelers follow their own senses without resorting to philosophy but as such modelers have not done a good job either, as this paper shows that they are oblivious to a new kind of entropy being instigated.

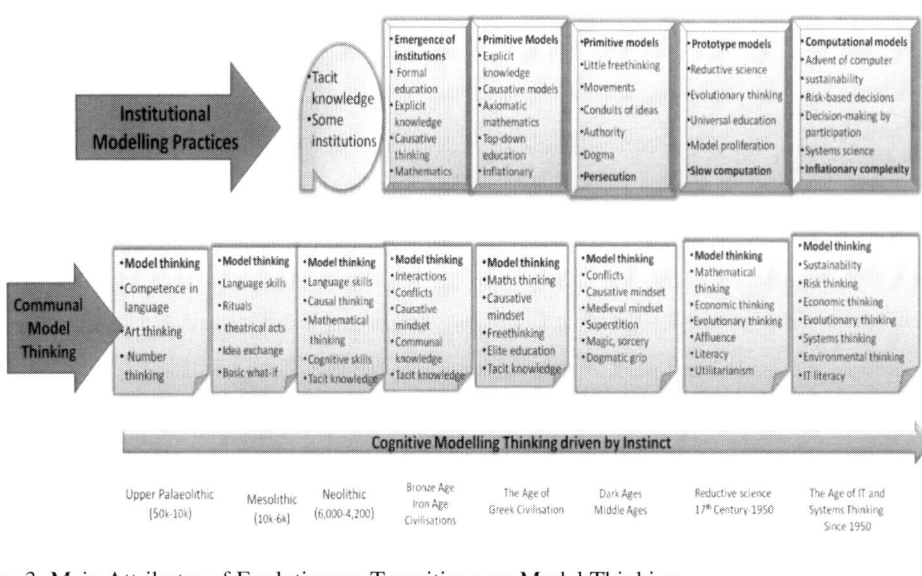

Figure 3. Main Attributes of Evolutionary Transitions on Model Thinking.

3. COGNITIVE MODEL THINKING OF INDIVIDUALS

Sebeok and Danesi (2000, Page 1) argue that human beings have a *modeling instinct* to produce *forms* to stand for objects, events, feelings, actions, situations, and ideas perceived to have some meaning, purpose, or useful function. The form can be *mental images* of externalized things called *representation* and Figure 4 depicts four types: (i) *signs* (words, gesture, etc.), *text* (stories, theories, etc.), (iii) *codes* (language, music, etc.), and (iv) *figural assemblage* (metaphors, metonyms, etc.).

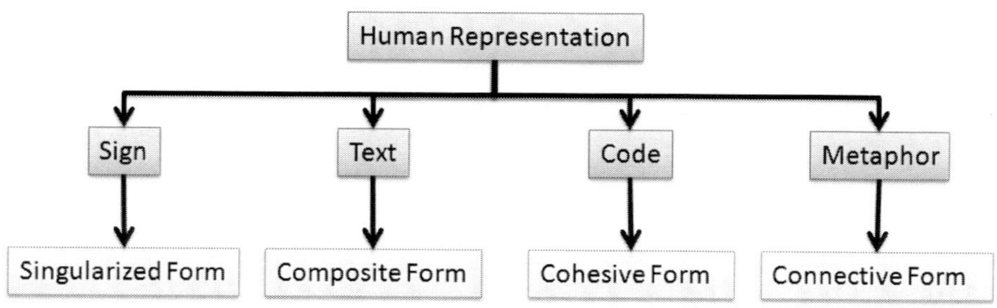

Figure 4. Linguistic Notion of Human Representation.

Instinctive model-making can ostensibly account for the origin of mathematical thinking and mathematical modeling. The evidence is provided by cognitive scientists through semiotic and abductive modeling of human beings and other animal species. Semiotics is the capacity for meaning-making and representation in many forms, e.g. text or signs, Chandler (2006, Pp 2-3). The term abductive inference is introduced by Charles Sanders Peirce (1839–1914) as follows: "When a chicken first emerges from the shell, it does not try fifty random ways of appeasing its hunger, but within five minutes is picking up food, choosing as it picks, and picking what it aims to pick. That is not reasoning, because it is not done deliberately; but" abductive inference, (cited in Magnani, 2007).

Magnani (2007) argues that many animals which are traditionally considered "mindless," in fact, make up a series of signs and are engaged in making, manifesting or reacting to a series of signs, which are fundamentally *model-based.* Their continuous process of "hypothesis generation" is (i) instinctual behavior, i.e. a kind of "wired" cognition, and (ii) representation-oriented behavior, i.e. non-linguistic pseudo-thoughts. This activity is at the root of a variety of *abductive* reasoning and is the process of inferring an explanation based on available information. The process of inference in this respect does not need to be more sophisticated than guessing. The author's interpretation of the literature on the subject in terms of feedback loops is that hardwired representation cognition is the active process of zero$^+$ feedback loop, where abductive reasoning means guessing the cause out for a given effect.

It is often thought that language is a measure of humans but what did make language in the first place? Magnani (2007) quotes Mithen (1996), Donald (2001) and Berm´udez (2002) and based on cognitive archaeology, he states that: it was not language that made cognition possible, rather cognition was rendered possible by the integration in social environments of pre-existent, separated, domain-specific modules in pre-linguistic hominids, e.g. complex motor skills learnt by imitation or created independently for the first time.

Based on the above, model thinking driven by cognition is ostensibly instinctive and manifests within individuals. Sebeok and Danesi (2000) remark that "Model-making constitutes a truly astonishing evolutionary attainment, without which it would be virtually impossible for modern man to carry out their daily life routines." Arguably, every individual is in possession of *cognitive model thinking*, which is a driver for cognition, language and culture. This mindset is captured in Figure 2. The language and culture elements ostensibly took off during the Upper Paleolithic Age (50,000 years ago). Arguably, the ability to "think" about a theme (say modeling, mathematics, risk, systems, the environment, economic, etc) does not require professional training but this is related to the cognitive capacity of individual human beings engaged in the process of abstraction to experiment with both causes and effects. As evolutionary systemic modeling stipulates four sequentially evolving feedback loops, the question is: what are the implications of feedback loops on cognitive model thinking and vice versa? The author's thoughts are that individuals do not undergo evolutionary transitions but the more they use their cognitive faculties (and nowadays their education and training), the greater is their development going through stages:

- Childhood, (zero$^+$ feedback loop), during which the child has a long journey of learning to connect causes with effects, ostensibly through their *instinctual* abductive reasoning capacity.
- Young ages (positive feedback loops), during which there is (i) an escalating process of cognition and learning of as many cause-and-effect instances as possible and (ii) cognition and learning of inherent discrepancies associated with cause-and-effect processes
- Middle ages (negative feedback), during which individuals mature by finding ways to get round discrepancies to create a steady pace of life through cognition, learning and trial-and-error
- Old ages (feedforward), during which one may develop foresight of life
- The author's knowledge of neuroscience is limited but can heuristically suggest the above feedback loops ought to be connected with appropriate connections in the neural system of the brain. Therefore, the instinctive stimulations are drivers following the birth for performing the most basic operations but then increasingly this capacity would be taken over by more sophisticated network of connection through the hand-in-hand working of instinctive stimulations and external cognitions, although the author is unable to specify them.
- In above heuristic account, d-loops cater for real-time activities, ℓ-loops for adaptations, and e-loops for the personal cognitive capacity.

The author's observation is that the above steps are quite distinct in everyone's life journey but there is no inevitability for their acquisition, as some individuals never go through all those steps; some are quick learners; and some are only good on certain matters by going through those steps but not other. The author is not quite sure whether these are internalized or are driven by external cues with internal mechanism. Ever since model thinking helped by language, new cognitive modeling activities have been priming including mathematical modeling, mathematics and risk. Cognitive model thinking is also engaged pluralistically with communal model thinking and, influences institutional modeling practices and is reshaped by them, which opens up new possibilities.

4. Communal Model Thinking

Community is the backdrop for everything related to the way of life since prehistory and is formed by cohesion of individuals. The author heuristically argues the following. (i) Changes in a community drive the collective mindset, which are slow and long-term, although strong autocratic or devoted individuals can mobilize communal mindsets. (ii) Communal mindsets are either passive or active and arguably, they do not undergo evolutionary transitions but co-evolve with institutional mindsets or become unstable. When communal mindsets are constrained, the cohesion between the individuals is transformed into resisting forces. (iii) Complex patterns of communal behavior may be studies by analogy with the concept of the Rankine's passive and active soil pressures but under normal condition, communal mindsets are most healthy when diversity is encouraged. Communal model thinking was not planned but emerged spontaneously in prehistory, as the culture was increasingly rife for new ways of doing things. The contextual roles of communal model thinking is outlined below, covering the time from the Upper Paleolithic Age until modern time and are captured in Figures 2-3. The actual dates are inevitable approximate.

The Upper Paleolithic Age (circa 50,000 – 10,000 years ago): Communal model thinking strengthened languages with simple grammars. Armstrong (2005) recounts one version of the building blocks of the cultural life for the Upper Paleolithic Age to be dominated by death, flux, the endless succession of events, and the cycle of the seasons, complemented by dreamtimes inhabited by the ancestors, archetypal beings, as well as by visions. One can almost detect the role of model thinking and linguistics in all these building blocks.

The Mesolithic period (circa 10,000 – 6,000 years ago): Farming emerged in Mesopotamia and the cultural life evolved away from archetypal beings of the past and embraced a holistic reality by having Gods, human beings, animals and plants sharing the same nature and invigorating each other. Ostensibly, this period also primed numbers giving rise to mathematical thinking. Armstrong (2005) states that during this age there was a realization that: *one should not expect to get something for nothing*. Although this is a deduction and possibly her deduction, it is an interesting way of transforming tacit knowledge inherent in the Mesolithic mindset to explicit knowledge.

The Neolithic period (6,000–4,200 years ago) gave rise to early civilizations, during which the people became more aware of the chain of cause-and-effect processes (causative mindset) with cultures springing up in the Fertile Crescent, Mesopotamia, Egypt, China. This period primed writing and thereby early written versions of history, causing mythos to retreat but not to its entire demise.

Greek civilization (circa 800BC–200AD) the communal model thinking primed philosophy and institutional modeling involving formal education during which freethinking flourished. Similar processes took place in China and India.

Middle Age–Modern time: freethinking fell victim under the stagnating grip of dogma but this triggered the selective advantage for the "right to challenge," as the essence of science or negative feedback – see Khatibi (2012a). This emerged in Italy in the 16th century and then was selected in Europe initially and throughout the world since the 20th century. The outcome is a more complex communal mindset coevolving with science. The attributes above are displayed in Figure 3 and can be considered as positive feedback ℓ-loops and e-loops.

5. INSTITUTIONALIZATION IN MODELING PRACTICES

Communal model thinking served as roots for countless areas of complexity in modern times and one such area is mathematics-based model thinking in institutions, as outlined below.

5.1. Priming of Institutionalization

The communal mindset and family structure are conduits to pass on some of the skills but passing on more specialist skills would have been haphazard by relying on family and community resources. Thus, the priming of institutions would have served the purpose of pooling together the required talent base from the community, e.g. formal education. Not surprisingly, methods of formal education up to modern times were often top-down and Guisepi (http://history-world.org/mesopotamia_a_place_to_start.htm) opens a window to the Sumerian in the following terms: "Formal education was practical and aimed to train scribes and priests. It was extended from basic reading, writing, and religion to higher learning in law, medicine, and astrology. Generally, youth of the upper classes were prepared to become scribes, who ranged from copyists to librarians and teachers. The schools for priests were said to be as numerous as temples."

A further example on formal education is provided by Marrou (1956) giving an insight to formal education in ancient Greece but the situation is not much different from the above. Generally, up to the 19th century, only the elite and some talented ones received education to become scribes and priests and by then there was no universal education. This was proportional to applying knowledge to practical problems then, as life was largely organized by cognitive knowledge and artisanship. The top-down nature of education then could be one of the factors to explain the slow pace of developments in mathematical modeling and other aspects of knowledge until the emergence of reductive science.

5.2 Building Blocks for Institutionalizing Mathematical Modeling

From the Mesolithic Age until the emergence of reductive science in the 17th century, the cognitive model thinking was increasingly within a rich culture and rife for priming new entities including: (i) the settled way of life, priming the focus on individuals, property and private ownership; (ii) urban way of life (in Sumerian Mesopotamia, Egypt and China in the Neolithic Age), primed the focus on understanding relationships beyond instinct and cognitive learning; (iii) primed the causative mindset to understand cause-and-effect relationships due to increased interactions, realignment and conflicts among different societies; (iv) primed a diversity of new activities creating dynamism for doing things differently by the emerging rich cultures. The attributes above are displayed in Figure 3 and can be considered as positive feedback ℓ-loops and e-loops. The emerging institutional mindsets primed new means of communication to preserve knowledge including: abductive reasoning, induction, deduction, analogy and metaphor, allegory, mathematical thinking, spatial thinking, homology, posit, hypothesis conjecture, theory/theorem/models. These terms and concepts have each their own history and definitions and serve as communication "currencies."

6. Primitive Modeling Activities

Figures 2–3 capture main features of *primitive* modeling, as outlined below.

6.1. Zero$^+$ Feedback Stage of Primitive Modeling

The Sumerian culture, going back more than 6,000 years, is the first one to leave behind written records on experimenting with different number systems. From Sumerian civilization until the emergence of Greek civilization, there were thriving cultural activities in many parts of the world, which led to the priming of mathematical thinking (see Khatibi, 2012b), itself through model thinking. Mathematical thinking primed a number of spontaneous outcomes including a counting system, observing the movement of heavenly bodies and calendars. The Sumerian calendar divisions have been transmitted into the global cultural life, who devised a system by dividing the year to 12 months, 4 seasons of 3 months, and each month to 4 weeks of 7 days; notably the religious significance of these divisions were unselected in time. The institutional modeling activities from the Sumerian time until the 17th century may be referred to as *primitive* modeling activities, which were largely unselected when challenged by the reductive science from the 17th century onwards.

6.2. Positive Feedback – Emergence of Philosophy

The propagation of ideas found new conduits by interactions and conflicts, or were constrained by them at a different scale during the periods through Bronze Age (5300–3200 years ago), Iron Age (3200-2800 years ago) the rise of the Greek civilization circa 800 BC and the emergence of ancient civilizations. The net gain was the emergence of the causative mindset and institutionalization (see Section 5), which culminated in the priming of a hereditary modeling activities and philosophy (positive feedback e-loop) driven by natural selection. Institutions became rather widespread as they are understood now. However, the institutions were for the elites but not for the universal participation; any change then was very slow but significant; and net gains were absorbed by the communal mindset over time (positive feedback ℓ-loops). By this time, the Sumerian, Babylonian and Egyptian cultures had lost their momentum; both Chinese and Indian cultures also primed comparable net gains in cultural activities through model thinking but the emergence of philosophy was their culmination.

6.3. Dark Ages and Middle Age

The significance of developments in applied mathematics during the Dark Ages (2nd-5th century) and the Middle Ages (5th-15th century) were proportional to the intellectual capacity of the time, which was retarded by the sweeping goal-fixation of the emerging dogmas instigated by religious dogmas, authoritarian rules of the time and the Aristotelian ontological mindset in the Christian and later in Islamic cultures. The outcome of this entropy was the collapse of the Greek philosophy and the stagnation of intellectual enquiries by the restrictions of Christian dogmas and the rise of rationalism in the Islamic cultures in the 8th century and its demise by the 14th century.

An example of using model thinking to enforce dogma is the debate between Abu Hanifa (699–767), a Muslim Imam, and an Atheist, although it is questionable if the Atheist would have dared to enter to an open debate in those times. The salient point of the debate is that in an exercise of reverse counting, the atheist agrees that there is nothing prior to *one* and Abu Hanifa exploits this as a proof that there is nothing before the "almighty." This debate shows: (i) the use (or the model) of mathematics in the Middle Ages in an analogous terms to non-mathematical entities; (ii) the debate was an analogy, because it expressed a similarity between two entities with no correspondence in their causes or in their governing laws; and (iii) actually there is 0.9999 prior to 1, or likewise there is zero prior to 1 and -1 prior to zero and this keeps going until $-\infty$; it may be noted that during the lifetime of Abu Hanifa the concept of negative numbers were not known but this was introduced by Alkhwarazmi (780-850) to the Islamic cultures, with a history that goes back to Chinese scholars more than 2,000 years ago.

6.4. Evolutionary Transitions

Prior to the 16th-17th century, mathematical modeling practices were proportional to the way of life then. It was barely and evolvable practice but its positive feedback e-loops survived with the following salient features:

- Primed entities selected, proliferated and went through evolutionary transitions: primitive model thinking did not trigger any significant area of institutional modeling activities (d-loops) prior to the emergence of reductive science in the 17th century. Those which emerged can be counted by fingers and include geometry, trigonometry, arithmetic, algebra and astronomy. Each of these endeavors flourished among the Sumerians, Babylonians, Egyptians, Chinese, Indians but geometry was formalized by the Greeks with contribution from the thinkers of the Islamic cultures during the Middle Ages and algebra was put on a firm base by the thinkers of the Islamic cultures. Each of these disciplines was institutionalized by these various cultures.
- Primed entities selected, proliferated but unselected in time: The geocentric model was institutionalized by observing the planets, which were going on for a long time before the Greek civilization. These observations were recorded particularly during the Islamic culture and gave rise to the first set of collected data, which were the very first seeds of mathematical modeling. However, mathematical modeling can reflect some of the inconsistencies in its fact engine but normally it does not have a self-

correction capability, although some genetic algorithm has such a capability. As the fundamental assumption of the geocentric hypothesis was never challenged, the anomalies shown by the data were subjected to increasingly sophisticated models overlooking the principle of parsimony. Evolutionary prototypes in nature are often parsimonious, implying the selective advantage of an approach with a minimum number of assumptions. Should parsimonious solutions give rise to complex models, this is probably a clue for inherent inconsistencies.

- Entropy Inherent in Primitive Modeling**:** The problem with the primitive models was that they all were based on commonsense, were observed as cause-and-effects, or were reasoned by logical arguments to be ontological entities. These models were spearheaded by the authorities of the time and any challenge to them was the challenge against the authority, which would have been very costly to the challenging thinker. So, entropy instigated by such ontological mindsets was kept under control by coercion and this stagnated intellectual domains and knowledge bases until the 17th century in Europe.

7. Prototype Mathematical Modeling (1600-1950)

Figures 2–3 capture features of *prototype* modeling activities, outlined below.

7.1. Priming Prototype Mathematical Modeling – Zero$^+$ Feedback

The fixating grip imposed by Christianity in the Middle Ages had triggered a selective advantage for alternative mindsets, particularly in Italy during the Middle Age times. The 15th century movements of humanism in Italy followed by some other Christian countries provided a rich cognitive/communal basis to challenge the fixation and then this spread to philosophy, literature and art. The impact was the emergence of a conduit to transfer ideas and this evolved by making way to Classicism in the 16th century and Romanticism in the 18th century. The subtle "right to challenge" primed by the mindset of Galileo Galilei (1564–1642) and others in Europe. Very similar conditions emerged strikingly in the Islamic cultures but the right to challenge was not primed. Notably, the priming process is spontaneous and even the subsequent selection is not guaranteed, subtleties that are captured by evolutionary systemics.

The humanist movement questioned dogmatic authorities by challenging the received ideas by "priming" the following main elements in the enquiries of natural philosophy (later referred to as reductive science) and these were: (i) reductionism: reducing the whole into parts, (ii) negative feedback: challenging the assumptions by negative feedback loops; (iii) data: data measurement and not relying on pure commonsense alone, and (iv) some sort of flexibility to reduce discrepancies, for more details see Khatibi (2012a). These are the foundation of the *scientific method.* The emergence of reductive science virtually "unselected" every model legacies since the Greek philosophy other than what was available through technology and mathematics.

The term *unselect* is pivotal, as it signifies that the causative mindsets are vulnerable to fictions of common sense, deception, misconstrue or misgivings by individuals. Reductive science became firmly established in the 17th–18th centuries.

7.2. Positive Feedback of Prototype Mathematical Modeling

The time span of 1600-1950 was a period of proliferation of applied mathematics with a flurry of applications to diverse fields of sciences. Reductive science and technology worked hand-in-hand and these came to coexist symbiotically with mathematics and applied mathematics, thanks to the impetus by the Galileo's optimism on the power of mathematics. It is attributed to Galileo e.g. Wigner (1991, P532) stating that "the laws of nature are written in the language of mathematics." Owing to the conduit of ideas created by humanist and classicist movements, the primed building blocks of reductive science were selected by newly emerging scholars of the 16th–18th centuries, who were often elites, often lived with the patronage of royal courts to add glory to their status.

The capability of applied mathematics during 1600-1950 was wide and made up of a diversity of mathematical equations derived by sound laws of nature or by well established empirical principles. The emerging capabilities penetrated gradually and gave rise to mathematical modeling as below:

i. Cohen and Stewart (1994) argue that "In other areas of science, especially those where really accurate measurements or repeatable experiments aren't possible, people nowadays tend to speak of 'models' rather than 'laws'.
ii. Stewart, 1990) argues that "The label 'analysis' is used today to describe calculus in its more rigorous form: the theory behind it rather than the computational technique. It acquired that connotation during the 18th century, when the theoretical side of calculus was being substantially extended." He adds that "The 18th century's main achievements were in setting up equations to model physical phenomena. It had much less success solving those equations".

The activities prior to 1950 did not warrant calling them as modeling but they were often formalized calculation procedures, analysis or applied mathematics. Since the emergence of reductive science, the number of practical problems receiving mathematical treatment underwent a phase of positive feedback d-loops and subsequent e-loops. The author's heuristic overview of the relevant activities is given here until the emergence of computation modeling in the 1950s and the following stages are recognized:

1. Rigorous explanations of natural phenomena at their classic macroscopic or microscopic scales were the main preoccupation of reductive science, e.g. physical science: the movements of heavenly bodies, gravity, light, heat, waves; chemical science: the periodic table, atomic structure, chemical reactions; and their working together: thermodynamics. There were very little data available then and enquiries provided an impetus to obtain data.
2. Deriving equations for practical problems was a customary preoccupation until the middle of the 20th century and most of the equations were derived even by the turn of

the 19th century. The activities in this period included the development of solutions by analytical, simplified, graphic or tabular techniques, which were often prestigious scientific endeavors (this stream of activities is still thriving in the form of developing *numerical modeling*).

3. The emergence of the *prototype modeling*, which took off from the 19th century to the middle of the 20th century to meet the needs of design works.

7.3. Institutionalization

Although some of the institutions of science (such as academies and royal societies) emerged in the 18th century, professional practices thrived by cognitively-driven traditional procedures, during which not much use was made of reductive science. In this period, reductive science and model-building were quite indistinguishable and confined to classical problems, e.g. modeling (describing) the celestial movements of the planet. However, science-driven professional procedures largely emerged in the 19th century owing to civil engineering, mechanical engineering and later on chemical and electrical engineering (20th century). The increasingly thriving culture of design practices then worked hand-in-hand with reductive science by developing working tools through deriving problem-oriented equations. This gave rise to a new mindset that the author refers to it as the *prototype modeling*. By the end of the 19th century, the various institutions accommodated communities of researchers, teachers, practitioners, consultants, contractors and policymakers and various authorities.

From the 19th century, reductive science overhauled cognitively learned artisanship to professional activities in terms of project works, many of which would have been handled on a piecemeal basis and were carried out by combining both traditional and scientific ways of design. However, owing to "objectivity" as the selective advantage of scientific design, there was an increasing tendency to use scientific principles.

7.4. Negative Feedback Loops

The key characteristic of prototype modeling was an implicit or explicit adoption of negative feedback through calibration. This is discussed further in Section 8.4 but its roots are traced here. For a mathematical modeler now, the issue of calibration and validation seems ubiquitous (timeless and space-less entities) but there is no such a thing as ubiquitous, as every habit of a practice has a source, selected at one time for a selective advantage and can be developed further or unselected upon losing the advantage. The author was therefore searching for the emergence of the concept of calibration within this culture of enquiry but yet to cite historic accounts to directly trace the emergence of this concept. An indirect understanding of the emergence of this concept is as follows.

The author's search for early examples of calibration is still in vain. One account, which seems plausible Hodgson (2008), attributes the origin of the term calibration to weaponry, where the size of ammunition fired by a gun is held to be directly related to the inside diameter of the barrel and both are called the *caliber*. When the idea of measuring an instrument for accuracy was first expressed as *calibration*, only about 150 years ago, the instrument in question was a thermometer. If the little glass tube that held the mercury or alcohol was not consistent in its inside diameter, the temperature gradients would not be regularly spaced. So a thermometer was measured in the same way that a gun barrel was. Note that callipers are used to establish the accuracy of something by referring to its standard. Both *caliber* and *calliper* are thought to be etymologically related and go back more than 400 years in English and come from French before that.

Hydraulics is one of the early branches of physical sciences, which can reflect early examples on incorporating empirical parameters and hence marking the need for early conceptions of calibration. The account given by Rouse and Ince (1963) on the emergence of the work by Antoine Chézy (1718-1798) on resistance to open channel flows is revealing and reproduced here as it also reflects on the emerging professional culture. "The circumstances which led to Chezy's resistance studies were the following. In 1760, the Acedémie des Sciences had been consulted by the Paris administration about the poor operation of its pumps and the resulting insufficiency of its water supply, and a committee was formed to study the problem. ... the administration decided that Perronet and Chezy should carry out the project. It was Chezy's task to determine the cross section of the canal and calculate the discharge. Since he could find nothing in the literature on this subject, he undertook his own investigation and submitted his recommendations to Perronet in the form of a report. This report was lost, but the files of the Ponts et Chaussées still contain his original manuscript."

Although Chezy derived the relationship for a particular river, it did not contain an empirical coefficient, now known as Chezy *C*. Rouse and Ince (1963) argue that Chezy appreciated that the formula should change from one river to another. This is significant, as prior to this, theoretical hydraulics had made significant advances by the emergence of the continuity, Bernoulli and Euler equations but applicable to idealized conditions. Leliavsky (1965) traces this subject from Galileo until the emergence of the modern concepts, reflecting on the up's and down's (or rather boom-and-busts) in the development of the modern concept of open channel friction parameter including a number of fallacies. The fact is that they were all resolved into a number of pluralistic solutions, a notion that is the second nature of science and mathematics, signifying the existence of a hereditary system, to be referred to as positive feedback e-loops.

7.5. Entropy

It was merely impractical for prototype modelers to obtain system-wide solutions, as their solutions were tedious. The strategies employed were: (i) devising a minimum analysis procedures by dividing a system into components and studying the most important components one-at-a-time; (ii) focusing on a design procedure for technical feasibilities and cost-effectiveness; (iii) ignoring impacts of real systems on the environment; and (iv) employing human calculators devoted to routine computations.

However, these solutions contained inherent entropy in the long-term, which included: (i) human calculators were not satisfactory solutions to solving the tedious equations; (ii) it was not clear how to account for discrepancies; (iii) a full cycle of manual calculation had to be repeated whenever any data had to change; (iv) gaining an insight into the system was not feasible due to the lack of a capability to carry out system-wide analysis and as such there was no perception of the impacts of the emerging design procedures on the environment.

These collectively acted as entropy and reduced the efficacy of prototype modeling by bring it to a stagnation point.

7.6. Overview

The prototype modeling activities may be captured by EvSyM as follows:

$$\boldsymbol{Prototype\ Modelling} = \boldsymbol{Positive\ Feedback}\left(\boldsymbol{d_{loop}} + \boldsymbol{\ell_{loop}} + \boldsymbol{e_{loop}}\right) + \boldsymbol{Negative\ Feedback}\left(\boldsymbol{d_{loop}} + \boldsymbol{\ell_{loop}} + \boldsymbol{e\_loop}\right)$$

8. Computational Modeling – Since 1950

Tedious manual computations acting as entropy triggered the evolutionary transition in the 1950s to exploit the emerging niche market by the advent of computers with no rethinking.

8.1. Emergence of Modeling Practices – Zero$^+$ Feedback Loop

There was no modeling in its modern sense prior to 1950. Chapman (1977) observes modeling in 1977 that: “In my days as an undergraduate the new paradigm of the model had just been launched: the take-off was sufficiently fascinating that the basic question of where the trajectory pointed was for the time forgotten.” Nonetheless, computational modeling built on earlier prototype modeling practices.

The evolutionary transition towards computational modeling was triggered by the advent of computers in the 1950s even though the initial computer programs were fragile and imposed the legacy of studying systems for one function, as multifunctional studies were not feasible due to software restrictions, see Khatibi (2002). In spite of their fragility, the emerging computational programs offered selective advantages: (i) models could be re-simulated, (ii) computational speed was radically different than that of manual calculations; (iii) system-wide models proved successful in providing an insight into the underlying processes and identifying the synergy among the components; (iv) it became possible to pool together the wealth of knowledge capital accumulated by reductive science in the 19th century up to 1950 and to transform them into working tools called models; (v) these tools had a potential to study blind spots, inter-component and inter-system synergies.

At the transition times, not much was unselected from the legacy of prototype modeling, as their design procedures were already scientific. However, in this transition, some of the past practices became obsolete, e.g. tabular and graphic solutions to some of the partial differential hydraulic equations. The fundamental rationale of the transition was to reinvent the past thinking in a new framework, without most of the past barriers, as in computer programs and later in software systems. The EvSyM description of the priming of computation modeling is:

$$\textbf{\textit{Computational Modelling}}$$
$$= \textbf{\textit{Primed due to the selective advantage of computational speed}}$$

8.2. Proliferation – Positive Feedback

The evolutionary transition in the 1950s was not a rethinking but an exploitation of a niche market. The proliferation of models since 1950 has been in pace with the rest of modern entities and like any newly emerging entity. The selective advantages gained impetus initially through various niche markets and then opportunistically populated virtually the landscape of science, *leading to the emergence of numerous islands of models*. The growth of modeling is organic and tedious to list its fields of modeling but consider one example from hydraulics:

- Modeling areas: modeling flows (low, operational, floods) groundwater, water quality, reticulation systems, erosion/sediment, water resources, draught, ecology, habitat; non-Newtonian liquids; waterhammer analysis.
- Modeling problems: in flood forecasting alone this includes (**i**) forecasting rainfall; (**ii**) forecasting flows in upper catchments, rivers, groundwater, canals and through pipes; (**iii**) forecasting soil moisture; and (**iv**) forecasting open sea water levels.

The initial computer programs in the 1950s and 1960s were one-off activities often implemented in research institutes and driven by opportunities created by computer programs. Initially, the data were embedded in computer codes and used sparingly. Implementations of computer programs required dedication for being user-hostile and requiring skills to deal with numerous runtime problems. When they reached professional environments in the 1970s, they were used sparingly. From 1980 onwards, the emerging mindset of software development had the ethos of proving user friendly facilities, so that the modelers largely guided by their intuition in preparing datasets, running the models and processing the results. Thus, legacy codes were incorporated into software applications and by 2000. The EvSyM description of the priming of computation modeling is:

$$\textbf{\textit{Computational Modelling}} = \textbf{\textit{Positive Feedback}}\left(\boldsymbol{d_{loop}} + \boldsymbol{\ell_{loop}} + \boldsymbol{e_{loop}}\right)$$

8.3. Directions of Information Flow

If information cannot flow in more than one direction, negative feedback is not feasible. The term information is used here in its generic sense and there are various signals propagated both in space and time. There are numerous examples from open channel hydraulics include: hydraulic information on river flows or the sea state; navigation information on minimum water level; control information for implementing control algorithms; environmental information on, say, constraining water quality parameters to within prescribed limits. A system can be operated by complying with any of these information settings. Of course, the more the directions of information flow, the more complex is the operation but increasingly the trend is to accommodate all these constraints. The strength of modeling is that, a model creates a ubiquitous environment where information can flow virtually in any direction. The only problem is to overcome barriers for their flows in software implementation and to find algorithms to cater for decision-making information.

8.4. Negative Feedback loops

The modeling community uses the term calibration or more generally "referencing," to refer to the concept of negative feedback loops. Without any form of referencing, a model is but a hypothesis at its best and a guess at its worst. One example is the hydrological routing technique, known as the Muskingum method Henderson (1966), requiring the estimation of three parameters before their applications (e.g. Chow, 1981, Henderson, 1966). Whilst the living organisms rely on trial-and-error in implementing negative feedback loops driven by fuzzy logic, calibration often employs visually-driven trial-and-error procedures but various algorithms are available, e.g. optimization or fuzzy logic.

The importance of negative feedback loops in mathematical modeling is due to underpinning the internal consistency, objectivity and defensibility of mathematical models. However, this role is subtle and has been badly overlooked. The key requirements for these loops stem from at least two major reasons: (i) to ensure that modeling assumptions are complied with, where they are often inherited from the theories or empirical rules of reductive science; and (ii) to identify the values of the parameters that are incorporated into the underlying mathematical formulations. Without the focus on the assumptions and parameter identification problems, mathematical models are hypothesis or systematized guesses but not scientific tolls. Thus, focusing on such essential requirements, mathematical modeling involves negative feedback loops, one way or another.

The signature of implementation of negative feedback is the capability for information to flow in more than one direction, as discussed in Section 8.3. This is necessary to estimate the values of the parameters embedded in the equations to close the gap between model performance and measured systems. According to the cybernetic/systems concept of negative feedback, a subsystem has to be assembled by a minimum of four control components. Three types of implementations are compared in Table 2 and illustrated in Figure 5. The implementation types consist of: (a) as in cybernetics/ systems science, (b) as in "scientific methodology, which is the first known explicit negative feedback loop, making up the essence of reductive science and (c) as in mathematical modeling through the process of calibration, although there are other procedures too.

Table 2. Comparing Three Implementations of Negative Feedback Loops

	Systems Science	Scientific Methodology	Model Calibration
Transmitter	• measure output data • obtain target data • transmit data to fact engine	• no formal transmitter • obtain observed data • obtain predicted data • pass data to fact engine	• no formal transmitter • obtain observed data • obtained modeled data • pass data to fact engine
Fact Engine	• Can be a PID algorithm, • Other algorithms ok • Decide on correction	• an objective comparison • a visual comparison • Decide on correction	• a visual comparison • use an optimization method • Decide on correction
Actuator	• Passes decision to real system	• Passes decision to real system	• Passes decision to real system
Flexible Interface	• System capable of modulation & sensitive to correction	• Modify assumptions and try again	• Reset system parameters

Fact engines are the core in implementing negative feedback loops signifying:

- Negative feedback is so called as the algorithms are implemented for *reducing* discrepancies in the model performances and the real system.
- Negative feedback is an externalization of the right to challenge.
- The importance of fact engines is extremely subtle and is not so much related to their actual algorithms but to the externalization of the authority exercised for judgment. It happens that even the simplest approach of visual comparison is often effective, since the very exercise of visual comparison demonstrates that the modeler has relinquished own authority and own interests for a common exercise of judgment.

Relinquishing the right to judge is not the act of altruism but a reciprocal act for the return of objectivity, transparency, openness and mutual trust and above all internal consistency. A lack of understanding of the interconnection of the externalization of judgment and its reciprocity with openness and trust explains a great deal in the postmodern discourse rejected by science.

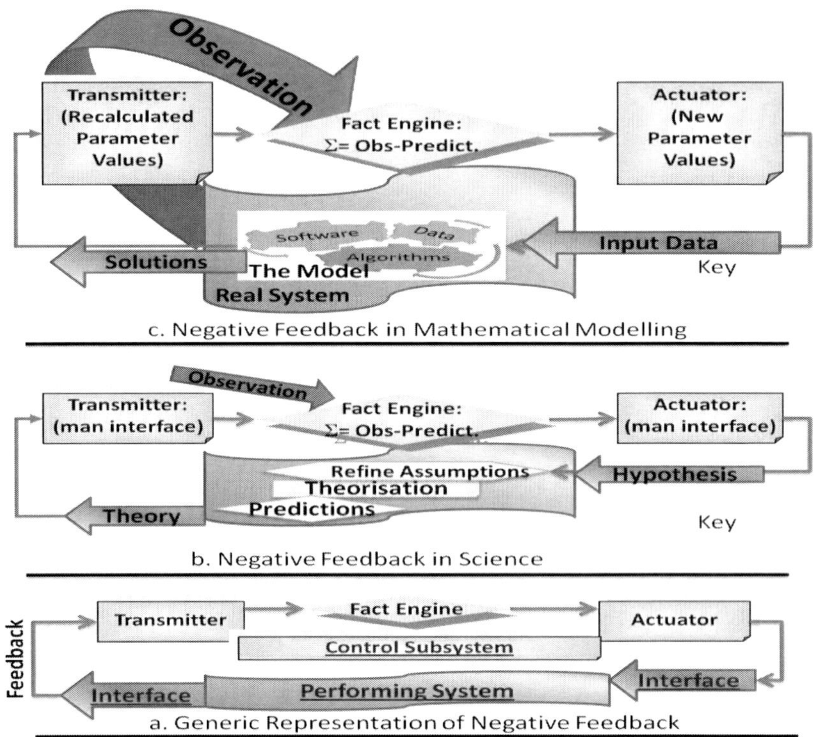

Figure 5. Architecture of Negative Feedback in Reductive Science.

Negative feedback loops underpin internal consistency and some of the ways for reinforcing it in mathematical models include: (i) *mathematical information* may be used to test the compliance of the models with a diversity of their assumptions; (ii) *historic data* are allowed into the models to calibrate (identify) the values of the parameters contained in the mathematical formulations; (iii) *fit-for-purpose* status of calibrated models have been tested from early days as it was realized that applying calibrated models as prediction tools would be a leap of imagination or a tremendous degree of faith and therefore they devised the validation procedure to test model performance before certifying them fit-for-purpose; (iv) *design information* may be fed to models as safety factors; (vi) *realtime data* are fed to forecasting problems using updating/assimilation techniques since the 1980s when the faith on validated models dwindled (see Goswami et al. (2005); (vii) *control information* may be fed into models using the PID algorithms (e.g. Shaw (2003), Willis (1999), which goes back to the late 19th century; and (viii) *noise reduction* information were may be fed into models using Kalman filters since the 1960s; (ix) *model variability* is an important information, which underlies uncertainty analysis and are a form of negative feedback.

Calibration problems were not transformed into a class of mathematical problems until 1950. Hadamard (1952) presented such problems as a category of inverse problems in mathematics leading to improperly-posed set of equations lacking unique solutions. Nonetheless, the conditions required for the quality of modeling solutions are often overlooked, see Khatibi et al (2001) for more details. The computational modeling activities may be captured by EvSyM as follows:

$$\boldsymbol{Prototype\ Modelling = Positive\ Feedback\left(d_{loop} + \ell_{loop} + e_{loop}\right) + Negative\ Feedback\left(d_{loop} + \ell_{loop} + e\_loop\right)}$$

The above account largely corresponds to negative feedback d-loop; ℓ-loops are wide within modeling practices, as practitioners keep adapting their best practice; e-loops cover performance criteria.

8.5. Entropy

Negative feedback loops in mathematical models do not cater for all sources instigating entropy reducing their efficacy. In fact, modeling practices encourage entropy either due to inconsistency between their internal building blocks or among different interacting models, as follows:

Proliferation of software applications: Modeling is an enterprise driven by niche markets, opportunism and abductive reasoning, which may seem as incoherent conglomerates. For the same set of modeling problems, there are diverse bespoke software packages, where professional organizations end up being their repositories. When raw data are processed by these systems, they are not easily transferrable from one system to another.

Multiplication of modeling datasets: Close architecture in software development gave rise to the emergence of "islands of models." The tendency to proliferate includes: (i) availability of different algorithms, this point was also made by Beven (2000) stating: "I gave up making a list of available models when I reached a count of 100 more than 20 years ago;" (ii) different versions, e.g. calibration/validation/application with a variety of scenarios; (iii) applications to different problems, e.g. low/high flows, water quality; (iv) there are different users; and (v) modeling datasets are reused across different organizations riddled with different levels of dedication.

Fragmentation in the modeling community: Modeling community is diverse and includes practitioners, software developers, academics, private organizations and competent authorities. They are severely fragmented and are all driven by different motivations and goals. For instance, academic goals are driven by research on topical issues are often measured by publishing journal papers but consolidating modeling practices fall short of such goals. Likewise, professionals are atoned to communicate successes of modeling but the vast experiences on model failures are not publicized. Fragmentation was seen so severe that it became a focal point during the 5th European Framework of research Program (FP5) in the years 2000-05 and the FP6 aimed to address this by encouraging more partnerships. Despite the emergence of new solutions, still improvements across different systems require the reinvention of the wheel.

There are a host of gaps in modeling practices and Khatibi (2008) lists some of the *technical gaps* including: (i) lack of tools to test the integrity of the models and to test the compliance with the assumptions; (ii) lack of institutionalization on the lifecycle management of modeling practices to treat *ad hoc* elements for ensuring their defensibility; (iii) lack of model pluralism, which makes it difficult to draw consensus from sets of results for similar scenarios. Some of *institutional gaps* are due to: (i) massive imbalance between the R&D budgets allocated to defense and civil sectors displaces goals and retards the rethinking on modeling practices; (ii) meager budgets allocated to R&D activities by commercial organizations are often for gaining commercial advantages and not for the consolidation of modeling practices.

These collectively reflect the absence strategic thinking and consolidation, as seemingly there is no prize for consolidating the emerging knowledge on modeling. Thus, the above factors encourage entropy in modeling enterprises.

8.6. Co-evolution and Adaptation over 50 Years:

Computational modeling was opportunistic and was fuelled by the optimism of Laplacian deterministic mindset during 1950-2000 by creating a perception of delivering truth. The arrival of the software venture in the 1980s opened up commercialism into modeling and academic interests receded to blue-sky type of problems on modeling, e.g. developing risk and uncertainty tools. The main intellectual movement in this period was raising alarms on the promise of truth by the deterministic mindset and exposing its misgivings, which was transformed into uncertainty movement and now in full swing. Notably, the uncertainty movement is not unselecting the deterministic capabilities but deterministic models are reused in probabilistic frameworks. This confirms that modeling with its inherent negative feedback loops is intrinsically sound but wrong-footed when driven by determinism.

8.7. Feedforward in Computational Models

Feedforward loops underpin external consistency of a product with the environment. Khatibi (2011) argues that the principle of sustainability and risk-based decision-making are feedforward loops already prevailed in science triggering a process of overhauling scientific practices and policymaking according to these loops. These loops have also found their way into mathematical modeling practices as follows.

To ensure sustainability and risk-based decision-making two streams of activities have been topical in recent years: (i) a diversity of capabilities are available on uncertainty analysis at their proof-of-concept stages; and (ii) open architecture was advocated to interconnect a host of modeling systems, e.g. Khatibi et al (2004) and OpenMI (http://www.openmi.org /reloaded/association/organisation.php). It is noted that there are other initiatives to ensure external consistency in modeling practices as policymakers promote best practice guidelines, specifications, modeling protocols to comply with a minimum standard particularly when models are issued to third-party users.

Although uncertainty analysis is a form of negative feedback in modeling, the ongoing uncertainty movement is a manifestation of feedforward in modeling practices. The movement has increasingly become strong since 1990, reflecting the failings of deterministic models. It is noted that there is no significant advocacy towards unselecting deterministic modeling tools but reinventing them in a new framework. None of the deterministic modeling techniques have been abandoned for good but they are increasingly being reused in stochastic frameworks to create masses of data, driven by stochastic analysis. The computational modeling activities may be captured by EvSyM as follows:

$$\begin{aligned} \textit{Prototype Modelling} = \textit{Positive Feedback}\left(d_{loop} + \ell_{loop} + e_{loop}\right) + \\ \textit{Negative Feedback}\left(d_{loop} + \ell_{loop} + e_{loop}\right) + \\ \textit{Feedforward}\left(d_{loop} + \ell_{loop} + e\_loop\right) \end{aligned}$$

9. Evolutionary Dynamism of Modeling Complexity – First Line of Enquiry

9.1. Definition of Models by Evolutionary Systemic Rationale

A model cannot be separated from its complexity context and therefore rather than defining it in a few words, it should be seen in terms of a hierarchy of attributes, some of which are summarized in Figure 6 and Table 3.

The main features are highlighted as follows: (i) Each modeling technique contains Orders 1-3 information, the proportions of which depend on feedback loops, i.e. the higher the loop, the greater is their reliability and less uncertainty. (ii) Evolving from models with zero$^+$ and positive feedback loops towards those with negative feedback and feedforward loops strengthen the proportions of Order 1 information contents and reduce Order 3 but this is not eliminated. (iii) Comparatively, some reciprocity is evident between feedback and uncertainty so that models with higher feedback reduce uncertainty but those with higher feedbacks are often complex and not parsimonious, may crash due to a host of runtime problems, and may *infrequently* be susceptible to unwarranted results with high consequences. It is flagged that the information content of a model (Order 1) can be hardened by feedback loops but at some expense for achieving better knowledge, as to be discussed later.

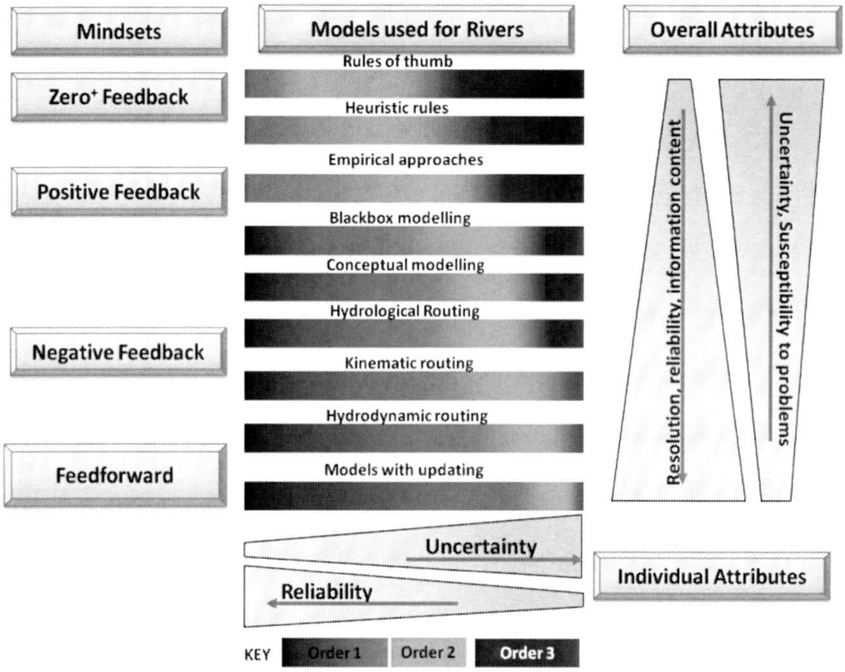

Figure 6. Indicative Proportions of 1st, 2nd and 3rd order Information and Attributes.

Table 3. Attributes of a Mathematical Model

Attribute	Description
Building blocks:	(i) mathematical formulations, (ii) datasets, (iii) modeling software. Note: each can be broken down to smaller elements.
Architecture: d-loops + ℓ-loops + e-loops	each of the four feedback loops may be present in a model: (i) *zero⁺ feedback* prevails in mathematical formulations based on rules of thumb, stochastic analysis of inputs our outputs or both or probabilistic; (ii) *positive feedback* prevails in mathematical formulations that are subject to gross errors and there is not any data to check them, e.g. black box and conceptual models and many rational or empirical formulae; (iii) *negative feedback* prevails in systems of equations based on laws of nature, timeseries analysis of inputs our outputs or both, where the internal consistency of the models are reinforced by negative feedback loops, (iv) *feedforward* prevails in models that their external consistency are continually checked feedforward loops.
Architect: e-loops	The law of natural selection is the architect but itself is reshaped by the feedback loops.
Functions	Models emulate the behavior of physical or social systems as a surrogate to aid understanding and decision-making.
Procedures	Modeling involves the use of some parameters, to link the respective worlds. Their values are determined by some sort of referencing (calibration or conditioning). The referenced model serves as a versatile replacement of the real system, but not its identical replica, to gain an insight into system behaviors.
Modeling truth	The author does not regard models as an expression of any truth, as these are contentious constructs of philosophy and logic and irrelevant in modeling.
Emergent properties	Models are tools to create information. Computational models achieve this by the modeling trio. Their information contents are depicted in Figure 6 in indicative terms for hydraulic modeling as an example, where the various techniques are categorized in terms of their increasing resolution.
Total Information = Order 1 + Order 2 + Order 3 + Order 4	
Order 1 Information	This concerns performing systems in term of their internal consistency Example 1: flow capacity in water treatment systems, through the system; Example 2: flow and level values in rivers throughout; Example 3: water level throughout navigational canals.

Attribute	Description
Order 2 Information	• This concerns the external consistency of a system in terms of overall cultural requirements, such as sustainability and reducing risk; • Example 1: In water treatment systems, this information is the containment of the pressure distribution in the system within an acceptable range and upgrading them for improved water quality by pre-ozonation, post-ozonation, dissolved air floatation counter current; • Example 2: in rivers this is maintenance of the system and reducing risks; • Example 3: in navigational canals, this information is operating the system within multi-functional uses by maintaining flows within a range, beyond which risks may prevail.
Order 3 Information	• This concerns assessing randomness or uncertainty an a measure of risk. • Example 1: In water treatment systems, this information caters for randomness in demand and supply and water security in the long-run; • Example 2: very large events, e.g. risk of tsunami or climate change; • Example 3: in navigational canals, this caters for climate change.
Order 4	• This acknowledges information not yet perceived/conceived.

Note: The above Information Orders are selections from a larger set of orders that the author is preparing for another paper

9.2. Insight into Consistency of Mathematical Modeling

Negative feedback loops in modeling practices underpins their internal consistency as they express the *right to challenge*. These loops characterize various attributes, as outlined in Table 3, including gaining an *insight* into mathematical models, *transparency*, *defensibility, interconnectivity*, *intelligibility of information* and *two-way flow of information* from the model to the real-world and vice versa. This mindset lends itself to pluralism and therefore can coexist with a model that has no negative feedback loop built into it. The full role of negative feedback loops may be explored by the following examples.

Example 1 – unselecting a model: Consider the Ptolemaic geocentric model, which was a geometric construction based on naked eye observations. The direction of information flow in this model was from man to man because man was a measure of everything. This model responded to anomalies by the growing sophistication of epicycles and in doing so, it failed: (i) to explore the internal consistency of the model; and (ii) it overlooked model parsimony.

Example 2: Hydraulic modeling is one of the fields, where pluralism has manifested in terms of the assemblage of different modeling techniques with different feedback loops. These models are categorized in terms of their resolutions (coarse, medium and fine) and depicted in Figure 7, which transforms the tacit knowledge on hydraulic models into explicit knowledge. These models are first considered on their individual merits in terms of their information content, uncertainty, diversity and data requirement, as follows. (i) Techniques with low information content (low resolution) can normally be expressed in diverse ways but the higher their resolution, the more well-defined they become. (ii) There seems to be some sort of reciprocity between model resolution and uncertainty, as well as between data requirement and the susceptibility of the models to diversification. (iii) There seems to be congruence between model resolution and data requirement or between uncertainty and diversification. It is flagged that reciprocity reflects the incurrence of some penalty for achieving better knowledge, as to be discussed later.

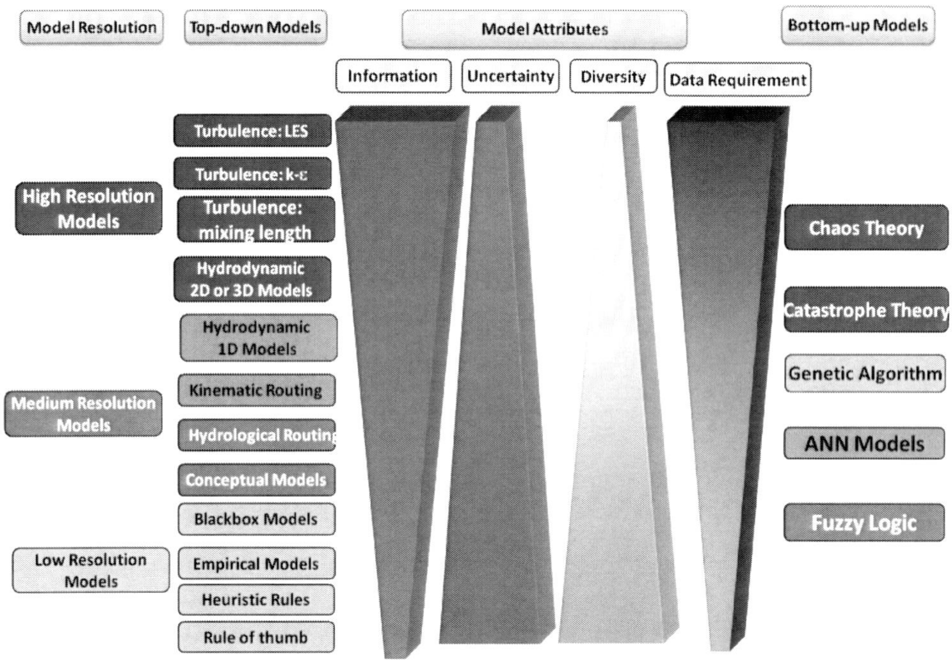

Figure 7. Categories of Modeling Techniques used in Hydraulics and their Attributes.

Owing to the culture of pluralism, an open channel hydraulic model is very likely to be the assemblage of a host of these models. Evolutionary systemic modeling unravels that each of these modeling techniques has different feedback mindsets. These are depicted in Figure 6 as follows:

- *Zero$^+$ feedback mindset*: This mindset may prevail in many models, such as those based on rule-of-thumb, e.g. modeling lateral inflows, which often involve gross approximations.
- *Positive feedback mindset*: This mindset prevails in a variety of ways, e.g. when modeling overflows from a river to its floodplains, in which case the modeler sets parametric values by selecting default values without calibrating them or testing compliance with the assumptions.
- *Negative feedback mindset*: This mindset prevails when the modeling includes referencing of the model performances by calibration-and-validation or model performance is continually maintained by updating.
- *Feedforward mindset*: This mindset prevails when the parameters of the hydraulic system are reset by PID algorithm, Kalman filtering or other control algorithms to ensure consistency of the results with external requirements, e.g. water demand in irrigation or water supply systems.
- Model management has not been standardized and it is not common practice even to have a case history of each model to keep their records, where decisions are made by a range of modelers across different organizations with different management practices, Khatibi (2008).

Each model is infested with default parameter values and they often remain unchallenged. So each model has a capacity to severely deviate from their true values as particular values of the parameters may cause delineation between mild and severe deviations. The underlying explanation for such behaviors is the possibility of the existence of low-order chaos driven by an attractor. For assembled models, each mindset may act as an attractor and the outcome may be complex. Certain models can push their performance to rapidly varying dynamic zones, in which the model performance can be quite different. The experience of such behavior happens from time to time but modelers have no way of explaining these seemingly "ghost" features. Thus, increased complexity in pluralistic models triggers possible inadvertent dynamic behaviors in modeling systems.

Overall: It is flagged that pluralism allows compound models to be affected by multiple mindsets at the expense of no insight into their internal consistency.

9.3. Foresight of Mathematical Modeling

A further insight into pluralism and the information represented in Figures 2 and 3 is that EvSyM makes it evident that mathematical modeling practitioners are "abductive" modelers in the sense they use these models as best available information to infer explanations. Modelers by their nature, make use of any information available without being deterred if its information content is falsifiable contrary to the Popperian sense or its structure is not simple contrary to Beven's (2002) modeling vision. Evidently there is life after falsification and parsimony but at a price, as explored below in terms of its external consistency.

The external consistency of a system or a practice with the environmental settings is underpinned by feedforward loops and this encourages articulating foresight. The problem is that a system often has many solutions, some of which are compatible with the external environment but the others are not. Traditional feedforward loops were implemented in terms of two design criteria: technical feasibility and economic cost-effectiveness. In recent years, there is a transition to overhaul traditional design practices to multi-dimensional criteria. Mathematical modeling has a formidable role on this.

The past performance criteria were driven by utilitarianism; whereas the multi-dimensional criteria are driven by the principle of sustainability, decision-making by participation and risk-based decision-making, Khatibi (2011). So there is a process of rethinking, where policies are being overhauled to comply with sustainability. The question is then what is the role of mathematical modeling in this rethinking? To the best knowledge of the author, the only questions asked on modeling, as a whole, have been by philosophers, as reviewed in Section 2 of this paper. The conclusion was that philosopher discourses were indulged with own issues without addressing real modeling issues. Thus, there is no vision for modeling complexity, as a whole, even though modeling acts as a sculptor and as a sculpture but yet without foresight of its own.

Notably, the multi-criteria approach depends on free-flow of information among the systems and goal-orientation systematizing interactions among the systems and with the environment. If a mathematical model is assembled in a rigid fashion, soon it could be obsolete and a complete waste of resources. It is flagged that foresight of mathematical models is not clear as models are required to maintain their external consistency with one another and with the environment. Goal-orientation is the key for maintaining external consistency through flexibility, modularity and the ability for free-flow of information, but these are not explicit in the agenda of the modeling community. It is questionable whether the emerging open architecture will be enough for goal-orientation.

9.4. Some of the Elements to Formulate a Vision

EvSy-modeling offers some of the elements for a vision, depicted in Figure 8.

9.4.1. Modeling complexity (Element 1 – see Figure 8)

Modeling complexity treated as amorphous is increasingly dynamic and in an inflationary state; it is a by-product of abductive reasoning, opportunism and pluralism. In the received view of modeling, there is little critical view of this complexity for being under the grip of ontology, i.e. models are often treated as timeless entities. Thus, their interconnectivities and evolutionary transitions are overlooked. Any critical thinking is philosophical but the subsequent doctrines retard the development of a scientific vision. Pragmatic visions are often blinkered. So, the formulation of a comprehensive vision for modeling practices as a whole is outstanding.

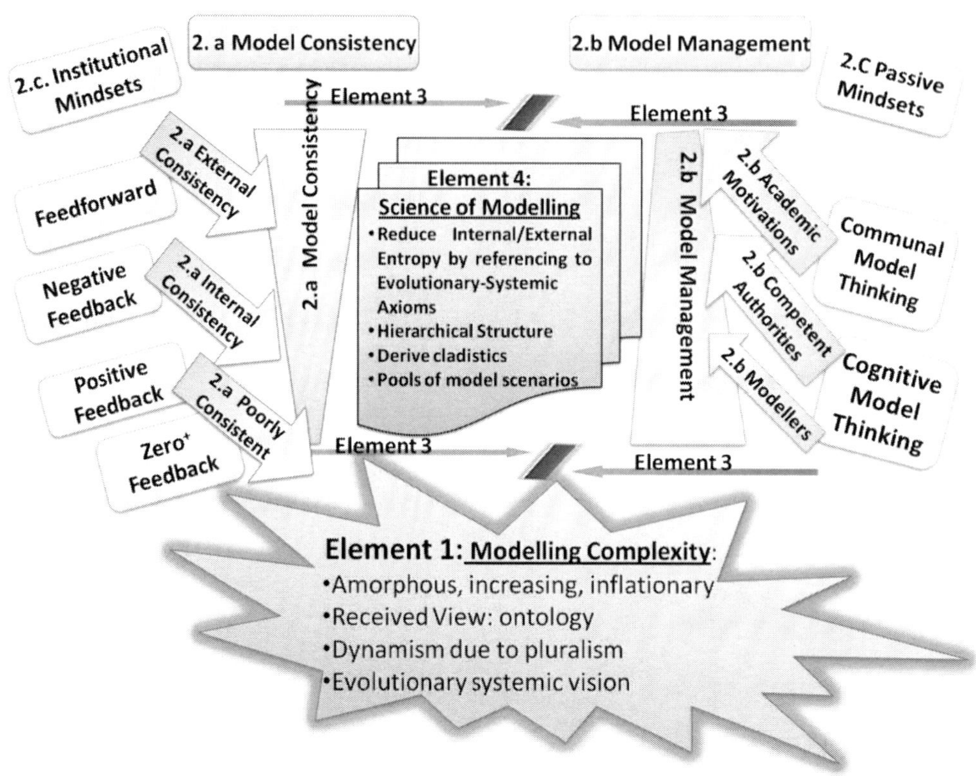

Figure 8. Key Drivers Triggering the Need for Forming the Science of Modelling.

9.4.2. Architecture of the Complexity (Elements 2.a-2.c – see Figure 8).

Arguably, modelling complexity is a by-product of feedback loops operated, at least, at three levels, as outlines below.

Model Consistency (2.a): Model performances depend on feedback/ feedforward loops:

- Zero$^+$ feedback and positive feedback loops signify poor consistency
- Negative feedback loops signify catering for internal consistencies
- Feedforward loops signify catering for external consistencies.

Model Management (2.b): The performance of any model depends on the improvement initiatives of the stakeholders including: (i) modelers are engaged in abductive reasoning and do not really pay much attention to fundamental issues; (ii) competent authorities commission formulations of modeling strategies, best practice procedures and modeling specifications, which are important to do the required job but in a blinkered way; (iii) consultants are employed to do modeling or to develop best practice guidelines but they are also keen to satisfy their clients; (iv) academia is motivated by reputation measured by the number of publications and least concerned with consolidating knowledge. The initiatives driven by these stakeholders fall far short of institutionalizing modeling complexity.

Modeling Cultures (2.c): Modeling performances may also depend on cultural/ professional mindsets due to different levels of professional care and by catering for public expectations.

9.4.3. Entropy (Element 3)

The paper provides evidence for increased entropy on the following basis: (i) the gain in reliability of models by their increased resolution is at the expense of increased complexity; (ii) increased complexity signifies that modelling complexity is in its inflationary stage and its entropy is increasing; and (iii) the pluralism of feedback loops at various strata are likely to instigate increased dynamism, as follows:

1. Each feedback loop can act as attractors/repellents in the sense of chaos/catastrophe theories, instigating dynamic behaviors;
2. Dynamic behaviors can be triggered by:
    a. Different types of feedback loops collected pluralistically in single models (Element 2.a),
    b. Interactions of model consistency (technical implementation of feedback loops) with model management (Elements 2.a-2.b), and
    c. Perceptions of wider communities, and conceptions of modelers on model capabilities (Element 2.c);
3. It is heuristically suggested that the increased dynamism is a possible expression of a new kind of entropy, which may be referred to as external entropy.
4. Currently, dynamic behaviors are overlooked but likely to overlook other feasible solutions, and discardnthem.

It should be noted that the interpretation of increased dynamism as entropy unravels entirely a different picture on modeling complexity. It reminds the expression that *you do not get something for nothing*, the Mesolithic mindset as discussed in Section 4. The wisdom for consistency has resurfaced in modern times that the benefits of pluralism have increased complexity and dynamism, instigating as a new type of entropy invoking the need for regulation, consolidation, and slowing down.

9.4.4. Referencing

EvSy-Modeling ofto modeling practices shows that the four feedback loops did not emerge at once but they were primed and selected one-by-one as followa:

- Communal model thinking: perception of zero$^+$, positive and negative feedback mindsets
- Primitive modeling: zero$^+$ and positive feedback loops
- Prototype modeling: zero$^+$, positive and negative feedback loops
- Computational modeling: zero$^+$, positive and negative feedback and feedforward loops

Elements 2 explain modeling complexity as it is now, but Elements 3 that of evolutionary transitions. By referencing Elements 2 with Elements 3, it is evident that the feedback loops in both cases have an architectural role but in modeling complexity, the role increases from model consistency (Element 2.a) to model management (Element 2.b) and modeling culture (Element 2.c). The collective outcome of these elements is there are massive holes in modeling practices but practitioners just ensure that they avoid them rather than treat them. Also due to the absence of cladistics, any investment in new modeling capabilities is not guided towards a common goal. The EvSyM vision of modeling practices is not blinkered but it is referenced (calibrated) by the history of past practices, and guided by a roadmap through its cladistics. As per Section 6, the substance of these arguments makes a research case for forming modeling science.

9.4.5. Drivers of the Vision (Element 5)

The amorphous image of modeling complexity is suggestive of a healthy culture, but in reality, it reflects a thriving opportunism and the absence of institutionalization. Today's status of modeling complexity reminds reductive science facing natural/social complexity in 17th century, at which time reductive science was not institutionalized and had the image of organically growing complexity. The institutionalized science unraveled its own hierarchical structure, which greatly simplified scientific studies. There is no institutionalization on modeling complexity as a whole and therefore its organically-growing image is of a jumble, which is not healthy. However, if the problem with modeling complexity is just a matter of image, it is likely to be tolerated without translating into any proactive action plans. This Paper argues that the increased dynamism and complexity shrouds in itself a new kind of entropy: external entropy.

10. STRUCTURE OF MODELING COMPLEXITY – SECOND LINE OF ENQUIRY

In the symbiosis of science, mathematical modeling and mathematics, science explains the complexity of life, which has acquired a hierarchical organization comprising physics, chemistry, biology, psychology, sociology and anthropology. Both mathematics and mathematical modeling are normally driven by science but without displaying a similar hierarchy or indeed any hierarchy. This is largely because mathematical modeling is often opportunistic and responds to local problems in science, and mathematics is a rigorous expression of the formulation used in the model, although both modeling and mathematics can be developed independent of science and a development in mathematics can remain abstract.

Local problems of science were often confined to its particular areas in one hierarchies but this restrictive level of abstraction has increasingly being seen anomalous. In certain disciplines, models are already integrated:

Example 1: weather forecasting models incorporating the atmospheric phase at various resolutions and ocean models, e.g. see Figure 9.a (after Beersma, 2002);

Example 2: open channel friction parameter, which currently a lumped value accounts for a diversity of contributing factors but there is an increasing concern that physical, chemical and biological and anthropological factors operate at different timescales and resolutions and without accounting for them in appropriate manners, uncertainties can undermine the defensibility of the models.

Example 3: Similar ideas are given by Saunders et al (2007), see Figure 9.b.

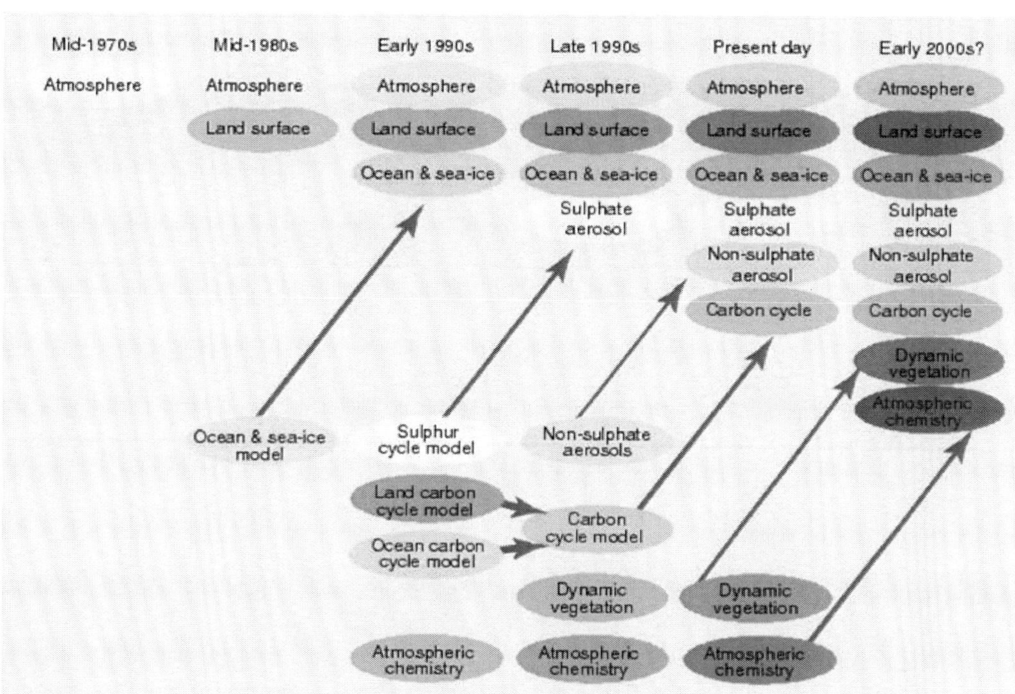

Figure 9.a. Integrating Global Climate Modeling System, Beersma (2001).

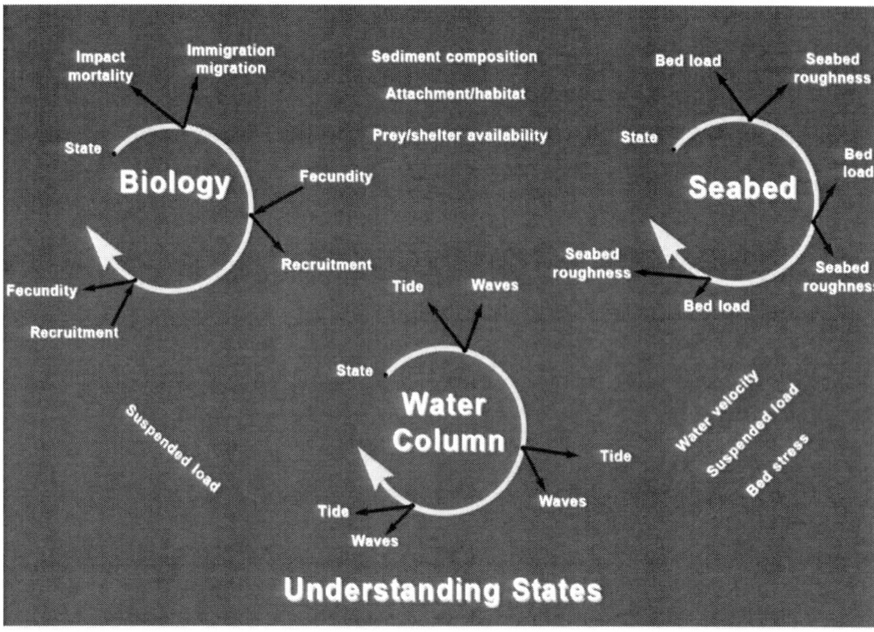

Figure 9.b. Coupling Physical and Ecological Models, Saunders et al. (2007).

The increasing tendency to integrate models creates new challenges on software development and opens up issues on their architecture and the availability of their source-codes and their ownership. Models of complex systems tend to indicate unanticipated outcomes. Without any scientific insight into inherent complexity, the current trend of amorphous growth is inevitable and this is not a sound economical proposition, as this requires the "reinvention of the wheel". The complexity of mathematical modeling is in an urgent need of consolidation. The author advocates generalization of cladistic approaches to each area of complexity and this is the second line of enquiry in this paper. The aim is to reconstruct modeling complexity paralleling the world of science to show its inherent hierarchical structure, as outlined below (using hydrodynamic examples):

- *Quantumistic* level: The backdrop for the formation of modeling complexity is presented in Sections 4-8, where a thriving culture is essential during which there is no perception of the future
- *Physicalistic/chemicalistic* levels: Identify appropriate building blocks (slope, velocity, time, space, sectional properties) to form compounds (mathematical expressions for forces and storage) each with emergent properties (force or volume) at *physicalistic/chemicalistic* hierarchies.
- *Biologistic/psychologistic* levels: The working of the compounds together (e.g hydrodynamic equations) gives rise to higher emergent properties (solved equations creates meaningful information) at the *biologistic* level when focused on its internal consistency; *psychologistic* level focuses on the ability to cope with the external world.
- *Sociologistic/anthropologistic* levels: There would be different ways of developing equations to solve the same problems. The application of one method is deemed

sufficient but there is already a growing culture of used ensemble solutions of seeking consensus from applying many different models. Under these conditions, the cohesion among different models would reflect a *sociologistic* approach. The emerging cultures would signify an *anthropologistic* hierarchy.

The problem of complexity would then be to transform mathematics and modeling capabilities into hierarchically organized complexity, similar to science. This may sound as a wish list or an ideological proposition, unless there is a good reason for the transformation. Below are some of the reasons:

- Currently, mathematics is formulated as a top-down discipline and it is very difficult to make sense of existing equations and theorems. A hierarchically-organized mathematics would create a bird'e eye view into mathematical equations and theorems.
- There is a good insight into conic sections to understand point, line, circle, ellipse, parabola and hyperbola but there is no similar insight for many other problems. Understanding hierarchical structure can open up the opportunity for deriving cladistics at the biologistic level.
- The four feedback loops can characterize the modeling complexity at psychologistic, sociologistic and anthropologic levels.
- This hierarchical approach creates transparency with science.
- Evolutionary systemics lead to a bottom-up approach, as nothing is imposed on the reality of the complexity but the aim is to study it as it is.

The author believes that transforming amorphous modeling complexity into a hierarchically-organized complexity is a feasible proposition. He has developed an analytical calculus for the sequences of natural numbers by factorizing a sequence into the sequence of building blocks (the architecture) and a sequence of counters (the rules or architect) developing new operators, which are capable of unraveling inner structure of number sequences. The author is optimistic that this calculus can serve as model of hierarchy and interconnectivity at each hierarchy. This is outlined by Khatibi (2012a, 2012b) and Khatibi et al (2012).

Conventional modeling scenarios overlook hierarchical factors affecting the subject of the studies. A systematic way to formulating scenarios is a challenge. It is suggested that a systematic approach can be devised by compiling a pool of scenarios and selecting intelligently from the pool based on the specific characteristic of a particular study. Some of the author's ideas are as follows:

1. Establish baseline models without any change
2. Identify areas of complexity relevant to a particular study
3. Identify hierarchies appropriate to each area of complexity
4. Identify building blocks at each hierarchical level and use the four feedback loops at each hierarchy (although the aim is not to catalogue them), as outlined in Table 3.

Table 3. Broad Scenario Attributes for Psychologistic–anthropologistic Levels

Hierarchies	**Zero^{+} Feedback Loops**	**Positive Feedback Loops**	**Negative Feedback Loops**	**Feedforward Loops**
Physicalistic Biologistic	The four feedback loops would operate differently at each of these hierarchical levels and different sets of problems are often handled at different resolutions. This is a vast subject and cannot easily be summarized in this table.			
Psycho-logistic (incidents)	To emulate change scenarios triggered spontaneously by natural incident acting spontaneously or by individuals instigating change (inputs or outputs) with no knowledge of impacts e.g. fly dumping	To emulate change scenarios triggered by natural acts with definable cause-and effects equations or policymakers setting the pace to change who guess the outcomes but are aware of discrepancies	Emulating change scenarios, in which natural acts are regulated or individuals attempt to reduce entropy by initiating negative feedback, e.g. recycle household rubbish, efficient energy use	To emulate change scenarios to modulate **proactively** the contributions or impacts of incident by nature or by individuals e.g. adapting to climate change
Socio-logistic (processes)	To emulate change scenarios triggered spontaneously by **sections of a society** (inputs or outputs) without knowing the impacts with switch between rationality or irrationality, e.g. profit motives of	To emulate change scenarios triggered by **communal, social or institutional consciousness** escalating causative processes with entropy, e.g. the actions of the individuals using fossil fuel and those identifying global warming	Emulating change scenarios, in which sections of the **society** attempt to reduce entropy by initiating negative feedback, e.g. organizations having Carbon Footprint policies	To prevent goal fixation or goal displacement by the virtue of foresight, e.g. the Green Movements in the 196s-1990s, until their ideals led to the Principle of Sustainability
Anthropo-logistic (culture)	To emulate change scenarios triggered spontaneously by **various cultures** (inputs or outputs) without knowing the impacts with switch between rationality or irrationality	To emulate change scenarios enshrined by **governance or cultures** escalating causative processes with entropy, e.g. the policies and philosophies of utilitarianism during the Industrial Revolution	Emulating change scenarios of reducing entropy is enshrined by the **governance and global cultures**, e.g. environmental impact assessments driven by sustainability	To prevent goal fixation or goal displacement by the virtue of foresight coded to them, e.g. the principle of sustainability as one such foresight
Pluralism	Each of the above types of scenarios is a mindset and can act on its own or the newly emerging mindset can create a pluralistic culture with the older mindsets. This is the strength of evolutionary systemic modeling, which emulates reality, where there is the pluralism of these mindsets associated with individuals, communities and institutions. Each mindset can act as an attractor or repellent in the sense of chaos and/or catastrophe theories and therefore in principle, it is possible to study the dynamics of complexity			

11. Discussion, Research Case for Modeling Science

11.1. Ontology Looming over Mathematical Modeling Cultures

Ontological mindset seems to overshadow modeling complexity but Figures 2 and 3 challenges such a perception as they transforming a tacit knowledge of evolution in modeling complexity into explicit knowledge by virtue of attributes at each epoch. These thesis and antithesis are scrutinized, as follows:

- Without the concept of evolution, biology was very convoluted to the extent that man arbitrarily granted his own species by a special status. Darwinism has successfully pushed ontology to extinction only in the biological science. However, ontology infests, modeling complexity, as well the rest of science and complexity but evolutionary systemics is able to pinpoint evolutionary processes throughout the man's cultural horizon even discerning communal model thinking, institutional modeling practices and their co-evolution.
- The difference between understanding evolutionary and ontological approaches is the difference between no parallax and a parallax error. The position of no parallax and concept of parallax errors are spatial but evolutionary transitions are temporal. Ignoring this difference, the argument is that evolutionary thinking is equivalent to the no parallax position, i.e. natural selection remains the driver but the driven entities vary as the attributes of the environment changes; whereas the ontological mindset is equivalent to parallax error by denying the epochal attributes in Figure 3 and by denying the role of natural selection.
- Ontological approaches are reminiscent of the situation prior to the publication of the law of natural selection in 1859 by Charles Darwin (1809–1882), by which time any conception of the origin of species/life was denied (i.e. there was a parallax error and ontologically-minded people refused to change their viewpoints to see their errors). The single law of natural selection provided an explanation for the origin of all the species that became the cornerstone of science by the virtue of relying on evidence. No matter which species is under consideration, each and all of them lead to the origin. Evolutionary systemics extends evolution to all areas of complexity, including modeling complexity.
- The culture becomes rife to ontological mindsets by overlooking evolution. It happens that an ontology is a single snapshot of complexity, once adopted at one time, it becomes a norm and sometimes it is coerced, e.g. the coercion of the Aristotelian ontology by Christianity. Thus, ontology is like a child that never grows.
- The culture for modeling complexity is subtly ontological, as masses of models are churned out with no incentive to establish their interconnectivity for the simple reason that the evolution in modeling complexity is overlooked. Evolutionary systemics can provide an insight into modeling complexity and unearth their generic variations.

11.2. Outline of a Vision for Mathematical Modeling Complexity

It is one thing to identify possible solutions and articulate a vision and another thing to produce solutions. Evidently, it is beyond the means of a single researcher to embark on the massive task of formulating a vision for modeling and implementing it. An international project is needed to investigate the feasibility of forming and delivering the *science of modeling*. Arguably, such a project would include a vision with quick win deliverables and a long term plan.

The Vision: The paper transformed tacit knowledge into explicit knowledge that models are increasing in complexity and dynamism and therefore this is regarded as a new kind of entropy, to be referred to as the external entropy. Thus, there is more in the seemingly amorphous modeling complexity than meets the eye but the author is not aware of any investigation on these identified issues. The vision suggested in this paper comprises: (i) the need to reduce entropy, as guided by evolutionary systemics; (ii) *the need for hierarchy in modeling complexity*: the paper suggests that the hierarchical structure of science can serve as model for modeling complexity; (iii) *the need for a model of interconnectivity at the chemicalistic and biologistic levels*: the paper suggests that equivalences of the Periodical Table and the cladistics can be derived by using a building block approach to restructure mathematics; (iv) *the need for a model of interconnectivity at psychologistic, sociologistic and anthropologistic levels*: the paper suggests the four feedback loops can be used for this purpose.

Short-term quick win deliverables can carry out knowledge mining and some of these activities are suggested as follows: (i) to identify possible patterns in modeling; (ii) to assess the feasibility of modularizing modeling practices and tools for developing user-designed capabilities by creating a pool of modules, encouraging goal-orientation in setting up models, and refining models according to the changing internal and external requirements; (iii) to give a particular attention to the exchange of information across interfaces; (iv) to express strength and weaknesses of the models in the sense that a model is not used indiscriminately but identify suitable models for each brand of modeling problems; and (v) to encourage running many models and gain consensus from their collective results; (vi) to create a capability that automatically challenges the integrity of the modeling results by developing appropriate feedback loops and ensures that the models are compatible with the goals; and (vii) to create an automatic capability for seeking dynamic behaviors in the models.

Long- and medium-term activities can identify gaps and barriers and can devise strategies to tackle them. The paper flags some of them as follows: (i) without deriving the cladistics of modeling complexity, its growth will remain organic implying the absence of a strategy; (ii) goals of modeling and their performance measures need to be institutionalized else the practices would suffer from haphazard approaches; (iii) modeling capabilities are catered for by commercial enterprises but science and mathematical modeling will suffer without an open source system for software development; (iv) cataloguing systematically the types of feedback loops in modeling practices and devising conditions for the integrity of the modules without negative feedback loops; and (v) although modeling is pluralistic by its nature, there are many factors retarding pluralism and this should be addressed.

CONCLUSION

At the dawn of the emergence of computational mathematical modeling in 1954, Einstein (1954) remarked that "as far as the laws of mathematics refer to reality, they are not certain; and as far as they are certain, they do not refer to reality." So, there is a gap between the real world and the world of mathematics, filled by models, where two factors are put face to face: (i) any model is a viable option, i.e. there is no right or wrong model, provided that a model is selected only after challenging its performance – this is an expression of pluralism; (ii) the gap is bridge by pluralism, natural selection and abductive reasoning. Abductive reasoning is an interesting concept and captures well the psychology of the modeling community, as models are information leading to explanation and modelers are not characteristically deterred by shortfalls inherent in their models but keen on inferring explanations by using information compiled within the modeling datasets and mathematical relationships.

The paper uses *the evolutionary systemics* (Khatibi, 2011 and Khatibi, 2012a) to model evolutionary transitions in mathematical modeling through the various evolutionary ages of the cultural *Homo sapiens*, within the time span of 190,000 years ago until now. The literature is suggestive that model thinking is instinctive. It ostensibly gave rise to language as its first fruit. Just as language uses words to represent abstract/concrete entities, modeling also uses one entity, say mathematics, as a surrogate for understanding another. Just as languages has evolved, so have models.

Evolutionary systemics integrates evolutionary thinking with systemic thinking using systemic/cybernetic concepts of *zero*$^+$ *feedback*, *positive feedback*, *negative feedback* and *feedforward loops*. These loops also explain evolutionary transitions in modeling practices and form a vision for rethinking. Their salient features are outlined below.

Evolutionary systemics transforms tacit knowledge into explicit knowledge, shedding light on the co-evolution between institutional and communal mindsets: (i) communal mindsets would encourage tacit knowledge, (ii) institutional mindsets would encourage explicit knowledge articulated by individuals or by institutions, which are either using tacit knowledge or priming new knowledge; (iii) the emerging explicit knowledge was either propped up by the authorities or proliferated by positive feedback loops and primed new areas of applied mathematics; (iv) channels of communication were primitive and included tablets, scroll and books, the printed form of which appeared only since the 15th century.

Evolutionary systemics unearths the evolutionary transitions suggesting that language is ostensibly the first fruit of model thinking and after its emergence, it amplified the *communal model thinking* and the subsequent evolutionary transitions were institutionalized comprising:- (i) *primitive modeling*; (ii) the right to challenge triggered the emergence of reductive science as an evolutionary transition, which triggered *prototype modeling* until 1950 (was driven by zero$^+$, positive and negative feedback loops); (iii) *computational modeling* emerged by the impetus from the advent of computers (was driven by zero$^+$, positive and negative feedback and feedforward loops). The paper unearths the retarding impacts of the ontological mindset over modeling complexity and seeks to understand its hierarchical structure by using the hierarchical structure of science as a model for modeling complexity for deriving possible cladistics.

In parallel with the above, this paper shows that the prevailing mindset over modeling complexity is ontological. When the attributes of pluralism, selection and evolutionary thinking are combined, the modeling culture in the past 60 years can be overarched in the following terms:

- Overview: Over the past 60 years of mathematical modeling, there was, and still there is, no grand plan but the organic growth driven by natural selection operating on modeling entities.
- Models: are made up of data (datasets), mathematical relationship (the modeling engine) and produce information (or solutions) by a set of algorithms deemed to be internally consistent and defensible but subject to uncertainty.
- Modeling complexity: Modeling complexity is seemingly amorphous, rife for an inflationary growth and riddled with increasingly dynamic.
- Modeling Cultures: Mathematical modeling practices have no insight into the complexity they are shaping and no foresight of the future.
- Greatest shortfalls: the absence of transparency among different areas of modeling encourages *ad hoc* practices, hampering knowledge transfer. This in turn, encourages the "reinvention of the wheel."

This paper shows that modeling complexity is poorly understood and if there are visions, they are fragmentary. The paper identifies various elements contributing to the formulation of a vision and argues that there is a research case for forming the *science of modeling* through an international project with short-term and long-term objectives toward mapping out modeling complexity and delivering a goal-orientated modeling capability. The driver is increasing complexity and dynamism in modeling complexity, tantamount to a new kind of entropy. If steps are taken, the return is substantial and includes: (i) transparency within various areas of modeling complexity, facilitating the transfer of knowledge between them; (ii) maximizing potential returns of investments by preventing the reinvention of the wheel and exploiting synergies between different capabilities; (iii) improving the defensibility of the models; (iv) enhancing the culture of modeling communities; (v) articulating goals for modeling complexity.

ACKNOWLEDGMENT

The author is grateful to Mr. John Little of New College, Swindon, UK, for thoroughly reviewing a draft version of the paper and offering his detailed constructive comments; and also due to Prof. Oleg Makarynskyy, of Asia-Pacific Applied Science.

REFERENCES

Abbott, M. B. (1991). "*Hydroinformatics.*" Avebury Technical, London, Britain

Anand, D.K. (1980), *Introduction to control systems*, Pergamon Unified Engineering Series

Armstrong, K. (2005), *A short history of mythology*, Pp 63-65, Publ. By Canongate, Edinburgh

Bailer-Jones, D. M. (2002), Models, Metaphors and Analogies, *The Blackwell Guide to the Philosophy of Science*, Edited by Machamer P. and Silberstein, M., Published by Blackwell Publishers Ltd 2002, USA

Barth, K. (1960), *Church dogmatics*, Vol. III: *the doctrine of creation*, Part 3, translation Bromiley G.W. and Ehrlich R.J. Clark, Edinburgh; cited in Abbott (1991)

Beersma, J.J., (2001), *"IPCC projections of future climate change,"* Electronic proceedings of MITCH Workshop I, 2001, 12-13 December, Delft, The Netherlands, *(http://www.knmi.nl/onderzk/klimscen/papers/Mitch_beer)*

Beven, K., (2000), *Rainfall-runoff modelling*, John Wiley and Sons Limited pp

Beven, K., (2002), "Towards a coherent philosophy for modelling the environment," *Proc. R. Soc. Lond.* A (2002) 458, 2465–2484

Berm´udez, J.L. (2003): *Thinking without Words*. Oxford University Press, Oxford

Brigandt, I., and Griffiths, P.E. (2007) "The Importance of Homology for Biology and Philosophy,"
(http://paul.representinggenes.org/webpdfs/Brigandt.Griff.07.ImportHomol.pdf)

Chandler, D. (2007), *Semiotics: the basics*, Published by Routlege, as an imprint of the Taylor and Francis group.

Cartwright, N. (1983), *How the Laws of Physics Lie*, Clarendon Paperbacks, Oxford University Press

Chapman, G. P., (1977), *Human and environmental systems*, Academic Press, New York

Checkland P.B, (1981) "*Science and the systems movement*"; printed in *systems behaviour*, Open University set book, edited by Open Systems Group, 3rd edition, 1981.

Chow, V. T., (1981), *Open channel hydraulic*,McGraw-Hill Kogakusha, Ltd.

Cohen J. and Stewart I., (1994), *The collapse of chaos – Discovering simplicity in a complex world*, Viking, published by the Penguin Group, Page 19

(Eds.), published by Springer-Verlag Berlin Heidelberg

Defra (May 2003), R&D Technical Report FD2302/TR1, *Risk, Performance and Uncertainty in Flood and Coastal Defence – A Review*; by Sayers, PB ., Gouldby, BP., Simm, JD., Meadowcroft, I. and Hall, J.

Donald, M. (2001): *A Mind So Rare. The Evolution of Human Consciousness*. W.W. Norton and Company, New York

Diller, A., (2002), "Constructing a Comprehensively Anti-Justificationist Position," Jarvie, I., Milford, K., and Miller, D. (2002); Karl Popper, A Centenary Assessment, Selected Papers from Karl Popper, Volume II, *Metaphysics and Epistemology*, Ashgate

Duhem, P., (1906), *Aim and Structure of Physical Theory*,Published by Princeton University Press, 1954, Part I, Chapter IV

Einstein, A. *Ideas and Opinions*. (New York, 1954)

Frigg, R. (2008), Models in Physics, Forthcoming in *Routledge Encyclopaedia of Philosophy* (web version).
http://www.romanfrigg.org/writings/Models_in_Physics_REP.pdf

Hadamard, J., (1952), *Lectures on Cauchy's problem in linear partial di_erential equations*, Dover, New York, 1952. MR 14:474f

Hartmann, S., Hoefer, C. And Bovens, L. *Nancy Cartwright's Philosophy of Science*, First published 2008 by Routledge, 270 Madison Ave, New York, NY 10016

Hoeffer, C., (2008), "Introducing Nancy Cartwright's Philosophy of Science," Pp. 1-13, in *Nancy Cartwright's Philosophy of Science*, Edited by Hartmann, S., Hoefer, C. And Bovens, L., First published 2008 by Routledge, 270 Madison Ave, New York, NY 10016

Goswami, M., O'Connor, K.M., Bhattari, K.P. and Shamseldin, A.Y. (2005), "Assessing the performance of eight real-time updating models and procedures for the Brosna River", *Hydrology and Earth Sciences*, 9(4), 394-411 (2005).

Henderson, F.M. (1966), *Open channel flow*, Macmillan Publishing Co. Inc.

Jarvie, I., Milford, K., and Miller, D. (2002); Karl Popper, A Centenary Assessment, Selected Papers from Karl Popper, Volume II, *Metaphysics and Epistemology*, Ashgate

Khatibi, R.H., Williams, J.J.R. and Wormleaton, P.R., (December 1997), "Identification problem of open channel friction parameters", *ASCE J. of Hyd. Eng.*, Vol. 123 No. 12.

Khatibi R. H. (2002), "Formulation of Modelling and Systems Strategy - End Users Initiatives on Improving Flood Forecasting", Paper presented at the second workshop of the MITCH workshop in Barcelona available at *http://www.mitch-ec.net.*

Khatibi R. H. (2003), "Systemic Knowledge Management in Hydraulic Systems - Paper 1: a postulate on paradigm shifts as a methodological tool", *Journal of Hydroinformatics*, Vol. 5.2, PP 127-140

Khatibi, R., Jackson, D., Curtin, J., Whitlow, C., Verwey, A., Samuels, P., (2003), "Vision statement on open architecture for hydraulic modelling software tools", the *J. of Hydroinformatic*

Khatibi, R., Lincoln, R., Jackson, D., Surendran, S., Whitlow, C., and Schellekens, J., (2004), "Systemic Data Management for Mathematical Modelling of Environmental Problems", *Journal of Management of Environmental Quality;* Vol. 15, No. 3, Pp. 318-330.

Khatibi, R., (2005) "Mutual inclusion in flood forecasting", *(http://www.actif-ec.net/conference2005/proceedings/ PDF_docs/Poster_Session/Khatibi_Rahman.pdf)*, Tromsø, Norway

Khatibi, R. (2008), "Model reuse in flood risk management," Pp 415-423, *Proc. of European Conference on Flood Risk Management*, FLOODrisk2008, Oxford, UK

Khatibi, R., (2011), "Evolutionary Systemic Modelling for Flood Risk Management Practices," *Journal of Hydrology* Vol. 401 Issue 1-2, Pp 36–52

Khatibi, R., (2012a), Learning from Natural Selection in Biology: Reinventing Existing Science to Generalize Theory of Evolution – *Evolutionary Systemics, Chapter X of this series*, Published by Nova Publishers

Khatibi, R., (2012b), *A study of mathematical practices by evolutionary systemic modelling; Chapter X of this series*, Published by Nova Publishers

Leliavsky, S. (1965), *River and canal hydraulics*, Vol IV, Chapman and Hall Ltd

Lilienfeld R., (1978), *The rise of systems theory*, Wiley, New York.

Linton, Christopher M. (2004), *From Eudoxus to Einstein—A History of Mathematical Astronomy*, Cambridge: Cambridge University Press, ISBN 978-0-521-82750-8

Magnani, L. (2007), Animal Abduction, From Mindless Organisms to Artifactual Mediators, In: *Model-Based Reasoning in Science, Technology, and Medicine* by Lorenzo Magnani and Ping Li

Marrou, H. I. (1956) *A history of education in antiquity*, University of Wisconsin Press

McCarthy, G. T. (1938), "The unit hydrograph and flood routing," unpublished manuscript, presented at a conference of the North Atlantic Division, U.S. Army, Corps of Engineers, June 24, 1938

McCarty, W., (2004) '*Modeling: A Study in Words and Meaning*', in A Companion to Digital Humanities pp. 254-272

Merton, R. K., (1969), *Social Theory and Social Structure*, Revised edition, Collier Macmillan

Mithen, S. (1996): *The Prehistory of the Mind. A Search for the Origins of Art, Religion and Science*. Thames and Hudson, London

Peirce, C.S. (1958), *Collected Papers of Charles Sanders Peirce*. Harvard University Press, Cambridge, MA (1931-1958) vols. 1-6, Hartshorne, C. and Weiss, P., eds.; vols. 7-8, Burks, A.W., ed.

Poli, R. (2006). The Ontology of What is Not There. In J. Malinowski, & A. Pietruszczak, *Essays in Logic and Ontology. Essays dedicated to Jerzy Perzanowski* (pp. 73-80). Amsterdam: Rodopi.

Popper, K., (1997) *The myth of the framework: in defence of science and rationality*, Printed by Routledge

Rivadulla, A. (2002), "Theories in physics, a rejoinder to Karl Popper's picture of Science *Discovery practices in natural sciences: from analogy to preduction*

Rouse, H., and Ince, S. (1963), *History of Hydraulics*, Pub. Dover Publications, Inc., New York, New York, U.S.A.

Saunders, J, Harris, JM , Bishop, SR, Schinaia, SA, Simons, R, Newell, RC, and Frost, NJ (2007) Coupling physical and ecological models: A new approach to predicting the impacts of aggregate extraction on biological recoverability. In: (Proceedings) *Change In Aquatic Ecosystems: natural and human influences*. (p. 30)

Schmuller, J., (2004), *Teach yourself UML in 24 hours*, Sams Publishing

Sebeok, T. A., and Danes, T. (2000), *The forms of meaning: modeling systems theory and semiotic analysis*, Walter de Gruyter, Gmbh and Co. Berlin.

Shaw, J.A. (2003), "The PID Control Algorithm, How it works, how to tune it, and how to use it," *(http://www.fadalti.com/other/PIDcontrolbook2.pdf, access on 12 March 2011)*

Simsion, G. () *Data Modeling: Theory and Practice*,

Stewart, I., (1990), *Does God play dice?*,Basil Blackwell, P. 108-109.

Taylor, I., (2003), *Galileo*, Hodder and Stoughton, P. 12.

von Bertalanffy, L. von, (1940) "The Organisation Considered as a Physical System"; reprinted in "*General Systems Theory*", Braziller, New York..

Wiener, N., *Cybernetics, or control and communication in the animal and the machine*. Cambridge, Massachusetts: The Technology Press; New York: John Wiley & Sons, Inc., 1948

Wigner, E., (1991)"The unreasonable effectiveness of mathematics in the natural sciences", *The world treasury of physics, astronomy and mathematics*, edited by Ferris, T., Published by Little Brown and Company

Willis, M.J. (1999), "Proportional-Integral-Derivative Control," *http://lorien.ncl.ac.uk/ming/pid/PID.pdf, accessed on 12 March 2011*

In: Natural Selection
Editors: J. R. Lynch and D. T. Williamson
ISBN: 978-1-62257-467-4

Chapter 6

NATURAL SELECTION APPLICATIONS FOR ALGORITHMIC COMPUTATION

Christos Dimitrakopoulos*[1]*, Konstantinos Theofilatos*[1]*,
***Spiros Likothanassis*[1] *and Seferina Mavroudi*[1,2]
[1]Department of Computer Engineering and Informatics,
University of Patras, Greece
[2]Department of Social Work, School of Sciences of Health and Care,
Technological Educational Institute of Patras, Greece

ABSTRACT

Natural selection is one of the cornerstones of modern biology and one of the great intellectual revolutions of human history. The term was first introduced by Charles Darwin who put forth a coherent theory of evolution and amassed a great body of evidence in support of this theory. In the last few decades, the continuing advance of modern technology has brought about something novel. Evolution is now producing practical benefits in a very different field, and this time, the creators can claim that their explanation fits the facts perfectly. This field is computer science, and the benefits come from a programming strategy which incorporates the notion of natural selection within algorithmic computation, and builds the so called evolutionary algorithms. The main goal of this survey is to summarize the previous research on two of the most important evolutionary algorithm techniques, which are Genetic Algorithms (GAs) and Differential Evolution (DE). Both algorithms are population-based, which means that a population of candidate solutions is optimized through the inspiration of biological motivations. GAs are efficient search methods based on principles of natural selection and population genetics. DE took a serious attempt to replace the classical crossover and mutation operators of genetic algorithms. We focus on these algorithms to guide us through a review of many of the most important innovations that they have provoked to science. In this paper, different variations of operations used by these algorithms such as fitness function, mutation, selection and crossover are discussed. We try to identify some of the problems of these methods that have not been studied systematically yet. We also describe the parallel potential of the evolutionary algorithms by focusing on the higher computational costs that big populations causes for some hard problems. As astonishing and counterintuitive as it may seem to some, evolutionary algorithms have proven to be an enormously powerful and successful problem-solving strategy, dramatically demonstrating the power of evolutionary principles.

Keywords: Natural Selection, Evolutionary algorithms, Genetic Algorithms, Differential Evolution, Adaptive Evolutionary Algorithms, Recombination, Crossover, Mutation, Selection, Population Representation, Fitness Evaluation, Multiobjective Optimization

1. Introduction

Mimicking biological evolution and harnessing its power for adaptation are problems that have intrigued computer scientists for at least four decades in order to approach solution finding in problems (Lawrence, 1991). Algorithms which have risen from this logic maintain a population of coded potential solutions of the problem that we are trying to solve, and apply on it the various procedures of natural selection. Hence, from generation to generation, these systems create new populations of potential solutions by evolving the previous populations and thus imitating the evolutionary process in nature.

Evolutionary Algorithms are based on a model of natural, biological evolution, which was formulated for the first time by Charles Darwin (Darwin, 1958). The Darwinian theory of evolution explains the adaptive change of species by the principle of natural selection, which favors those species for survival and further evolution that are best adapted to their environmental conditions (the well-known "survival of the fittest"). In addition to selection, the other important factor for evolution recognized by Darwin is the occurrence of small, apparently random and undirected variations between the phenotypes, i.e., the manner of response and physical embodiment of parents and their offspring. These mutations prevail through selection, if they prove their worth in light of the current environment, otherwise, they perish. Under advantageous environmental conditions, population size grows exponentially, a process which is generally limited by finite resources. This point of view is presently generally accepted as the correct macroscopic explanation of evolution (Back, 1996).

Evolutionary Algorithms (EAs), is an interdisciplinary research field with a relationship to biology, Artificial Intelligence, numerical optimization and decision support in almost any engineering discipline (Back, 1996). EAs are based on models of organic evolution, i.e. natural selection. They model the collective learning process within a population of individuals, with each of them representing not only a search point in the space of potential solutions to a given problem, but also a potential temporal container of current knowledge about the rules of the environment. The starting population is initialized and evolves towards successively better regions of the search space by means of randomized processes of recombination, mutation, and selection. The environment delivers a quality information (fitness value) for new search points, and the selection process favors those individuals of higher quality to reproduce more often than worse individuals. Recombination mechanism allows for mixing of parental information while passing it to their descendants, and mutation introduces innovation into the population. This process is currently used by three different mainstreams of Evolutionary Algorithms, i.e. Evolution Strategies (ESs), Genetic Algorithms (GAs), and Evolutionary Programming (EP). In this paper, we focus on analyzing the properties of GAs, the most famous sub-field of evolutionary algorithms, as well as on another contemporary evolutionary optimization technique, which is the Differential Evolution (DE).

2. Natural-Selection Based Optimization Techniques

2.1. Genetic Algorithms

Genetic algorithms (GAs), invented by John Holland in the 1960s, are the most widely used approaches to computational evolution. Holland presented GAs in a general theoretical framework for adaptation in nature. Holland's motivation was largely scientific as he was attempting to understand and link diverse types of natural phenomena but he also proposed potential engineering applications of GAs. Since the publication of Holland's book, the field of GAs has grown into a significant sub-area of artificial intelligence and machine learning (Lawrence, 1991). Research on GAs has spread from computer science to engineering and, more recently, to fields such as molecular biology, immunology, economics, and physics (Holland, 1995).

GAs are search meta-heuristic algorithms inspired by the principle of natural selection. As most meta-heuristic algorithms, they are useful and efficient if the search space is big and complicated or there does not exist any available mathematical analysis of the problem.

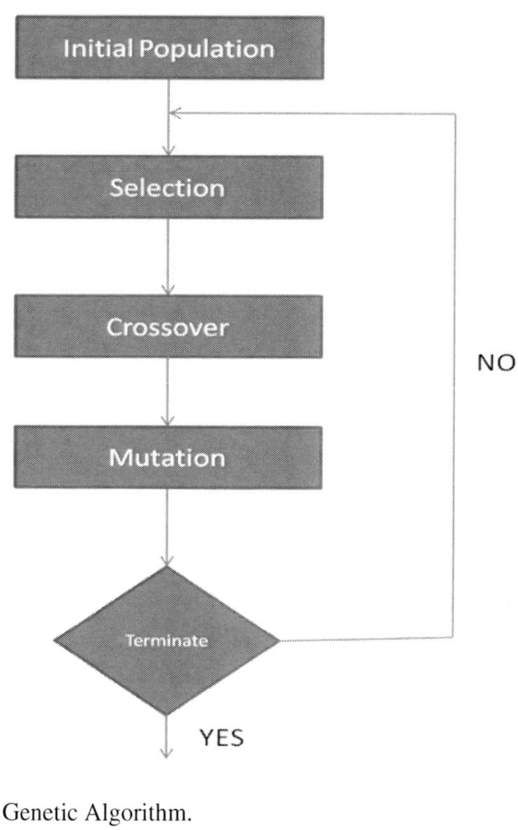

Figure 1. Flowchart of a Genetic Algorithm.

A population of candidate solutions, called chromosomes, is optimized via a number of evolutionary cycles and genetic operations, such as *crossovers* or *mutations*. Chromosomes consist of genes, which are the optimizing parameters.

At each iteration (*generation*), a fitness function is used to evaluate each chromosome, measuring the quality of the corresponding solution, and the fittest chromosomes are selected to survive. This evolutionary process is repeated until some termination criteria are met. It has been shown that GAs can deal with large search spaces and do not get trapped in local optimal solutions like other search algorithms (Holland, 1995).

2.1.1. Fitness Evaluation and Representation

Traditional GAs work on bit strings of fixed lengths λ, i.e., $I = \{0,1\}^{\lambda}$. To apply canonical GAs to continuous parameter optimization problems of the form $f : \prod_{i=1}^{n}[u_i, v_i] \rightarrow R(u_i < v_i)$, the bit string is logically divided into n segments of (in most cases) equal length λ_x (i.e., $\lambda = n*\lambda_x$), and each segment is interpreted as the binary code of the corresponding object variable $x_i \in [u_i, v_i]$. A segment decoding function $G^i : \{0,1\}^{\lambda x} \rightarrow [u_i, v_i]$ typically looks like

$$G^i(a_{i1} \dots a_{i\lambda x}) = u_i + \frac{v_i + u_i}{2^{\lambda x} - 1} * \left(\sum_{j=1}^{\lambda x} \alpha_{ij} 2^{j-1}\right), \quad (1)$$

where $(a_{i1} \dots a_{i\lambda x})$ denotes the ith segment of an individual $\alpha = (a_{11} \dots a_{n\lambda x}) \in I^{n*\lambda x} = I^{\lambda}$.

In terms of molecular genetics, there is an analogy to the discrete alphabet of nucleotide bases, double strands of which form the DNA. This information chain (the *genotype)* is decoded in several steps to amino acids (by means of the genetic code), to proteins, and finally to the phenotype by means of the epigenetic apparatus. Therefore, the GA representation scheme can be interpreted to be slightly closer to the natural model than both other algorithms discussed here. However, the impact of this additional representational level on the suitability for parameter optimization has still to be investigated further. The genetic code as well as the epigenetic apparatus has been completely ignored until now (Bäck and Schwefel, 1993).

2.1.2. Selection

Selection in canonical GAs emphasizes a probabilistic survival rule mixed with a fitness dependent chance to have (different) partners for producing more or less offspring. By deriving an analogy to the game-theoretic *multi-armed bandit problem,* Holland identifies a necessity to use *proportional selection* in order to optimize the trade-off between further exploiting promising regions of the search space while at the same time also exploring other regions. For proportional selection $s : I^{\mu} \rightarrow I^{\mu}$, the reproduction probabilities of individuals a_i are given by their relative fitness, i.e. (for every $i \in \{1, \dots, \mu\}$):

$$p_s(a_i) = \frac{\Phi(a_i)}{\sum_{j=1}^{\mu} \Phi(\alpha_j)} \quad (2)$$

Sampling μ individuals according to this probability distribution yields the next generation of parents. Obviously, this mechanism fails in the case of negative fitness or minimization tasks. Researchers usually add a constant value to all fitnesses (maybe the maximum positive fitness value), in order to avoid the existence of negative fitnesses. In minimization problems, they usually put their fitness function in the denominator of a fraction or multiply the fitness function with -1 and add a large enough constant.

The selection pressure which is the process of selecting the best individuals for the next generation is a critical parameter to be improved in GAs. If it is set too low, then the rate of convergence towards the optimum solution is too low. If the selection pressure is set too high, the system is likely to be stuck in a local optimum due to the loss of diversity in the population. Thus the selection methods controls the selection pressure, which in turn determines how fast the algorithms converge. The speed of convergence of different selection schemes was first studied by Goldberg and Deb (Goldberg and Kalyanmoy, 1991). The selection mechanism should be chosen such that it converges to the global optimum solution by without being caught into local optimum and also should encompass the knowledge of the existing data. There are many different selection techniques proposed in the literature (Sivaraj, 2011), with the most famous of them been the Tournament selection and the rank-based selection.

In tournament's selection most famous variant, two individuals are chosen at random and the better of the two individuals is selected with fixed probability p, $0.5 < p < 1$. In linear ranking selection (most famous variant of rank-based selection), first the individuals are ranked according to the fitness values. Those with high fitness values will be ranked high and those with low fitness values will eventually have lower ranks. Then the individuals are selected with a probability that is linearly proportional to the rank of the individuals in the population.

2.1.3. Recombination/Crossover

In canonical GAs, emphasis is mainly concentrated on *crossover,* the recombination operator of *GAs,* as the main variation operator that recombines useful segments from different individuals. Crossover $r'_{\{pc\}} : I^{\mu} \rightarrow I$ is again an operator working entirely on the bit representation, completely ignoring the genetic code and the epigenetic apparatus. It does not respect the semantic boundaries λ_x of the encoded variables. An exogenous parameter p_c (crossover rate) indicates the probability per individual to undergo recombination. Typical values for p_c are in the range [0.6, 1.0]. When two parent individuals $s = (s_1, \ldots, s_\lambda)$, $v = (v_1, \ldots, v_\lambda)$ have been selected (at random) from the population, crossover forms two offspring individuals s', v' according to the following scheme:

$$s' = (s_1, \ldots, s_{x-1}, s_x, v_{x+1}, \ldots v_l) \tag{3}$$

$$v' = (v_1, \ldots, v_{x-1}, v_x, s_{x+1}, \ldots s_l) \tag{4}$$

As before, $x \in \{I, \ldots, \lambda\}$ denotes a uniform random variable, and one of the two offspring individuals is randomly selected to be the overall result of crossover. This *one-point crossover* can be extended naturally to a generalized m-point crossover by sampling more than one breakpoint and alternately exchanging each second resulting segment. The *uniform crossover* operator drives the number of crossover points to an extreme by performing the random decision whether to exchange information between parents for each bit of the genotype anew. On the genotype level this operator corresponds well to discrete recombination in ESs. Actually, however, neither clear theoretical nor empirical evidence exists to decide which crossover operator is most appropriate, although several investigations tried to shed light on these questions.

Through the population, GAs implicitly maintain the statistics about the search space. This implicit statistics can be used explicitly to enhance GA's performance. Inspired by this idea, a statistics-based adaptive non-uniform crossover (SANUX) has been proposed. SANUX uses the statistics information of the alleles in each locus to adaptively calculate the swapping probability of that locus for crossover operation. A simple triangular function has been used to calculate the swapping probability. In this paper two new functions, the trapezoid and exponential functions, are proposed for SANUX instead of the triangular function. Experiment results show that both functions further improve the performance of SANUX (Zhang and Chung, 2005).

In (Yang, 2002) the basic conceptual features and specific characteristics of various crossover operators in the context of the Traveling Salesman Problem (TSP) are discussed. The results of experimental comparison of more than six different crossover operators for the TSP are presented.

2.1.3.1. Uniform Crossover Operator

The child is formed by an alternating randomly between the two parents.

2.1.3.2. Cycle Crossover

The Cycle Crossover (CX) builds offspring in such a way that each city (and its position) comes from one of the parents.

2.1.3.3. Partially-Mapped Crossover (PMX)

Partially matched crossover PMX, is made by randomly choosing two crossover points XP1 and XP2 which break the two parents in three sections. S1 and S3 the sequences of Parent1 are copied to the Child1, the sequence S2 of the Child1 is formed by the genes of Parent2, beginning with the start of its part S2 and leaping the genes that are already established.

2.1.3.4. The Uniform Partially-Mapped Crossover (UPMX)

The Uniform Partially Matched Crossover, uses the technique of PMX. Any times, it does not use the crossover points; instead, it uses a probability of correspondence for each iteration.

2.1.3.5. Non-Wrapping Ordered Crossover (NWOX)

Non-Wrapping Ordered Crossover (NWOX) operator is based upon the principle of creating and filling holes, while keeping the absolute order of genes of individuals. The holes are created at the retranscription of the genotype, if $x_{j,i} \in \{x_{k,a}, \ldots, x_{k,b}\}$ then $x_{j,i}$ is a hole.

2.1.3.6. Ordered Crossover (OX)

The Ordered Crossover method, is used when the problem is of order based, for example in U-shaped assembly line balancing etc. Given two parent chromosomes, two random crossover points are selected partitioning them into a left, middle and right portion. The ordered two-point crossover behaves in the following way: child1 inherits its left and right section from parent1, and its middle section is determined.

2.1.3.7. Crossover with Reduced Surrogate

The reduced surrogate operator constrains crossover to always produce new individuals wherever possible. This is implemented by restricting the location of crossover points such that crossover points only occur where gene values differ.

2.1.3.8. Shuffle Crossover

Shuffle crossover is related to uniform crossover. A single crossover position (as in single-point crossover) is selected. But before the variables are exchanged, they are randomly shuffled in both parents. After recombination, the variables in the offspring are unstuffed. This removes positional bias as the variables are randomly reassigned each time crossover is performed.

The experiment results demonstrated that OX operator enables to achieve a better solutions than other operators tested.

2.1.4. Mutation

Mutation in canonical *GAs* works on the bit string level and is traditionally referred to as a "background operator" (Holland 1975). It works by occasionally inverting single bits of individuals, with the probability p_m of this event usually being very small ($p_m \sim 1*10^{-3}$ per bit). Often, p_m depends neither on the number n of object variables nor on the total length λ of the bit string, This kind of mutation is ruled by pure randomness like the Monte Carlo method; one cannot speak of a (mean) step size. On a single individual, mutation $m'_{\{pm\}}(s_1, \ldots, s_\lambda) = (s_1', \ldots, s_\lambda')$ works as follows (for every $i \in \{1, \ldots, \lambda\}$):

$$s_i' = \begin{cases} s_i & , \quad x_i > p_m \\ 1 - s_i & , \quad x_i \le p_m \end{cases} \tag{5}$$

Here $x_i \in [0,1]$ is a uniform random variable, sampled new for each bit. The reader should be very careful about mutation rate values given in the literature, because originally mutation was defined as a substitution of a bit by a random element from the alphabet {0,1} (Holland 1975), such that mutation rates according to our definition are twice as large as mutation rates according to Holland's original one. However, because it is more appropriate to interpret mutation as a real change (instead of a random toss with probability one-half), it is used as an inversion event by most GA researchers.

2.1.5. Adaptive Crossover and Mutation Schemes

Mutation and crossover rate parameter are considered to be two of the most sensitive of the parameters that a genetic algorithm works with. For example, it has been shown that through using a mutation rate variation scheme that adapts the mutation rate parameter during the run of the algorithm, the time to find the optimum is decreased (Uyar and Eryigit, 2004).

The choice of p_c and p_m (crossover and mutation probability) is known to critically affect the behavior and the performance of the GA, and a number of guidelines exist in the literature for choosing them (De Jong, 1975). The generalized guidelines are inadequate because the choice of the optimal p_c and p_m becomes specific to the problem under consideration. Instead of using fixed p_c and p_m, some adaptive parameter control schemes that can relieve the burden of specifying the values of p_c and p_m, have been proposed. In (Grefenstette, 1986), a second level GA is used to select p_c and p_m. Although this method can select p_c, and p_m, according to the solution distribution, it is computationally expensive. In (Srinivas and Palnaik, 1994), an adaptive GA is proposed. P_c and p_m vary depending on the fitness values of the solutions. Although the procedures of tuning p_c and p_m are computationally efficient, the distribution factor that gives strong indication on p_c and p_m is lost.

Parameter control techniques are classified mainly into three groups based on the type of change they introduce:

- deterministic: the parameter value is updated according to some deterministic rule,
- adaptive: the parameter value is updated based on some feedback taken from the population
- self-adaptive: the parameter is evaluated and updated by the evolutionary algorithm itself.

2.1.5.1. Deterministic Approach

The deterministic mutation rate schedule provides the mutation rate to decrease from a value (generally, 0.5) to the optimum mutation rate (generally 1/L) without using any feedback from the population. The deterministic mutation rate schedule implementation proposed in (Smith and Fogarty, 1997) was reported in (Thierens, 2002) as having the most successful results for hard combinatorial problems. Based on this method, the time-varying mutation rate is calculated using the formula given in Eq. 6. In this formula, t is the current generation number and T is the maximum number of generations. In the original proposal for Eq. 6, the k value is chosen as 1.

$$p_t = (2 + \frac{L-2}{T-1} * t)^{-k} \tag{6}$$

2.1.5.2. Individually Adaptive Approach

In (Srinivas and Patnaik, 1994) an individually adaptive GA method (AGA) is chosen for the comparisons. In this method, the probabilities of crossover and mutation are adapted depending on the fitness values of the individuals. The adaptation of the p_c and p_m allows the individuals having fitness values of over-average to maintain their genetic material, while forcing the individuals with sub-average fitness values to disrupt. The mutation rate adaptation rule is given in Eq. 7. In this equation, f denotes the fitness value of the individual, f_{max} denotes the best fitness value of the current generation, and f_{avg} denotes the average fitness value of the current generation. The constants k_2 and the k_4 are chosen as 0.5.

$$p_m = k_2(f_{max} - f)/(f_{max} - f_{avg}), f \geq f_{avg}$$
$$p_m = k_4, f < f_{avg} \quad (7)$$

2.1.5.3. Self-Adaptive Approach

In the self-adaptive approach, the parameters are encoded into the chromosomes and undergo mutation and recombination. The basic idea is that better parameter values lead to better individuals and these parameter values will survive in the population since they belong to the surviving individuals. (Back, 1993) refer to this approach also as online learning. In their work, they propose a self-adaptation mechanism of a single mutation rate per individual. The mutation of this mutation rate value gives the new mutation rate through Eq. 8. In this equation, γ is the learning rate and controls the adaptation speed. It is taken as 0.22 in (Goldberg, 1989) and also in this study.

$$p' = (1 + \frac{1-p}{p} * \exp(-\gamma * \mathrm{N}(0,1)))^{-1} \quad (8)$$

In (Stanhope and Daida, 1997), a Gene Based Adaptive Mutation (GBAM) method is proposed. This approach experiments with varying mutation rate values during the run, using feedback from the population. Different from other known mutation adaptation strategies, in GBAM each locus has its own mutation rate value. An adaptive approach for adjusting mutation rates for the gene locations based on the feedback obtained by observing the relative success or failure of the individuals in the population is used.

In GBAM, there are two different mutation rates defined for each locus: a mutation rate value pm_1 for those genes that have an allele value of "1" at that locus and another mutation rate value pm_0 for those that have a "0". In the reproduction phase, the appropriate mutation rate is applied based on the gene allele value. Initially all mutation rates are set to a default value in the specified boundaries. During the GA run, the mutation probabilities pm_1 and pm_0 for all loci are updated at each generation using the feedback taken from the relative success or failures of those individuals having a "1" or "0" at that locus respectively. For a maximization problem, the update rule for the mutation rate subunits for one gene location can be seen in Eq.9. This update rule is applied separately for each locus.

$$p_{m0}^{+} = \begin{cases} p_{m0} + \gamma, (S_{avg} / P_{avg}) > 1 \\ p_{m0} - \gamma, (S_{avg} / P_{avg}) \leq 1 \end{cases}$$
$$p_{m1}^{+} = \begin{cases} p_{m1} - \gamma, (S_{avg} / P_{avg}) > 1 \\ p_{m1} + \gamma, (S_{avg} / P_{avg}) \leq 1 \end{cases} \quad (9)$$

It is expected that GBAM fastens the convergence of the GA to the optimum. This has been confirmed through tests performed on a simple one-max problem (Stanhope and Daida, 1997). Since this problem is a unimodal function fast convergence to the optimum is not a problem. However, extra convergence tests should be applied for multimodal fitness landscapes to escape from local optima, and necessary precautions should be taken to restore diversity.

2.2. Differential Evolution

Differential evolution (DE) is arguably one of the most powerful stochastic real-parameter optimization algorithms in current use. DE operates through similar computational steps as employed by a standard evolutionary algorithm (EA). However, unlike traditional EAs, the DE-variants perturb the current-generation population members with the scaled differences of randomly selected and distinct population members. Therefore, no separate probability distribution has to be used for generating the offspring. Since its inception in 1995, DE has drawn the attention of many researchers all over the world resulting in a lot of variants of the basic algorithm with improved performance.

The Differential Evolution algorithm is currently one of the most powerful and promising stochastic real parameter optimization algorithms (Das and Suganthan, 2011). As an evolutionary algorithm, it iteratively applies selection, mutation and crossover operators until some termination criteria are reached. However, in opposite to traditional evolutionary algorithms, DE is mainly based on a specific mutation operator which perturbs population individuals with the scaled differences of randomly selected and distinct population members. DE is able to handle continuous gene representation (Brest, 2006).

An iteration of the classical DE algorithm consists of the four basic steps—initialization of a population of search variable vectors, mutation, crossover or recombination, and finally selection. After having illustrated these stages, we now formally present the whole of the algorithm in a pseudo-code below. The terminating condition can be defined in a few ways like: 1) by a fixed number of iterations G_{max}, with a suitably large value of G_{max} depending upon the complexity of the objective function; 2) when best fitness of the population does not change appreciably over successive iterations; and alternatively 3) attaining a pre-specified objective function value (Qin and Huang 2009).

```
Procedure DE
Input: Randomly initialized position and velocity of the particles: xi(0)
Output: Position of the approximate global optima X*.
Begin
        Initialize Population
        Evaluate fitness
        For i = 0 to max_iteration do
                Begin
                        Create Difference-Offspring
                        Evaluate Fitness
                        If an offspring is better than its parent
                        Then replace the parent by offspring in the next generation
                        End if
        End For
End
```

2.2.1. Initialization of the Parameter Vectors

DE searches for a global optimum point in D-dimensional real parameter space R^D. It begins with a randomly initiated population of NP D dimensional real-valued parameter vectors. Each vector, also known as genome/chromosome, forms a candidate solution to the multidimensional optimization problem. Since the parameter vectors are likely to be changed over different generations, we may adopt the following notation for representing the ith vector of the population at the current generation:

$$X_{i,G} = [x_{1,i,G}, x_{2,i,G}, x_{3,i,G}, \ldots, x_{D,i,G}] \quad (10)$$

The initial population (at $G = 0$) should cover this range as much as possible by uniformly randomizing individuals within the search space constrained by the prescribed minimum and maximum bounds: $X_{min} = \{x_{1,min}, x_{2,min}, \ldots, x_{D,min}\}$ and $X_{max} = \{x_{1,max}, x_{2,max}, \ldots, x_{D,max}\}$. Hence, we may initialize the jth component of the ith vector as

$$x_{j,i,0} = x_{j,min} + rand_{i,j}[0,1] * (x_{j,max} - x_{j,min}) \quad (11)$$

where $rand_{i,j}[0,1]$ is a uniformly distributed random number lying between 0 and 1 and is instantiated independently for each component of the ith vector.

2.2.2. Mutation with Difference Vectors

In DE-literature, a parent vector from the current generation is called *target* vector, a mutant vector obtained through the differential mutation operation is known as *donor* vector and finally an offspring formed by recombining the donor with the target vector is called *trial* vector. In one of the simplest forms of DE-mutation, to create the donor vector for each ith target vector from the current population, three other distinct parameter vectors, say $X_{r_1^i}, X_{r_2^i} and\ X_{r_3^i}$ are sampled randomly from the current population. The indices r_1^i, r_2^i $and\ r_3^i$ are mutually exclusive integers randomly chosen from the range [1, NP] which is also different from the base vector index i. These indices are randomly generated once for each mutant vector. Now the difference of any two of these three vectors is scaled by a scalar number *F* (that typically lies in the interval [0.4, 1]) and the scaled difference is added to the third one whence we obtain the donor vector $V_{i,G}$. The process can be expressed as

$$V_{i,G} = X_{r_1^i,G} + F * (X_{r_2^i,G} - X_{r_3^i,G}) \quad (12)$$

2.2.3. Crossover

The DE family of algorithms can use two kinds of crossover methods—*exponential* (or two-point modulo) and *binomial* (or uniform). In exponential crossover, we first choose an integer *n* randomly among the numbers [1, D]. This integer acts as a starting point in the target vector, from where the crossover or exchange of components with the donor vector starts. We also choose another integer *L* from the interval [1, D]. *L* denotes the number of components the donor vector actually contributes to the target vector. After choosing *n* and *L* the trial vector is obtained as

$$u_{j,i,G} = v_{j,i,G} for\ j = \langle n\rangle_D\ \langle n+1\rangle_D, \ldots, \langle n+L-1\rangle_D \quad (13)$$

$$x_{j,i,G} = for\ all\ other\ j \in [1, D]$$

An efficient technique for adapting control parameter settings associated with differential evolution was described in (Thierens, 2002). DE has only a few control parameters, which are kept fixed throughout the entire evolutionary process. However, it is not an easy task to properly set control parameters in DE. (Thierens et. al., 2002) present an algorithm—a new version of the DE algorithm—for obtaining self-adaptive control parameter settings that show good performance on numerical benchmark problems. The results demonstrated that the specific algorithm with self-adaptive control parameter settings is better than, or at least comparable to, the standard DE algorithm and evolutionary algorithms from literature when considering the quality of the solutions obtained.

2.2.4. Selection

To keep the population size constant over subsequent generations, the next step of the algorithm calls for *selection* to determine whether the target or the trial vector survives to the next generation, i.e., at $G = G + 1$. The selection operation is described as

$$X_{i,G+1} = U_{i,G} \text{ if } \mathrm{f}(U_{i,G}) <= \mathrm{f}(X_{i,G}) \tag{14}$$

$$= X_{i,G} \text{ if } \mathrm{f}(U_{i,G}) > \mathrm{f}(X_{i,G})$$

where the angular brackets $\langle\rangle_D$ denote a modulo function with modulus D.

On the other hand, binomial crossover is performed on each of the D variables whenever a randomly generated number between 0 and 1 is less than or equal to the Cr value. In this case, the number of parameters inherited from the donor has a (nearly) binomial distribution. The scheme may be outlined as

$$u_{j,i,G} = u_{j,i,G} \; if \; \left(rand_{i,j}[0,1] \leq Cr \; or \; j = j_{rand}\right) \tag{15}$$

$$= x_{j,i,G} \; otherwise$$

2.2.5. DE with Adaptive Selection of Mutation Strategies

DE can encompass a number of trial vector generation strategies, each of which may be effective over certain problems but poorly perform over the others (Feoktistov and Janaqi, 2004). In (Qin, 2009), for the first time Qin *et al.* proposed a self-adaptive variant of DE (SaDE), where along with the control parameter values the trial vector generation strategies are also gradually self-adapted by learning from their previous experiences in generating promising solutions. Consequently, it is possible to determine a more suitable generation strategy along with its parameter settings adaptively to match different phases of the search process/evolution.

In SaDE, four effective trial vector generation strategies namely the DE/rand/1/bin, DE/rand-to-best/2/bin, DE/rand/2/bin and finally DE/current-to-rand/1 were chosen to constitute a strategy candidate pool. The first three DE-variants are equipped with binomial type crossover while the last one uses arithmetic recombination. In the SaDE algorithm, for each target vector in the current population, one trial vector generation strategy is selected from the candidate pool according to the probability learned from its success rate in generating improved solutions (that can survive to the next generation) within a certain number of previous generations, called the learning period (LP). The selected strategy is subsequently applied to the corresponding target vector to generate a trial vector. More specifically, at each generation, the probabilities of choosing each strategy in the candidate pool are summed to 1. These probabilities are initially equal ($1/K$ for K total number of strategies in the pool) and are then gradually adapted during evolution, based on the *Success* and *Failure Rates* (Qin, 2009) over the previous *LP* generations. The adaptations of the probabilities take place in such a fashion that, the larger the success rate for the kth strategy in the pool within the previous *LP* generations, the larger is the probability of applying it to generate trial vectors at the current generation. The performance of SaDE was compared with the conventional DE and three adaptive DE-variants: ADE (Huang, 2007), SDE (Omran, 2005), and jDE (Brest, 2006) over a suite of 26 bound constrained numerical optimization problems, and the authors reported that, SaDE was more effective in obtaining better quality solutions, with the relatively smaller standard deviations, and higher success rates.

2.2.6. Adaptive DE with DE/Current-to-pbest Mutation

In order to avoid the need for problem specific parameter tuning and also to improve the convergence characteristics of DE, an adaptive DE-variant, called JADE, was recently proposed (Zhang and Sanderson, 2009). The algorithm implements a new mutation strategy, referred by the authors as DE/current-to-*p*best and uses an optional external archive to track the previous history of success and failure. It also updates the control parameters in an adaptive manner with generations. The DE/current-to-*p*best strategy is a less greedy generalization of the DE/current-tobest/strategy. Instead of only adopting the best individual in the DE/current-to-best/1 strategy, the current-to-pbest/1 strategy utilizes the information of other good solutions. Moreover, the recently explored inferior solutions are also incorporated in this strategy.

3. DISCUSSION

During the last years research in evolutionary computation and the application of EAs to real-world problems, have steadily and significantly expanded. The following paragraphs summarize some most remarkable trends from the author's perspective.

3.1. Parallel Potential of Evolutionary Algorithms

Evolutionary algorithms are generally able to find good solutions in reasonable amounts of time, but as they are applied to harder and bigger problems there is an increase in the time required to find adequate solutions. As a consequence, there have been multiple efforts to make EAs faster, and one of the most promising choices is to use parallel implementations (Berlich and Kunze, 2003).

The basic idea behind most parallel programs is to divide a task into chunks and to solve the chunks simultaneously using multiple processors. This divide-and-conquer approach can be applied to EAs in many different ways, and the literature contains many examples of successful parallel implementations. Some parallelization methods use a single population, while others divide the population into several relatively isolated subpopulations. Some methods can exploit massively parallel computer architectures, while others are better suited to multicomputers with fewer and more powerful processing elements.

There are three main types of parallel EAs : (1) global single-population masterslave EAs, (2) single-population fine-grained, and (3) multiple-population coarse-grained EAs (Berlich and Kunze, 2003). In a master-slave EA there is a single panmictic population (just as in a simple GA), but the evaluation of fitness is distributed among several processors. In this type of parallel EA, selection and crossover consider the entire population (known as global parallel EAs). Fine-grained parallel EAs are suited for massively parallel computers and consist of one spatially-structured population. Selection and mating are restricted to a small neighborhood, but neighborhoods overlap permitting some interaction among all the individuals. The ideal case is to have only one individual for every processing element available.

Multiple-population EAs are more sophisticated, as they consist on several subpopulations which exchange individuals occasionally. This exchange of individuals is called migration and it is controlled by several parameters. Multiple-population EAs are very popular, but also are the class of parallel EAs which is most difficult to understand, because the effects of migration are not fully understood. Multiple-population parallel EAs introduce fundamental changes in the operation of the EA and have a different behavior than simple EAs. Multiple-population parallel EAs are known with different names. Sometimes they are known as “distributed” EAs, because they are usually implemented on distributed memory MIMD computers. Since the computation to communication ratio is usually high, they are occasionally called coarse-grained EAs. Finally, Multiple-population EAs resemble the “island model” in Population Genetics which considers relatively isolated population, so the parallel EAs are also known as “island” parallel EAs. Since the size of the population is smaller than the population used by a serial EA, we would expect that the parallel EA converges faster. However, when we compare the performance of the serial and the parallel algorithms, we must also consider the quality of the solutions found in each case. Therefore, while it is true that smaller population converge faster, it is also true that the quality of the solution might be poorer. It is important to emphasize that while the master-slave parallelization method does not affect the behavior of the algorithm, the last two methods change the way the EA works. For example, in master-slave parallel EAs, selection takes into account all the population, but in the other two parallel EAs, selection only considers a subset of individuals. Also, in the master-slave any two individuals in the population can mate (i.e., there is random mating), but in the other methods mating is restricted to a subset of individuals. The final method to parallelize EAs combines multiple population with master-slave or fine-grained EAs. We call this class of algorithms hierarchical parallel EAs, because at a higher level they are multiple-population algorithms with single-population parallel EAs (either master-slave or fine-grained) at the lower level. A hierarchical parallel EAs combines the benefits of its components, and it promises better performance than any of them alone (Berlich and Kunze, 2003).

3.2. The Evaluation of Genetic Algorithms

When should an evaluation algorithm be used? For example, when an optimization problem is encountered, when should one consider the use of an evolutionary algorithm? Evolutionary algorithms are known as weak methods in the Machine Learning community. Weak are called the methods that do not exploit domain specific knowledge. Methods that exploit domain knowledge will almost always out-perform methods that are blind. This leads to the following observation: if one is interested in trying some form of evolutionary computation, try to add domain knowledge into the evolutionary algorithm.

One of the simplest tests to do before applying an evolutionary algorithm is to try some form of local search. In local search, a neighborhood structure is defined around every point in the search space. Search then proceeds from a point by testing all of the neighbors for an improving move. Any point where all of the neighbors are inferior is a *local optimum*.

An easy way to do local search is to apply a bit climber. This is especially true if a genetic algorithm is going to be used that also utilizes a bit encoding. In this case, the neighborhood is defined by flipping the L bits of the string representing the current point in the search space. This neighborhood is also known as the Hamming Distance-1 neighborhood. The entire search space is then *Hamming Space*. Dave Davis' algorithm called Random Bit Climbing (RBC) is a local search algorithm that climbs in Hamming Space (Lawrence, 1991). A random permutation is generated that determines the order in which bits are flipped Each improving move is accepted. After every bit has been tested, a new permutation is generated for the next pass. If RBC has checked every bit in the string and no improvement is found, a local optimum has been found and RBC is restarted from a new random point in the search space.

Other methods that might be used include such simple methods as forms of line search and the Nelder Mead simplex methods. None of these methods requires gradient information. If gradient information is available, then some form of non-linear gradient-based search should be attempted. (Whitley and Mathias, 1996) provide a more in-depth discussion of the evaluation of evolutionary algorithms for optimization and search problems (Whitley, 2001).

3.3. Multiobjective Optimization

In real-world optimization applications, it is often hard to formulate the optimization goal as a scalar function. Typically, there are several criteria or objectives, and not unusually, these objectives stay in conflict with each other. As an example, in the design of an automobile an engineer may wish to maximize crash resistance for safety and minimize weight for fuel economy. Simply combining the different associated objective functions in a linear way is usually unsatisfactory. Instead, one is interested in a so-called Pareto optimal set of solutions, i.e., any solution that cannot be improved with respect to one objective without worsening the situation with respect to the other objectives. Special strategies are therefore needed to deal with such multiobjective optimization problems (Fonseca and Fleming, 1993). Since EAs work on populations of candidate solutions, they represent a promising basic framework for multiobjective optimization. In the last few years many variants and extensions of classical EAs have been developed and were demonstrated to be superior techniques for this purpose (Coello and Veldhuizen, 2002).

3.4. Extended Hybrid Architectures

In particular with respect to solving discrete optimization problems, classical simple EAs are known to be typically less efficient than other state-of-the-art algorithms that are specifically tailored to the problems. As already mentioned above, EAs are therefore often combined with problem-specific algorithms. Beside this, also hybrids of EAs and other metaheuristics such as local search, simulated annealing, tabu search, guided local search, variable neighborhood search, and variable depth search are nowadays widely used, and for many hard combinatorial optimization problems, such approaches are today's leading algorithms for finding superior solutions to large problem instances in limited time. Such combinations are also called *memetic algorithms* (Lawrence, 1991). Typically, the EA's primary responsibility is to take care of diversification and coverage of all major areas of the search space, while the embedded other metaheuristic is mainly responsible for intensification of the search, thus, the fine-tuning of the most promising candidate solutions. Some highly promising work has also been done in combining EAs with exact optimization techniques including dynamic programming, branch-and-bound, and the large palette of (integer) linear programming techniques. In these hybrid systems, the exact technique may work as a subordinate of the EA in order to solve smaller subproblems, the EA may act as subordinate of the exact technique (e.g., to obtain better bounds), or both strategies can run in a parallel or interleaved way, exchanging information and stimulating each other. In the latter two strategies, the exact approach's property of being, in principle, able to prove optimality or to provide a quality guarantee can be retained (Puchinger and Raidl, 2005).

3.5. Theoretical Background

Research in the theory of evolutionary computation has also made significant progress. Today, various theoretical models and results help to understand under which conditions an EA works well or is supposed to perform only poorly. The spectrum of algorithms which could be rigorously analyzed, including the derivation of tight bounds for the expected run-times and the expected quality of obtained solutions, has significantly increased. Nevertheless, there still exist many open questions, and despite the achievements, theory is today unfortunately still far away from precisely predicting or fully describing the expected behavior of a more complex hybrid EA when applied to a hard problem from real world. See e.g. (Wright et. al., 2005) for recent theoretical results.

CONCLUSION

The main objective of the present paper was to throw some light on the theoretical aspect of two of the most important Evolutionary Algorithms, which are the Genetic Algorithms and the Differential Evolution. We reviewed the basic components of the algorithms, named selection, crossover/recombination, mutation and fitness evaluation and representation. We also focused on their adaptive approaches, which vary the mutation and crossover rates during the run of the algorithms. The main ultimate outcome of adapting an evolutionary algorithm may be the decrease of the time needed to find the optimum solution. This effect of adaptive schemes of evolutionary algorithms can be extremely meaningful in real-time problems where it is necessary for the algorithm to be able to find the solution (or a solution near the optimal one) of the problem sufficiently fast. Research in both algorithms is progressing very fast and numerous researchers from various fields are focusing on developing some workable schemes. Different companies are also working on getting commercial products. Hence, we hope that some commercial and effective schemes will be available in future.

REFERENCES

Bäck, T., Schwefel, H. (1993) *An Overview of Evolutionary Algorithms for Parameter Optimization*, Vol. 1, Issue 1, pp.1-23.

Bäck T., (1993) "Optimal Mutation Rates in Genetic Search", Proceedings of 5th International Conference on Genetic Algorithms, Morgan Kaufmann.

Back, T. (1996) Evolutionary algorithms in theory and practice: evolution strategies, evolutionary programming, genetic algorithms, Oxford University Press, USA; 1 edition.

Berlich, R., Kunze, M., (2003) Parallel evolutionary algorithms, *Nuclear Instruments and Methods in Physics Research Section A: Accelerators, Spectrometers, Detectors and Associated Equipment,* Vol. 502, Issues 2–3, 21 April 2003, pp. 467–470 .

Brest, J., Greiner, S., Boskovic, B., Mernik, M., and Zumer, V., (2006) "Self-adapting control parameters in differential evolution: A comparative study on numerical benchmark problems," *IEEE Trans. Evol. Comput.*, vol. 10, no. 6, pp. 646–657.

Coello, C.A., Veldhuizen, D.A., and Lamont, G.B. (2002) *Evolutionary Algorithms for Solving Multi-Objective Problems*, Kluwer Academic Publishers.

Darwin, C. (1958) *On the Origin of Species by Means of Natural Selection, or, the Preservation of Favoured Races in the Struggle for Life*. London: J. Murray

Das, S., and Suganthan, P. (2011) Differential Evolution: A survey of the State-of-the-Art. IEEE Transactions on evolutionary computation. vol.15(Issue 1), pp.4-30.

De Jong, K.A., (1975) "An analysis of the behavior of a class of genetic adaptive systems," PhD. Dissertation, University of Michigan.

Feoktistov, V., and Janaqi, S., (2004) "Generalization of the strategies in differential evolution," in *Proc. 18th IPDPS*, pp. 165a.

Fonseca, C.M. and Fleming, P.J. (1993) Genetic Algorithms for Multiobjective Optimization: Formulation, Discussion and Generalization, In: Proceedings of the 5th International Conference on Genetic Algorithms San Francisco, CA, USA: Morgan Kaufmann Publishers Inc., p. 416--423.

Goldberg, D.E., (1989) *Genetic Algorithms in Search Optimization and Machine Learning*, Addison Wesley.

Goldberg, D.E., Kalyanmoy D., (1991) A comparative analysis of selection schemes used in genetic algorithms, Foundations of Genetic Algorithms, Morgan Kaufmann.

Grefenstette, J. (1986) "Optimization of control parameters for genetic algorithms," *IEEE Trans. Syst. Man. and Cybernetics,* vol. 16. no. I , pp. 122-128.

Holland, J. (1995) Adaptation in natural and artificial systems: an introductory analysis with applications to biology, control, and artificial intelligence. A Bradford Book, Cambridge: Mass ;MIT Press.

Huang, V. L., Qin, A.K., Suganthan, P.N., and Tasgetiren, M.F. (2007) "Multiobjective optimization based on self-adaptive differential evolution algorithm," in *Proc. Congr. Evol. Comput.*, pp. 3601–3608.

Lawrence, D.D., (1991) Bit-Climbing, Representation Bias, and Test Suite Design. In L. Booker and R. Belew, editors, Proc. of the 4th Int'l. Conf. on GAs, pp. 18-23. Morgan Kaufmann.

Lawrence, D.D., (1991), Handbook of Genetic Algorithms, Van Nostrand Reinhold; 1st edition

Omran, M. G., Salman, A., and Engelbrecht, A.P. (2005) "Self-adaptive differential evolution," in *Proc. Comput. Intell. Security*, Lecture Notes in Artificial Intelligence 3801, pp. 192–199.

Puchinger, J., and Raidl, G.R., (2005) Combining metaheuristics and exact algorithms in combinatorial optimization: A survey and classification. In *Proceedings of the First International Work-Conference on the Interplay Between Natural and Artificial Computation*, LNCS. Springer.

Qin, A. K., Huang, V. L. and Suganthan, P.N., (2009) "Differential evolution algorithm with strategy adaptation for global numerical optimization, "*IEEE Trans. Evol. Comput.*, vol. 13, no. 2, pp. 398–417.

Sivaraj, R., Ravichandran, T., (2011) A review of selection methods in genetic algorithm, *International Journal of Engineering Science and Technology* (IJEST), vol. 3(Issue 5), pp. 3792 - 3797.

Smith J. E., Fogarty T. C., (1997). "Operator and Parameter Adaptation in Genetic Algorithms", *Soft Computing* 1, pp. 81-87, Springer-Verlag.

Srinivas, M., Patnaik, L. M., (1994) "Adaptive Probabilities of Crossover and Mutation in Genetic Algorithms", *IEEE Transactions on Systems, Man and Cybernetics,* Vol. 24, No. 4., pp. 656-667, IEEE.

Stanhope S.A., Daida J.M., (1997) "An Individually Variable Mutation-Rate Strategy", *Proceedings of the Sixth Annual Conference on Evolutionary Programming.*

Thierens D., (2002) "Adaptive Mutation Control Schemes in Genetic Algorithms", *Proceedings of Congress on Evolutionary Computing*, IEEE.

Uyar A.S., Eryigit G., Sariel S. (2004) An Adaptive Mutation Scheme in Genetic Algorithms Fastening Convergence to the Optimum, In Proceedings of 3rd APIS: Asian Pacific International Symposium on Information Technologies.

Whitley, D., (2001) An overview of evolutionary algorithms: practical issues and common pitfalls, *Information and Software Technology* , Vol. 43, Issue 14, pp. 817–831.

Whitley, D., Mathias, K., Rana, S., and Dzubera, J. (1996) Evaluating Evolutionary Algorithms. *Artificial Intelligence Journal*, vol.85, pp.1-32.

Wright, A., Vose, M., De Jong, K., and Schmitt, L., (2005) *Foundations of Genetic Algorithms Conference 2005 (FOGA-8)*, LNCS. Springer.

Yang, S., (2002) Adaptive crossover in genetic algorithms using statistics mechanism, Proceedings of the 8th International conference on Artificial life, pp. 182-185

Zhang, J., Chung, H. and Zhong, J. (2005) Adaptive Probabilities of Crossover and Mutation in Genetic Algorithms Based on Clustering Technique, Proceedings of the 2005 conference on Genetic and evolutionary computation, GECCO '05, pp. 1577-1578, ACM New York, NY, USA

Zhang, J., and Sanderson, A.C., (2009) "JADE: Adaptive differential evolution with optional external archive," *IEEE Trans. Evol. Comput.*, vol. 13, no. 5, pp. 945–958.

In: Natural Selection
Editors: J. R. Lynch and D. T. Williamson

ISBN: 978-1-62257-467-4

Chapter 7

ALLOZYMES, DNA AND NATURAL SELECTION: A MOLLUSKS TALE

E. P. Silva[*]
Laboratório de Genética Marinha e Evolução, Departamento de Biologia Marinha, Instituto de Biologia, Universidade Federal Fluminense, Brazil

ABSTRACT

Mollusks are a group in which some interesting examples have been described over the years concerning effects of natural selection over populations. A classical example is the observed clines of *Lap* alleles and salinity gradients in mussels. With the introduction of DNA techniques in the field of population genetics, it has been proposed that comparing the amount of geographic variation in allele frequencies for protein and DNA polymorphisms is a powerful method for detecting the effects of natural selection. Once more, mollusks are important models in these studies. These issues will be reviewed in this chapter and the importance of the molecular markers (allozymes and DNA) for studies on natural selection discussed in the light of the results obtained from mollusks which includes the Brazilian cockle *Anomalocardia brasiliana*, the brown mussel *Perna perna* and the blue mussel *Mytilus edulis*.

INTRODUCTION

The study of genetic evolution at the species level demands at first a description of the variation within and between populations. A second and more difficult task is to detect the role of natural selection in shaping the observed pattern of distribution of genetic variation. When the application of allozyme electrophoresis became widely used to describe genetic variation in natural populations (Lewontin, 1974), many studies have tried to relate different allelemorphs with environmental factors, especially temperature and salinity. Mollusks are a group in which some interesting examples have been described over the years. A classical

[*] E-mail: gbmedson@vm.uff.br.

example is the observed clines of *Lap* alleles and salinity gradients in mussels (Koehn, 1978, 1983, 1985; Hilbish and Koehn, 1985a, 1985b).

Currently, DNA techniques gained terrain over allozymes in the field of population genetics (Silva and Russo, 2000) and, following its development in the end of 1980s (Avise, 2004), a growing number of studies have been showing incongruence between estimates of genetic structure based on allozymes and DNA markers (Karl and Avise, 1992; Marchant and Shaw, 1993; Riginos *et al.*, 2002; Hurwood *et al.*, 2003; André *et al.*, 2012). Furthermore, it has been proposed that comparing the amount of geographic variation in allele frequencies for protein and DNA polymorphisms is a powerful method for detecting the effects of natural selection (McDonald, 1994). Drift and migration should cause the same pattern of geographical variation at two sets of neutral markers. On the other hand, discordant patterns between different sets of markers would be an indication of the effects of natural selection on at least one of these different marker sets. Once more, mollusks are important models in these studies (Karl and Avise, 1992; Riginos *et al.*, 2002; Bierne *et al.*, 2003; Riginos and Cunningham, 2005).

This review aims to examine the importance of the molecular markers (allozymes and DNA) for studies on natural selection discussed in the light of the results obtained from mollusks which includes the Brazilian cockle *Anomalocardia brasiliana* (Silva and Solé-Cava, 1994), the brown mussel *Perna perna* (Weber and Silva, 2008) and the blue mussel *Mytilus edulis* (Silva and Skibinski, 2009). Initially, to assist the reader, a short overview is presented of the molecular markers use in population genetics.

Molecular Markers

Since 1966, when Lewontin and Hubby published their historic papers (Hubby and Lewontin, 1966; Lewontin and Hubby, 1966), allozyme electrophoresis has been used widely among almost all groups of animals and plants, and every sort of problem in population genetics and its branches (Murphy *et al.*, 1996). However, such intense use of the method have revealed some limitations for some groups of living organisms and for some specific problems. Such limitations of the allozyme electrophoresis technique for special cases do not reflect the exhaustion of the method. By contrast it continues to produce invaluable knowledge which is more and more problem oriented (Silva and Russo, 2000; Gharbi and Said, 2011; Chung and Chung, 2012; Mayoll *et al.*, 2012).

However, what represents the limits of the allozyme approach? Sometimes the use of allozyme electrophoresis is limited by the amount of tissue present in the organism studied; for example, it is very difficult to study larvae with this technique, since it requires moderate amount of tissues more than present in the organism in the larval phase (Côrte-Real *et al.*, 1994; Boeger *et al.*, 2007). The need for fresh, or carefully frozen tissue, is another requirement of the technique, which sometimes cannot be met (Thorpe and Solé-Cava, 1994; Murphy *et al.*, 1996; Silva and Russo, 2000). Some species and groups show low levels of allozyme variation, which makes studies of population structure difficult, and this problem can be dramatic in cases of stock definition for fish or definition of strategies of conservationism for birds (Avise, 2004). Because an allozyme represents a phenotype its neutrality is not completely certain, which causes problems in interpretations based on this assumption (Karl and Avise, 1992; Begun and Aquadro, 1993; Riginos *et al.*, 2002; Silva and Skibinski, 2009; André *et al.*, 2012). Finally, in some cases differences between allozymes data and other techniques are found, which makes an alternative technique desirable (Avise, 2004). Such limitations for some specific problems bring DNA techniques forward as a good alternative.

DNA techniques can almost use all kinds of tissues (alcohol fixed, air dried and even fossils!) (Brown and Brown, 1994; Ravasi *et al.*, 2012). The development of PCR methods opens the possibility (theoretically) of working even with only one molecule or sequence of DNA (Saiki *et al.*, 1988). Most of the nucleotide changes can be detected by DNA techniques and many of them probably reflect silent substitutions or other genetic changes likely to be selectively neutral.

It is a mistake to allow that the introduction of DNA techniques prevent the use of allozymes as a molecular marker. Both approaches have their advantages and disadvantages, their limits and their idiosyncrasies. The future should find each of the techniques properly problem-oriented for solving evolutionary questions about the life and history of living organisms. In the next sections are presented examples which show this problem oriented approach in the case of the use of allozymes as well as the use of nuclear DNA markers couple with allozymes to underline the possible operation of natural selection in marine mollusks populations.

Brazilian Cockle *Anomalocardia Brasiliana* (Gmelin, 1791)

Anomalocardia brasiliana is a tropical marine bivalve that lives in sandy and muddy bottoms in the intertidal zone in sheltered waters of bays and lagoons from the Caribbean Sea to Uruguay (Rios, 1994, 2009). A. brasiliana can reach very high populations densities, and it is consumed and exploited commercially (Pezzuto et al., 2010). It is a dioecious and reproduces sexually through external fertilization. The larvae produced can remain in the plankton up to four weeks until settlement (Narchi, 1976; Luz and Boehs, 2011).

A very interesting aspect of the biology of A. brasiliana is its broad euryhalinity (Leonel et al., 1983). Populations of this species are common both in metahaline (up to 65) and hypohalyne (down to 20) environments (Silva et al., 2005).

One of the rare examples of an adaptative cline for a molecular marker in natural populations is the correlation usually observed between salinity and gene frequencies at the leucine aminopepetidase (Lap) locus of some euryhaline species of the mussel genus Mytilus. These mussels experience Lap genotype-dependent mortality. This differential mortality arises from the higher reaction rate of the Lap^{94} allele compared with the other Lap alleles. Mussels carrying copies of the Lap^{94} allele can accumulate cellular free amino acids at a faster rate than mussels with other alleles, but these same mussels also lose cellular nitrogen at faster rates (Hilbish et al., 1982). The ecological consequence of this genetically-based difference in individual physiology is that in low-salinity regions mussels with the Lap^{94} allele are at a selective disadvantage compared with other genotypes.

Along the shores of Long Island Sound (USA), there exists a cline in Lap allele frequencies which results directly from the biochemical and hence physiological differences which exist among the Lap genotypes. At high-salinity locations, the Lap^{94} allele exists at highest frequency (around 0.55), but decreases in frequency (to around 0.12) as salinity decreases over a distance of less than 30 km (Koehn et al., 1976). An annual cycle of selection occurs along the length of this salinity gradient. As annual pulses of immigrants move into the lower-salinity waters of the Sound, they experience differential selection which reduces the frequency of the Lap^{94} allele. The Lap cline is therefore unstable with regard to new recruits, but stable with regard to adult residents remaining after genotype-dependent spat mortality (Koehn et al., 1980). Selection thus occurs in a relatively brief temporal period among the newly settled mussel spat.

In studying seven localities along 2000 km of Brazilian coast with salinities ranging from 27 to 68, Silva and Solé-Cava (1994) demonstrated the presence of a significant (p<0,003) Lap cline for this species. This cline was strong in hypohaline conditions, but very week in salinities above that of normal oceanic seawater (34). This seems to indicate that, if natural selection is responsible for the maintenance of this cline, it is acting more effectively at salinities less than 34. This conclusion is supported by physiological studies on other species which show osmoregulation at low salinities, but do not control their cell volume at high salinities, simply losing water to the medium (Pierce, 1971; Garton and Berg, 1989). Figure 1 shows the significant correlation between C-allele from the Lap-1 locus with salinity, according to an exponential model. Therefore, the results obtained for A. brasiliana is well described by the Long Island Sound model of selection which describes changes in Lap allele frequencies due to a salinity gradient.

$$\text{ArcSin}\sqrt{f_{lap\text{-}1\text{-}c}} = e^{(2.97(\pm 0.25)\ -\ 0.025(\pm 0.006)\text{ppt S})}$$

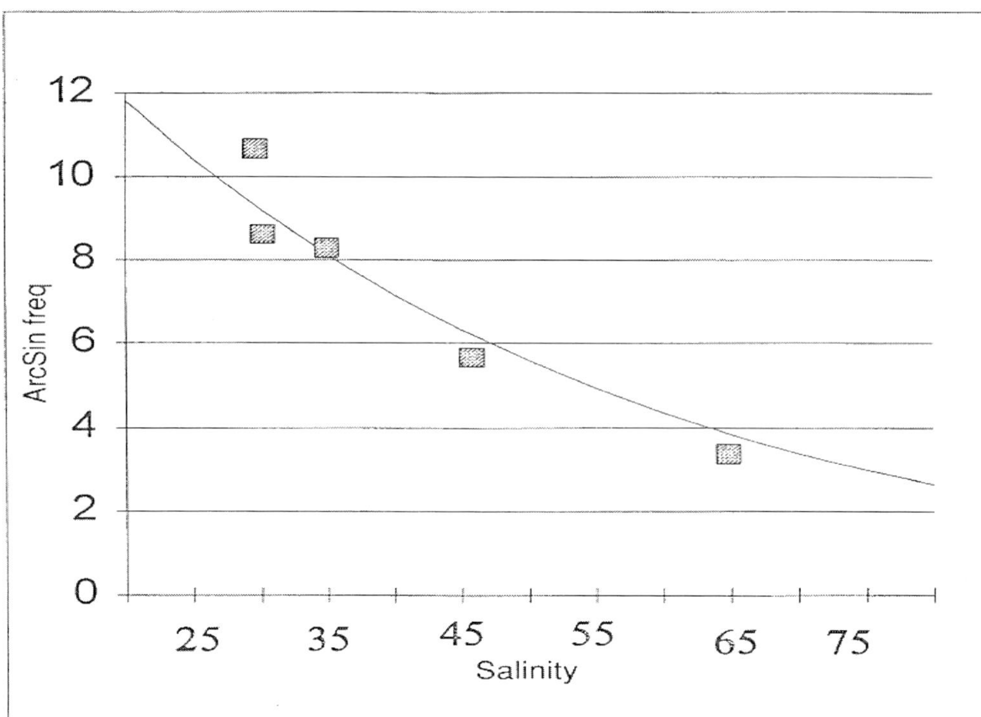

Figure 1. Regression analysis of the frequency of the *Lap-1* C-allele of *Anomalocardia brasiliana* in relation to salinty. The significant correlation ((p<0,003) shows accordance to an exponential model.

Brown Mussel *Perna Perna* (Linnaeus, 1758)

Perna perna occurs throughout the tropical and subtropical regions of the Atlantic and Indian Oceans (Hicks et al., 2001), and it was recorded for the Gulf of Mexico (Hicks and Tunnel, 1995). In South America, it is found on the coast of Venezuela (Bravo et al., 1998) and from the town of Vitória, in southeastern Brazil, to Uruguay (Souza et al., 2004), but does not occur between those points.

The species is dioecious and reproduces sexually through external fertilization. In Brazil reproduces several times a year, with high spawning capability in certain seasons and high larval density in the plankton (Beduschi et al., 2009). P. perna is of traditional importance as a food resource both in Brazil and Africa and great aquaculture potential (Metri et al., 2002).

In studying *P. perna* populations at three sites on the coast of Rio de Janeiro State (Arraial do Cabo, Itaipu and Búzios), for both subtidal and intertidal levels, results (Weber and Silva, 2008) showed significant differences in A-allele frequencies at the catalase (*Cat*) locus between intertidal and subtidal populations both at Arraial do Cabo and Búzios. The tests revealed significant differences both between populations at the same site as between different sites. For the superoxide dismutase (*Sod-2*) locus, significant differences were found in the B-allele frequencies between intertidal and subtidal populations at Arraial do Cabo and Itaipu. This allele is fixed in the subtidal populations, whereas the intertidal populations present three segregating alleles (A, B and C). Table 1 resumes these results.

The differences in allele frequencies obtained (for the A-allele of *Cat* and the B-allele of *Sod-2*, between the intertidal and subtidal organisms) may be plausibly explained by natural selection. In this case, natural selection would be acting on a microscale (environmental variation between the intertidal and subtidal regions) and, consequently, leading to differential mortality against carriers of those alleles.

Table 1. Contingency tables of A-allele frequencies at the *Cat* locus and B-allele frequencies at the *Sod-2* locus between subtidal and intertidal populations at Arraial do Cabo, Búzios and Itaipu

Sampling sites	Locus / Allele	
	Cat /A	*Sod-2* /B
ACI/ ACS	S	S
BZI/ BZS	S	NS
ITI/ ITS	NS (boxed)	S

S - Significant at 5% level.
NS- Non significant.
□- Significant only for comparisons between sampling sites.
ACI- Arraial do Cabo Intertidal.
ACS- Arraial do Cabo Subtidal.
BZI- Búzios Intertidal.
BZS- Búzios Subtidal.
ITI- Itaipu Intertidal.
ITS- Itaipu Subtidal.
Four out of six results are significant at 5% level within the same sampling sites and other (IT-Itaipu) is significant only for comparisons between sampling sites.

This mortality may be due to the intense environmental stress to which individuals on the intertidal zone are subjected, such as that caused by desiccation periods (heat stress), strong wave action and variation in oxygen availability (oxidation stress) (Storey and Churchill, 1995). The tide-influenced frequent immersion/emersion process causes intertidal organisms to undergo rapid reoxygenation which, after long periods of hypoxia, induces overproduction of highly toxic radicals (O2- and H_2O_2) (Richier *et al.*, 2003). Enzymes constitute part of the antioxidant system developed by organisms to counter possible damage due to rapid reoxygenation (Ross *et al.*, 2001). The selective mechanism involved in *P. perna* mortality on the intertidal zone may thus be related to the differential activity of the A-allele of *Cat* and the B-allele of *Sod-2*, enzymes which are directly involved in the physiological mechanisms of antioxidant defense. It may be that these alleles exhibit low efficiency under intense periods of stress and are being removed from the population for this reason.

Blue Mussel *Mytilus Edulis* (Linnaeus, 1758)

Mytilus edulis is a marine mussel which inhabits littoral and shallow sublitoral waters, occurring in both open water and brackish estuaries (Seed, 1992). It lives on any substrata which provide a secure anchorage, such as rock, stones, shingle, dead shells and even compacted mud or sand (Seed, 1976). Factors limiting the upper shore distribution limits of mussels are mainly related to physiological intolerance to temperature. Lower limits of distribution are more determined by biological factors, especially predators (Seed and Suchanek, 1992). Mytilus species have a worldwide distribution inhabiting sea waters in Europe, America, Oceania and Asia (Gosling, 1992a; Seed, 1992).

Mussels from this species are generally dioecious, although hermaphroditism is known (Seed, 1976). Fertilization is external and fertilized eggs develop into a ciliated trocophore larvae in approximately 24-48 hours (Lutz and Kennish, 1992). Marine mussels are molluscs of economic relevance being cultivated in over 20 countries in the world (Hickman, 1992).

In Mytilus species, many years were devoted to the study of population genetics using protein electrophoresis (Levinton and Koehn, 1976; Koehn, 1991; Gosling, 1992b). In Britain, the results of this effort have revealed little geographic differentiation (Skibinski et al., 1983) for Mytilus edulis suggesting that gene flow is extensive for this species. More recently, some studies have shown contrasting patterns of differentiation among allozymes and DNA markers across zones of hybridization between the species M. edulis and M. trossulus (Riginos et al., 2002; Bierne et al., 2003; Riginos and Cunningham, 2005).

Studies which compare patterns of geographic variation shown by different markers, especially at allozyme and nonallozyme loci, have been suggesting that the DNA and allozymes polymorphisms have different evolutionary dynamics, and that balancing selection may be acting on the allozyme variation whereas variation for the DNA markers is neutral (Karl and Avise, 1992; Hurwood et al., 2003). Concerning the Mytilus problem Riginos et al. (2002) argue that allozymes show high differentiation between the two species, in contrast to nDNA and mtDNA markers, due to natural selection acting on allozymes in the different environments where the species live. However, Bierne et al. (2003), in expanding the genetic dataset from 5 to 18 allozymes loci and adding 2 extra nDNA markers to the five analysed by Riginos et al. (2002), question this reported difference between the two sets of markers and suggest that the conclusion of selection acting on the allozyme markers is not warranted.

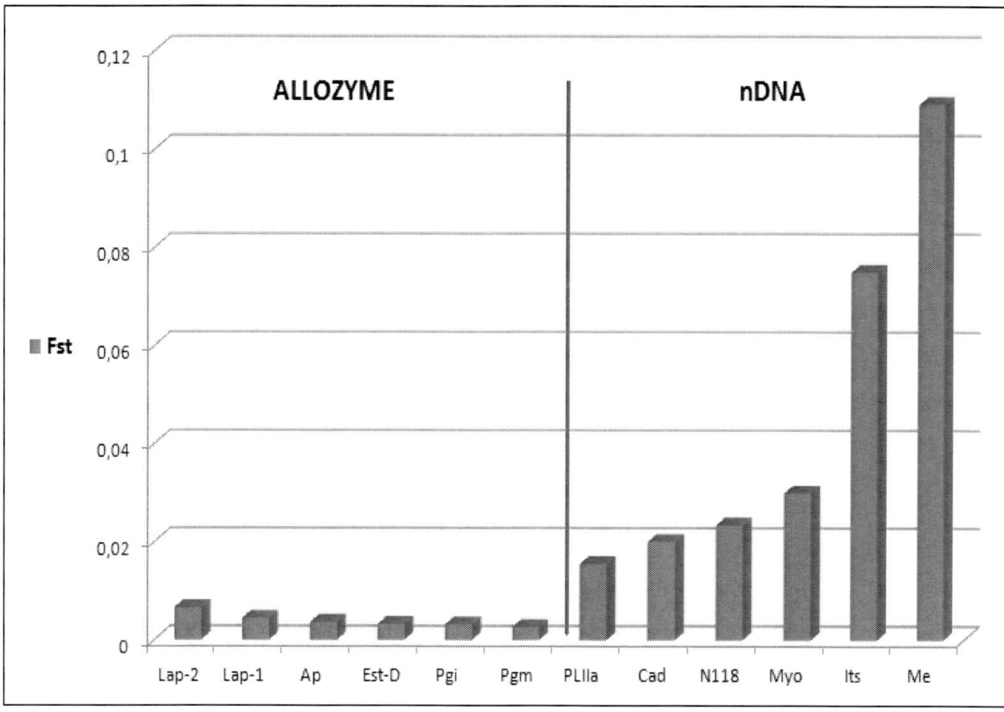

Figure 2. Estimates of F_{ST} for populations of *Mytilus edulis* using allozymes and nuclear DNA markers shows greater amount of geographic differentiation for nDNA markers than for allozymes. The Mann-Whitney U-test applied to these data confirms that F_{ST} is significantly different between the allozymes and nDNA polymorphisms at the 1% level suggesting that at least one class of polymorphisms is affected by selection.

The different interpretations here refer to the causes of patterns of geographic variation shown by different markers in hybrid zones between Mytilus species. An alternative approach, which has been applied to many organisms, is to study patterns of differentiation within a mussel species (Begun and Aquadro, 1993; Marchant and Shaw, 1993; Burton and Lee, 1994; Pogson et al., 1995; Piel and Nutt, 2000; Krafsur, 2002). This was the approach used by Silva and Skibinski (2009) to compare nuclear DNA (nDNA) population structure in 6 Mytilus edulis populations around the British Isles with data from a previous allozymes study on the same (two sites) or very close localities (the other 4 sites) (Skibinski et al., 1983). Comparison of the amount of geographic differentiation in allele frequencies for nDNA and allozymes shows F_{ST} values for M. edulis populations for nDNA markers which are significantly greater than those for allozymes (Figure 2). This result for nDNA provides evidence for restricted gene flow among the six Mytilus edulis populations studied. Moreover, the difference between levels of geographic variation for allozyme and nDNA reported support the hypothesis that balancing selection may be reducing the variation of allozyme allele frequencies among different geographic regions.

Although these results shows support to the hypothesis of natural selection on allozymes markers, they are qualitatively different from those of Riginos et al. (2002) and Riginos and Cunningham (2005) where zones of hybridization between the species M. edulis and M. trossulus were studied. These authors provided evidence for directional selection driving differences in allozyme frequencies between species while the results of Silva and Skibinski

(2009) support the hypothesis of balancing selection within a species. The two results are not necessarily contradictory. For example although balancing selection might maintain levels of variation within species, the selection might have a directional component in periods when species do adapt to new environments.

CONCLUSION

The strategy employed to underline natural selection operation on allozymes loci in wild population is basically a trial to relate different allelemorphs with environmental factors, especially temperature and salinity. An example of these attempts are offered by the *Adh* polymorphisms in *Drosophila* (Day *et al.*, 1974a, 1974b), *Ldh* in fishes (Powers and Place, 1978; Place and Powers, 1984), *Gpi* in copepods (Burton and Feldman, 1983) and *Lap* in mussels as reviewed here. Although, these studies supports a role for natural selection (Gillespie, 1991), providing fascinating examples of biochemical adaptation, evidence for the selective maintenance of a few well-characterized enzyme polymorphisms does not generalize the case of natural selection maintenance of observed allozyme variation. Furthermore, it is a widespread belief that adaptive processes are infrequent in protein evolution.

Since the development of allozyme electrophoresis as a method for detection of genetic variation in natural populations (Lewontin and Hubby, 1966), the genetic structure of natural populations has been inferred from data on allele frequencies at multiple allozyme loci. Although several exceptions can be suggested, the general conclusion coming from these studies is that gene flow is a powerful homogenising force that acts to prevent genetic divergence, especially for species with planktonic larvae (Waples, 1987). Mussel species like *Mytilus* represent a good example confirming this general conclusion (Skibinski *et al.*, 1983). The main assumption behind these studies was, and still is, the neutrality of allozyme polymorphisms (Nei, 1987). More recently however, studies on mitochondrial and nuclear DNA markers have been showing significant geographic structure for populations previously regarded as homogenous, both in the marine (Karl and Avise, 1992; Burton and Lee, 1994; Pogson *et al.*, 1995) and in the terrestrial environment (Begun and Aquadro, 1993; Marchant and Shaw, 1993). Two conclusions are provided by these studies. The first is the reevaluation of the conclusion that gene flow is able to bridge long geographic distances as once supposed and the second concerns the contradiction of the assumption of neutrality for allozymes. The results offered in this chapter for *Anomalocardia brasiliana*, *Perna perna* and *Mytilus edulis*, are in line with these conclusions.

Nuclear DNA markers showed levels of geographic variation which were significantly higher than allozymes. Because gene flow is a process that affects all loci equally, the contrasting patterns observed for nDNA markers and allozymes cannot be explained by gene flow. Therefore, the results implicate a role for balancing selection in affecting the homogeneity of the protein variation among natural populations of *Mytilus*.

In the same way that in the last 25 years the neutrality of allozyme loci was assumed when studying the population structure of natural populations, the assumption now behind comparisons between patterns of geographic variation shown by allozymes and DNA polymorphisms, is the neutrality of the DNA polymorphisms (McDonald, 1994). However, studies have been showing significant deviations from a strictly neutral model also to DNA markers (Silva and Skibinski, 2009; Quintela *et al.*, 2010; Herrera *et al.*, 2011). The results suggest the effects of purifying selection acting on DNA loci what suggests that early underlying assumptions of the neutrality for DNA polymorphisms must also be taken with caution.

Departures from the neutral mutation model can be attributed to several sources such as the Wahlund effect, population expansion, bottleneck effects and natural selection. What the new data are revealing is that the question of what causes protein polymorphism and divergence is still debatable, but evidence suggests it deserves a new level of appreciation.

The sequence of DNA molecules is perhaps the best method to reveal genetic variation, and therefore, in the testing of neutral theory, this should be the option to be pursued. This follow up would put results regarding comparison between allozymes and DNA molecular markers beyond simple surveying of population structure and straight into the scientific framework of the debate between neutralism and selectionism.

REFERENCES

Andre C.; Larsson L.C.; Laikre L.; Bekkevold, D.; Brigham, J.; Carvalho, G.R.; Dahlgren, T.G.; Hutchinson, W.F.; Mariani, S.; Mudde, K.; Ruzzante, D.E.; Ryman, N. 2012. Detecting population structure in a high gene-flow species, Atlantic herring (*Clupea harengus*): direct, simultaneous evaluation of neutral vs putatively selected loci. *Heredity* 106(2):270-280.

Avise, J.C. 2004. *Molecular Markers, Natural History and Evolution*. 2° Edition. Chapman and Hall, London.

Beduschi, P.; Melo, C.M.R. and Ferreira, J.F. 2009. The influence of techniques of larvae rearing and seed collectors on the survival rate and recovery efficiency of the brown mussel *Perna perna* (L.) in laboratory. *Braz. Arch. Biol. Technol.* 52(1): 145-152.

Begun, D.J. and Aquadro, C.F. 1993. African and North American populations of *Drosophila melanogaster* are very different at the DNA level. *Nature* 365:548-550.

Bierne, N.; Daguin, C.; Bonhomme, F.; David, P. and Borsa, P. 2003. Direct selection on allozymes is not required to explain heterogeneity among marker loci across a *Mytilus* hybrid zone. *Molecular Ecology* 12: 2505-2510.

Boeger, W.A.; Pie, M.R.; Falleiros, R.M.; Ostrensky, A.; Darrigran, G.; Mansur, M.C.D. and Belz, C.E. 2007. Testing a molecular protocol to monitor the presence of golden mussel larvae (*Limnoperna fortunei*) in plankton samples. *Journal of Plankton Research* 29(11):1015-1019.

Bravo, M.I.S.; Chung, K.S. and Perez, J.E. 1998. Salinity and temperature tolerances of the green and brown mussels, *Perna perna* and *Perna viridis* (Bivalvia: Mytilidae). *Rev. Biol. Trop.* 46(Suppl. 5):121-125.

Brown, T.A. and Brown, K.A. 1994. Ancient DNA: Using molecular biology to explore the past. *BioEssays* 16:719-726.

Burton, R.S. and Feldman, M.W. 1983. Physiological effects of an allozyme polymorphism: glutamate-pyruvate transaminase and response to hyperosmotic stress in the copepod *Tigriopus californicus*. *Biochem. Genet.* 21:238-251.

Burton, R.S. and Lee, B.-N. 1994. Nuclear and mitochondrial gene genealogies and allozyme polymorphism across a major phylogeographic break in the copepod *Tigriopus californicus*. *Proceedings of Natural Academy of Sciences USA* 91:5197-5201.

Chung, M.Y. and Chung, M.G. 2012. A review of the use of genetic markers in orchid systematics with emphasis on allozymes. *Biochemical Systematics and Ecology* 41:62-73.

Côrte-Real, H.B.S.M.; Holland, P.W.H. and Dixon, D.R. 1994. Inheritance of a nuclear DNA polymorphism assayed in a single bivalve larvae. *Mar. Biol.* 120:415-420.

Day, T.H.; Hillier, P.C. and Clarke, B. 1974a. The properties of genetically polymorphic isozymes of alcohol dehydrogenase in *Drosophila melanogaster*. *Biochem. Genet.* 11:141-153.

Day, T.H.; Hillier, P.C. and Clarke, B. 1974b. The relative quantities and catalytic activities of enzymes produced by alleles at the alcohol dehydrogenase locus in *Drosophila melanogaster*. *Biochem. Genet.* 11:155-165.

Garton, D.W. and Berg, D.J. 1989. Genetic variation at the Lap locus and ammonia excretion following salinity transfer in an estuarine snail. *Comp. Biochem. Physiol.* 92A:71-74.

Gharbi, A. and Said, K. 2011. Genetic variation and population structure of *Holothuria polii* from the eastern and western Mediterranean coasts in Tunisia. *Journal of the Marine Biological Association of the United Kingdom* 91(8):1599–1606.

Gillespie, J.H. 1991. *The Causes of Molecular Evolution*. University Press, Oxford.

Gosling, E.M. 1992a. Systematics and geographic distribution of *Mytilus*. In: Gosling, E.M. (ed), *The mussel Mytilus: Ecology, physiology, genetics and culture*, Elsevier, London. Developments in aquaculture and fisheries science 25, pp. 1-20.

Gosling, E.M. 1992b. Genetics of *Mytilus*. In: Gosling, E.M. (ed), *The mussel Mytilus: Ecology, physiology, genetics and culture*, Elsevier, London. Developments in aquaculture and fisheries science 25, pp. 309-382.

Herrera, C.M.; Pozo, M.I. and Bazaga, P. 2011. Clonality, genetic diversity and support for the diversifying selection hypothesis in natural populations of a flower-living yeast. *Molecular Ecology* 20(21):4395-4407.

Hilbish, T.J.; Deaton, L.E. and Koehn, R.K. 1982. Effect of an allozyme polymorphism on regulation of cell volume. *Nature* 298:688-689.

Hilbish, T.J. and Koehn, R.K. 1985a. The physiological basis of natural selection at the LAP locus. *Evolution* 39:1302-1317.

Hilbish, T.J. and Koehn, R.K. 1985b. Dominance in physiological phenotypes and fitness at an enzyme locus. *Science* 229:52-54.

Hickman, R.W. 1992. Mussel cultivation. In: Gosling, E.M. (ed.), *The mussel Mytilus: Ecology, physiology, genetics and culture*, Developments in aquaculture and fisheries science, vol. 25, Elsevier, London, pp. 465-510.

Hicks, D.W.; McMahon, R.F. and Ingrao, D.A. 2001. Two invasive mussels in the genus *Perna* in the Gulf of Mexico. In: *Virtual Proceedings for the State of the Bay Symposium V*. Texas Natural Resource Conservation Commission, Austin, Texas, pp. 159-170.

Hicks, D.W. and Tunnell Jr., J.W. 1955. Ecological notes and patterns of dispersal in the recently introduced mussel, *Perna perna* (Linné, 1758), in the Gulf of Mexico. *Am. Malac. Bull.* 11(2):203-206.

Hubby, J.L. and Lewontin, R.C. 1966. A molecular approach to the study of genic heterozygosity in natural populations. I- The number of alleles at different locos in *Drosophila pseudoobscura*. *Genetics* 54: 577-594.

Hurwood, D.A.; Hughes, J.M.; Bunn, S.E. and Cleary, C. 2003. Population structure in the freshwater shrimp (*Paratya australiensis*) inferred from allozymes and mitochondrial DNA. *Heredity* 90: 64-70.

Karl, S.A. and Avise, J.C. 1992. Balancing selection at allozyme loci in oysters-implications from nuclear RFLPs. *Science* 256:100-102.

Koehn, R.K. 1978. Biochemical aspects of genetic variation at the LAP locus in *Mytilus edulis*. In: Battaglia, B. and Beardmore, J.A. (eds), *Marine Organisms: Genetics, Ecology and Evolution*, Plenum Press, London, pp. 211-227.

Koehn, R.K. 1983. Biochemical genetics and adaptation in molluscs. In: Hochachka, P.W. (eds), *The Mollusca. Volume 2: Environmental biochemistry and physiology*, Academic Press, London, pp. 305-330.

Koehn, R.K. 1985. Adaptative aspects of biochemical and physiological variability. In: Gibbs, P. E. (ed.), *Proc. 19th European Marine Biology Symposium*, Cambridge University Press, London, pp. 425-441.

Koehn, R.K. 1991. The genetics and taxonomy of species in the genus *Mytilus*. *Aquaculture* 94: 125-145.

Koehn, R.K.; Milkman, R. and Mitton, J.B. 1976. Population genetics of marine pelecypods. IV. Selection, migration, and genetic differentiation in the blue mussel *Mytilus edulis*. *Evolution* 30: 2-32.

Koehn, R.K.; Newell, R.I.E. and Immermann, F. 1980. Maintenance of an aminopeptidase allele frequency cline by natural selection. *Proc. Natn. Acad. Sci.* USA 77: 5385-5389.

Krafsur, E.S. 2002. Population structure of the tsetse fly *Glossina pallidipes* estimated by allozyme, microsatellite and mitochondrial gene diversities. *Insects Molecular Biology* 11:37-45.

Leonel, R.M.V.; Magalhães, A.R.M. and Lunetta, J.E. 1983. Sobrevivência de Anomalocardia brasiliana (Gmelin,1791) (Mollusca: Bivalvia), em diferentes salinidades. *Boletim de Fisiologia Animal*, Universidade de São Paulo 7:63-72.

Levinton, J.S. and Koehn, R.K. 1976. Population Genetics. In: Bayne, B.L. (ed), *Marine mussels: Their ecology and physiology*, Cambridge University Press, Cambridge. International Biology Program 10, pp. 357-384.

Lewontin, R.C. and Hubby, J.L. 1966. A molecular approach to the study genic heterozygosity in natural populations. II- Amount of variation and degree of heterozygosity in natural populations of *Drosophila pseudoobscura*. *Genetics* 54: 595-609.

Lewontin, R.C. 1974. *The Genetic Basis of Evolutionary Change*. Columbia University Press, EUA.

Lutz, R.A. and Kennish, M.J. 1992. Ecology and morphology of larval and early postlarval mussels. In: Gosling, E.M. (ed.), *The mussel Mytilus: Ecology, physiology, genetics and culture*, Developments in aquaculture and fisheries science, vol. 25. Elsevier, London, pp. 53-85.

Luz, JR. and Boehs, G. 2011. Reproductive cycle of *Anomalocardia brasiliana* (Mollusca: Bivalvia: Veneridae) in the estuary of the Cachoeira River, Ilhéus, Bahia. *Braz. J. Biol.* 71(3):679-686.

Marchant A.D. and Shaw, D.D. 1993. Contrasting patterns of geographic variation shown by mtDNA and karyotype organisation in two subspecies of *Caledia captiva* (Orthoptera). *Molecular Biology and Evolution* 10:855-872.

Mayol1, M.; Palau1, C.; Rossello, J.A.; Gonzalez-Martynez, S.C.; Molins, A. and Riba1, M. 2012. Patterns of genetic variability and habitat occupancy in *Crepis triasii* (Asteraceae) at different spatial scales: insights on evolutionary processes leading to diversification in continental islands. *Annals of Botany* 109: 429-441.

McDonald, J. H. 1994. Detecting natural selection by comparing geographic variation in protein and DNA polymorphisms. In: Golding, B. (ed.), *Non-Neutral Evolution: Theories and Molecular Data*, Chapman and Hall, London, pp. 88-100.

Metri, R.; Rocha, R.M. and Marenzi, A. 2002. Epibiosis reduction on productivity in a mussel culture of *Perna perna* (Linné, 1758). *Braz. Arch. Biol. Technol.* 45(3):325-331.

Murphy, R.W.; Sites, J.W., Jr.; Buth, D.G. and Haufler, C.H. 1996. Proteins: Isozyme electrophoresis. In: Hillis, D.M.; Moritz, C. and Mable, B.K. (eds), *Molecular Systematics*, Sinauer Associates, Inc. Sunderland, Massachusetts, USA, pp. 51-120.

Narchi, W. 1976. Ciclo anual de gametogênese de *Anomalocardia brasiliana* (Gmelin, 1791) (Mollusca: Bivalvia). *Boletim de Zoologia da Universidade de São Paulo* 1: 331-349.

Nei, M. 1987. *Molecular evolutionary genetics*. Columbia University Press, New York.

Pezzuto, P.R.; Schio, C. and Almeida, T.C.M. 2010. Efficiency and selectivity of the *Anomalocardia brasiliana* (Mollusca: Veneridae) hand dredge used in southern Brazil. *Journal of the Marine Biological Association of the United Kingdom* 90(7):1455-1464.

Piel, W.H. and Nutt, K.J. 2000. One species or several? Discordant patterns of geographic variation between allozymes and mtDNA sequences among spiders in the Genus *Metepeira* (Araneae: Araneidae). *Molecular Phylogenetics and Evolution* 15:414-418.

Pierce, S.K. 1971. Volume regulation and valve movements by marine mussels. *Comp. Biochem. Physiol.* 39A:103-117.

Place, A.R. and Powers, D.A. 1984. The *LDH-B* allozymes of *Fundulus heteroclitus*: II. Kinetic analysis. *J. Biol. Chem.* 259:1309-1318.

Pogson, G.H.; Mesa, K.A. and Boutilier, R.G. 1995. Genetic population structure and gene flow in the atlantic cod *Gadus morhua*: A comparison of allozyme and nuclear RFLP loci. *Genetics* 139:375-385.

Powers, D.A. and Place, A.R. 1978. Biochemical genetics of *Fundulus heteroclitus*. I-Temporal and spatial variation in gene frequencies of *LDH-B, MDH-A, GPI-B* and *PGM-A*. *Biochem. Genet.* 16:593-607.

Quintela, M.; Berlin, S.; Wang, B. and Höglund, J. 2010. Genetic diversity and differentiation among *Lagopus lagopus* populations in Scandinavia and Scotland: evolutionary significant units confirmed by SNP markers. *Molecular Ecology* 19:2380-2393.

Ravasi, D.F.; Peduzzi, S.; Guidi, V.; Peduzzi, R.; Wirth, S.B.; Gilli, A. and Tonolla, M. 2012. Development of a real-time PCR method for the detection of fossil 16S rDNA fragments of phototrophic sulfur bacteria in the sediments of Lake Cadagno. *Geobiology* 10(3): 196-204.

Richier, S.; Merle, P.L.; Furla, P.; Pigozzi, D.; Sola, F. and Allemand, D. 2003. Characterization of superoxide dismutases in anoxia- and hyperoxia-tolerant symbiotic cnidarians. *BBA-Gen. Subjects.* 1621:84-91.

Riginos, C. and Cunningham, C. W. 2005. Local adaptation and species segregation in two mussel (*Mytilus edulis* x *Mytilus trossulus*) hybrid zones. *Molecular Ecology* 14: 381-400.

Riginos, C.; Sukhdeo, K. and Cunningham, C. W. 2002. Evidence for selection at multiple allozyme loci across a mussel hybrid zone. *Molecular Biology and Evolution* 19: 347-351.

Rios, E. C. 1994. *Seashells of Brazil.* Fundação Universidade Federal Rio Grande, Rio Grande do Sul, Brazil.

Rios, E. C. 2009. *Compendium of Brazilian sea shells.* Rio Grande: Evangraf, Rio Grande, Brazil.

Ross, S.W.; Dalton, D.A.; Kramer, S. and Christensen, B.L. 2001. Physiological (antioxidant) responses of estuarine fishes to variability in dissolved oxygen. *Comp. Biochem. Phys. C* 130:289-303.

Saiki, R., Gelfand, D.H., Stoffel, S., Scharf, S., Higuchi, R., Horn, G.T., Mullis, K.B. and Erlich, H.A. 1988. Primer directed enzymatic amplification of DNA with a thermostable DNA polymerase. *Science* 239: 487-491.

Seed, R. 1976. Ecology. In: Bayne, B.L. (ed.), *Marine mussels: Their ecology and physiology*, Intern. Biol. Program 10 Cambridge University Press, Cambridge, pp. 13-65.

Seed, R. 1992. Systematics evolution and distribution of mussels belonging to the genus Mytilus: An overview. *Am. Malacol. Bull.* 9:123-137.

Seed, R. and Suchanek, T.H. 1992. Population and community ecology of Mytilus. In: Gosling, E.M. (ed.), *The mussel Mytilus: Ecology, physiology, genetics and culture*, Developments in aquaculture and fisheries science, vol. 25, Elsevier, London, pp. 87-169.

Silva, E.P.; Fernandes, F.C.; Soares-Gomes, A. and Abreu, C.M. 2005. Sandy beach macrobenthos assemblages at a hypersaline coastal lagoon, Lagoa de Araruama, RJ, Brasil. *Journal of Coastal Research* 42(SI):265-270.

Silva, E.P. and Russo, C. 2000. Techniques and statistical data analysis in molecular population genetics. *Hydrobiologia* 420:119-135.

Silva, E.P. and Skibinski, D.O.F. 2009. Allozymes and nDNA markers show different levels of population differentiation in the mussel *Mytilus edulis* on British coasts. *Hydrobiologia* 620:25-33.

Silva, E. P. and Solé-Cava, A. M. 1994. Genetic variation and population structure in the tropical marine bivalve *Anomalocardia brasiliana* (Gmelin, 1791) (Veneridae). In: Beaumont, A. (ed.), *Genetics and Evolution of Aquatic Organisms*, Chapman and Hall, London, pp. 159-168.

Skibinski, D.O.F.; Beardmore, J.A. and Cross, T.F. 1983. Aspects of the population genetics of *Mytilus* (Mytilidae: Mollusca) in the British Isles. *Biological Journal of the Linnean Society* 19:137-183.

Souza, R.C.C.L.; Fernandes, F.C. and Silva, E.P. 2004. Distribuição atual do mexilhão *Perna perna* no mundo: um caso recente de bioinvasão. In: Silva, J.S.L.V. and Souza, R.C.C.L. (eds.), *Água de lastro e bioinvasão*, Editora Interciência, Rio de Janeiro, pp.157-172.

Storey, K.B. and Churchill, T.A. 1995. Metabolic responses to anoxia and freezing by the freeze tolerant marine mussel *Geukensia demissus*. *J. Exp. Mar. Biol. Ecol.* 188:99-114.

Thorpe, J.P. and Solé-Cava, A.M. 1994. The use of electrophoresis in invertebrate systematics. *Zoologica Scripta* 23:3-18.

Waples, R.S. 1987. A multispecies approach to the analysis of gene flow in marine shore fishes. *Evolution* 41:385-400.

Weber, L.I. and Silva, E.P. 2008. Sistemática molecular e genética de populações. In: Resgalla Jr., C.; Weber, L.I. and Conceição, M.B. (orgs), *O Mexilhão* Perna perna (L.): *Biologia, Ecologia e Aplicações*, Editora Interciência, Rio de Janeiro, pp. 121-149.

In: Natural Selection
Editors: J. R. Lynch and D. T. Williamson
ISBN: 978-1-62257-467-4

Chapter 8

DIFFERENCES IN MICROBIAL FUNCTIONS IN SOILS OF SEMI-NATURAL AND IMPROVED GRASSLANDS

Zabed Hossain *** *and Shu-ichi Sugiyama***
Faculty of Agriculture and Life Science,
Hirosaki University, Japan

ABSTRACT

Soil microbes are important component of the terrestrial ecosystems that play the critical role in decomposition of organic substances and mineralization of nutrients. Since soil microbes influence the nutrition of plants in the nitrogen-limited areas there has been an increased interest on using soil microbes as an important indicator of soil quality.

It is also widely recognized that changes in soil microbial biomass and community composition can influence decomposition of organic matter and nitrogen cycling in soil. Grasslands constitute about one thirds of the earth´s area and are regarded as one of the most productive ecosystems on earth.

Since management measures are often taken in order to optimize the productivity of this system, enhanced knowledge about the functional differences of the soil microbes under various management activities is relevant for proper management of these systems. This chapter attempts to review studies on the structural and functional differences of soil microbial communities in various grassland ecosystems. This paper also presented data on the differential responses of soil microbial functions such as decomposition rate and nitrogen mineralization rate to litter inputs in order to reveal the characteristic differences of the semi-natural and improved grasslands with special references to grasslands of Shiriyazaki in the northern Honshu island, Japan.

Keywords: Grasslands, microbial function, phospholipid fatty-acids (PLFA), soil microbial community

*Department of Botany, University of Dhaka, Dhaka 1000, Bangladesh, Phone: 880-2-9661920/7555 (Office), 880-01727736087 (Mobile), E-mail: zabed@du.ac.bd

INTRODUCTION

Soil microbes play the critical role in decomposition of organic matter and mineralization of nutrients in the terrestrial ecosystems (Wardle 2002). Soil microbes (bacteria and fungi) are the most abundant group constituting more than 80% of the total biomass of the biotic community in soil (Davet 2004). It is also recognized that changes in soil microbial biomass and community composition can potentially influence the decomposition of organic matter and nitrogen cycling in soil (Bardgett et al., 2003; Lipson and Schmidt, 2004). Because of these manifold properties, there is a growing interest on using microbial communities as a parameter for soil quality.

Understanding the structural and functional characteristics of the soil microbes is therefore important for the enhancement of our knowledge about the various components as well as the functions of the terrestrial ecosystems. Grasslands constitute about one thirds of the earth´s area and are regarded as one of the most productive ecosystems on earth (Schlesinger 1997, Bardgett and Cook, 1998). Grasslands are complex terrestrial ecosystems having the components of producers, consumers and decomposers.

A number of studies have shown that grazing intensity has significant impact on the soil microbial structure and functions through a number of ways including the changes of vegetation aboveground (Sugiyama et al. 2008, Bardgett *et al.* 1999, 1996, Grayston *et al.* 2001, 2004, Kojima *et al.* 2007). The results of these studies reveal that unimproved semi-natural grassland is dominated by the fungal community while improved grassland is by bacterial community.

The differences in soil microbial community between these grasslands have been attributed to the differences in the plant derived resources to the soil. Wardle *et al.* (2004) suggested that fast-growing plant species in the nutrient-rich sites select bacteria-dominated decomposer community for rapid nutrient cycling that help dominance of the fast-growing plants and the slow-growing plant species in the nutrient-poor sites select fungi-dominated microbial community for the slower turnover of nutrients.

However, there has been limited experimental data to examine this hypothesis. Shiriyazaki has developed about 5-km^2 area of semi-natural grasslands along the shoreline of the Pacific Ocean as a result of horse grazing for over 300 years.

During this time no human management activities such as fertilization and sowing of pasture species have been applied. The semi-natural grasslands of Shiriyazaki area in the northern Honshu island of Japan consist of two floristically distinct vegetations; short-grass and tall-grass vegetations (Sugiyama et al. 2008).

The short-grass vegetation is characterized by the dominance of C_4 grass, *Zoysia japonica* and low height of vegetation resulting from frequent grazing by horses, while the tall-grass vegetation, which is subject to less frequent grazing, is characterized by the dominance of C_4 tall grass *Miscanthus sinensis,* C_3 grass *Bromus japonicus* and forbs. Improved grassland, located 1 km away from semi-natural grassland, was established about 30 years ago by sowing of herbage species and has been subjected to fertilization and grazing by cows.

The differences in plant vegetation among the grasslands and the management activities have caused differences in soil microbial community structure between these grasslands (Sugiyama et al. 2008, Hossain et al. 2007). The grasslands of Shiriyazaki thus provide suitable sites to examine the functional response of soil microbial communities to litter inputs. Since modification of plant species composition is commonly practiced in the grasslands in order to increase the plant productivity (Pearson and Ison 1997), study on the structural and functional responses of the soil microbial communities of the grasslands with various disturbance regimes is relevant not only for the proper management but also for the enhancement of our understanding about functions of the grassland ecosystems.

The objectives of the present study, therefore, were to examine the functional characteristics of the soil microbes of the semi-natural tall-type and the improved grassland by applying repeated inputs of litter of contrasting quality (high nutrients but low defense chemicals versus low nutrients but high defense chemicals) to the grassland soils.

1. Materials and Methods

1.1. Collection of Soil and Leaf Litter

In this study, the soil and leaf litter were collected from the grasslands of Shiriyazaki (41°25′N and 141°26′E) located in the northern Honshu island of Japan. Soils (0-10 cm) were collected from two sites, each one for semi-natural tall-grass and improved grasslands of Shiriyazaki. Soils were sieved (2 mm) to homogenize and remove roots, macro fauna, and coarse sands. Leaf litter was collected from *Miscanthus sinensis* and *Trifolium repens* because of the previous report having contrasting leaf litter quality (Hossain and Sugiyama 2008).

Mature leaves instead of senesced litter were collected, because litter chemical composition largely varied with the progression of decomposition and thus sufficient amount of litter with uniform quality was not collected in fields.

Further, it is reported that plant species did not differ in nutrient resorption efficiency from senescing leaves (Aerts et al. 1999). Immediately after collection, leaf litter was dried at 60 °C for 48 hours to make constant weight.

1.2. Analysis of Soil Physico-Chemical Properties and Leaf Litter Chemistry

Nitrogen was extracted from 5.0 g fresh soil in 50 ml 2N KCl. We followed the Kjeldahl method to determine available nitrogen content; the amount was expressed as mg N 100 g^{-1} dry soil. Soil extractable phosphorus (P_2O_5) content was determined using colorimetric method after extraction from 1.0 g dry soil in 0.002 N H_2SO_4 solution with ammonium-molybdate-ascorbic acid as the reagent. Absorbance was determined using a spectrophotometer (U 1000; Hitachi Ltd., Japan) at 710 nm. Measurement of soil pH was done by using distilled water (1:2, using water suspension). Soil moisture content was determined by weight loss after drying 10.0 g soil at 60 °C for 24 h. Soil total N (%) and total C (%) contents were analyzed by Automatic High Sensitive N-C analyzer (Sumigraph NC-90A, Sumika-bunseki, Japan). We determined leaf nitrogen and phosphorus as nutrient elements and phenol and tannin as defense chemicals. Nitrogen content of the leaf litter was determined by following Kjeldahl method after wet digestion in conc. H_2SO_4. The same extract was used for the determination of phosphorus using spectrophotometer at 710 nm absorbance.

Phenol and tannin contents were determined following slightly modified procedures as described by Bärlocher and Graça (2005) and by Gessener and Steiner (2005), respectively. Detailed procedure has been described elsewhere (Hossain and Sugiyama 2008).

1.3. Measurement of Soil Respiration and Nitrogen Mineralization Rates

We measured soil respiration rate (CO_2 flux) using an automatic CO_2 analyzer (6400; Li-Cor Biosciences, Lincoln, NE) equipped with a soil CO_2 emission chamber (LI-6400-09; Li-Cor). The CO_2 concentration inside the pot was maintained at 400 μmol during measurements. We measured the CO_2 flux at 7 days interval after litter input to the soil. The procedure has been described in other study (Hossain and Sugiyama 2008).

Available soil nitrogen ($N0_3^-$ and NH_4^+) was analyzed before and after incubation for determination of the mineralization rate. Nitrogen mineralization rate was calculated as the amount of inorganic nitrogen increased per day per 100 g dry soil during the incubation period.

1.4. Microbial Community Analysis by Phospholipid Fatty Acids (PLFA) Profiling

The two soils used in the study were characterized by analyzing PLFA profiles as described by Hossain et al. (2007) based on the slightly modification by Frostegård *et al.* (1993). Fresh soil of 7.0 g was used to extract PLFA. Fatty acid nomenclature accords to that of Frostegård *et al.* (1993).

The fatty acids, i15:0, a15:0, 15:0, i16:0, 16:1ω7, i17:0, a17:0, cy17:0, 2OH-16:0, 18:1ω7, 18:1ω5 and cy19:0 were used for bacterial biomarkers. Those of 16:1ω5c for mycorrhizae and 10Me16:0 and 10Me18:0 were used for actinomycetes, and 18:2ω6 was used for saprophytic fungi (Bossio and Scow 1998).

1.5. Experimental Design

A microcosm study was conducted by following a reciprocal experimental design that allowed incubation of each litter type (*T. repens* and *M. sinensis*) with each of the two soil types (semi-natural tall-type and the improved grassland soils). Fresh soil equivalent to 20.0 g dry weight was taken into a polyvinyl plastic pot of 19.6 cm^2 area and 3.5 cm depth. A total of 4 cycles of incubation, each one for four weeks, thus 16 consecutive weeks in total, was conducted to examine microbial functions (soil respiration rate and nitrogen mineralization rate) at different cycles.

For this, a total of 72 (2 soils × 3 litter types [including control soil, without litter] × 3 replication × 4 cycles) samples were made to collect samples destructively after completion of the period of each cycle.

After completion of each cycle and collection of samples, the undecomposed litters in the remaining treatments were replaced by adding new litter. Collection of samples and replacement of new litter were done in the similar way successively to complete four cycles of incubation. Control soils were also stirred with a forceps to mimic the disturbances experienced by the samples of leaf litter treatment while mixing litter with soil. Litter of 0.15 g in weight that was cut into small sizes of 0.25 cm in length was added with soil. Prior to incorporation of litter into soils, soils were made of equal in moisture content of 40% by adding required amounts of sterilized distilled water. The samples were incubated at 27 °C throughout the experiment. Samples were observed regularly to monitor the moisture content and, if necessary, water was added gently to maintain the moisture content stable throughout the experiment.

1.6. Data Analysis

A three-way ANOVA was applied to determine the effects of the factors of soil, litter, and cycle as well as their interaction on soil properties. Further, % SS (sum of square) explained by the factors was determined to elucidate the relative importance of the effects.

One-way ANOVA was also conducted to elucidate the differences among the variables. Data analyses were conducted using computer software (JMP 4.0 SAS Institute Inc., Cary, NC, USA).

2. RESULTS

2.1. Chemical Properties of Leaf Litter Used in the Experiment

Chemical properties of the leaf litter of the two plant species used in the present study is shown in Table 1. Leaf nitrogen and phosphorus contents were significantly higher in *T. repens* than those in the *M. sinensis*. On the other hand, *M. sinensis* leaf contained significantly higher abundance of tannin than that in the *T. repens*.

Although phenol content did not differ significantly, it was relatively higher in *M. sinensis* leaf. These results reflect that *T. repens* litter was of high quality and that of the *M. sinensis* was low quality.

Table 1. Properties of leaf litter used in the experiment; N (mg N $100mg^{-1}$ leaf), P (mgP_2O_5 $100mg^{-1}$ leaf litter), phenol (mg phenol $100mg^{-1}$ leaf litter) and tannin (mg tannin $100mg^{-1}$ leaf litter). Values within parentheses are SE

	Miscanthus sinensis	*Trifolium repens*	F	P
N	4.25 (0.05)	7.32 (0.20)	216.83	0.0001
P	1.34 (0.10)	1.78 (0.03)	16.69	0.0150
Phenol	0.09 (0.01)	0.07 (0.01)	4.49	0.1000
Tannin	0.07 (0.01)	0.05 (0.00)	11.67	0.0300

2.2. Physico-Chemical Properties of Soils Used in the Experiment

Available nitrogen and phosphorus contents were significantly higher in the soils of improved grassland than the semi-natural grassland (Table 2). Soil nitrogen content was almost double (15.75 μg/100 g soil) in the improved grassland compared to that in the semi-natural grassland (7.65 μg/100 g soil). Soil total carbon (%) was significantly higher in semi-natural grassland than that in the improved grassland, however, soil total nitrogen (%) and pH did not differ statistically between the two soils.

2.3. Microbial Community Composition of Soils Used in the Experiment

PLFA analysis showed that although the semi-natural tall-type and the improved grasslands did not differ significantly in total microbial PLFA and bacterial PLFA they differed significantly in mycorrhizal PLFA and saprophytic fungal PLFA (Table 3). Mycorrhizal PLFA was significantly higher in the improved grassland soil than that in the semi-natural grassland soil. Saprophytic fungal PLFA, on the other hand was more than four times higher in the semi-natural tall-type grasslands (4.57% nmol PLFA g^{-1} soil) than that in the improved grassland soil (1.01% nmol PLFA g^{-1} soil).

Table 2. Properties of soil used in the experiment; soil properties: N (mgN 100g^{-1} soil), P (mgP_2O_5 100 g^{-1} soil), total N (%), total C (%), and pH. Values within parentheses are SE

	Seminatural tall-grass	Improved grassland	F	P
Available N	7.65 (0.00)	15.75 (1.01)	4	0.001
P	14.98 (0.78)	18.47 (0.78)	10.13	0.034
Total N	0.83 (0.02)	0.73 (0.07)	1.93	0.237
Total C	10.34 (0.31)	8.07 (0.73)	8.18	0.046
pH	6.74 (0.14)	7.13 (0.19)	2.84	0.167

Table 3. Soil microbial communities as determined by PLFA profiling of the soil used in the present study; total PLFA (μg PLFA g^{-1} soil), bacterial, branched, cyclopropyl, hydroxyl, methyl, mycorrhizal, and fungal PLFA (%nmol PLFA g^{-1} soil). Values within parentheses are SE

PLFA	Semi-natural tall-grass	Improved grassland	F	P
Total	34.71 (0.60)	36.82 (1.75)	1.3	0.320
Total bacterial	47.29 (0.84)	47.18 (0.99)	0.008	0.933
Branched	29.34 (1.13)	30.26 (1.20)	0.31	0.606
Cyclopropyl	4.88 (0.26)	5.31 (0.32)	1.08	0.357
Hydroxy	4.40 (1.34)	2.43 (0.04)	2.18	0.214
Methyl	7.98 (0.15)	7.98 (0.32)	0.00	0.986
Mycorrrhizal	3.54 (0.09)	5.13 (0.07)	187.43	0.000
Fungal	4.57 (0.59)	1.01 (0.04)	35.93	0.004

2.4. Soil Respiration and N Mineralization Rates

Microbial functions (soil respiration rate and nitrogen mineralization rate) in response to the litter input to the soils are shown in Table 4. Soil respiration rate was significantly affected by soil types, litter types, and cycles, as well as by interactions among these factors, except that between litter and cycle and among soil, litter, and cycle (Table 4, Figure 1).

Nevertheless, soil effect was higher than the other effects and respiration rate was significantly higher in improved grassland soil than the semi-natural soil. Cycle effect on soil respiration rate was clearly observed in improved soil; it gradually decreased from first cycle to fourth cycle irrespective of litter types. However, in semi-natural soils, there was no such clear trend between cycles. Soil respiration rate, in general, was higher in improved soil treated with *T. repens* litter than that with the *M. sinensis* litter. Nitrogen mineralization rate was significantly affected by soil type, litter type, and cycle as well as by interactions among these factors except that between soil and cycle and among soil, litter, and cycle (Table 4, Figure 1). Nevertheless, litter effect was higher than that of other factors. Nitrogen mineralization rate, in generally, was significantly higher in *T. repens* litter in both soils across cycles. Overall, nitrogen mineralization rate was significantly higher in semi-natural grassland soil than improved grassland soil.

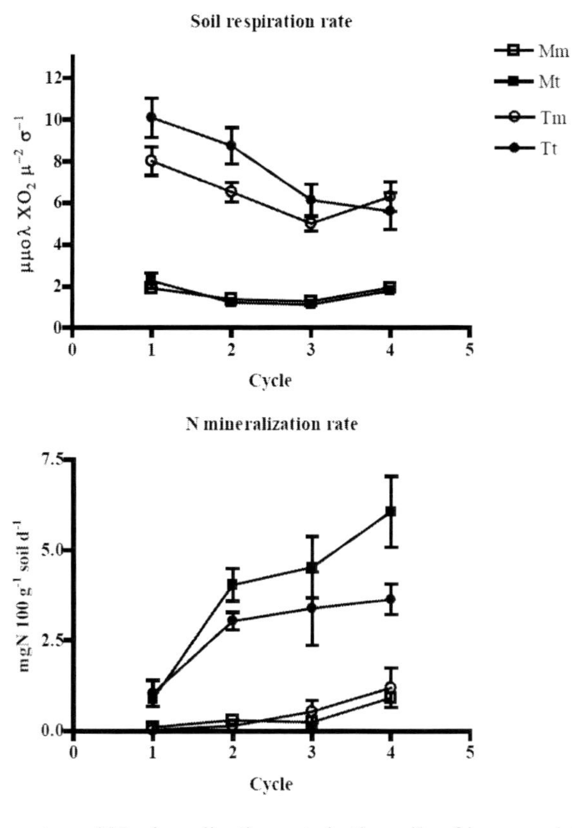

Figure 1. Soil respiration rate and N mineralization rate in the soils of improved (T) and semi-natural tall (M) grasslands treated with leaf litter of *Trifolium repens* (t) and *Miscanthus sinensis* (m) after completion of four cycles (each one for one month) of incubation.

Table 4. Overall effects (%SS, sum of squares) of soil, litter, cycle, and their interactions on soil respiration rate (µmol CO_2 m^{-2} s^{-1}) and N mineralization rate (mgN $100g^{-1}$ soil d^{-1})

		Soil respiration rate		N mineralization rate	
Source	df	%SS	P	%SS	P
Soil	1	79.01	<0.0001	1.66	0.0469
Litter	1	0.90	0.0477	53.06	<.0001
Cycle	3	7.04	<0.0001	19.71	<.0001
Soil×Litter	1	0.98	0.0321	2.26	0.0241
Litter×Cycle	3	1.10	0.1751	8.60	0.0010
Soil×Cycle	3	4.09	0.0015	1.02	0.4992
Soil×Litter×Cycle	3	0.79	0.2714	1.79	0.2315

3. Discussion

Data of the present study revealed that microbial communities of semi-natural tall-type and improved grassland soils differed in their functions (soil respiration rate and the nitrogen mineralization rate) in response to the repeated inputs of litter. Results demonstrated that the soil microbial respiration in the semi-natural tall type grasslands seemed to be resistant while that in the improved grasslands seemed to be responsive to litter inputs. Further, the results showed that microbial communities in the semi-natural grasslands were more efficient in nitrogen mineralization than those in the improved grasslands. Data also revealed that soil respiration and the nitrogen mineralization rates in the improved and semi-natural grasslands were not correlated. The difference in the pattern of soil microbial communities in their respiration rate in the semi-natural and improved grasslands in response to the repeated litter inputs could be related with the differences in the soil microbial communities between the two soils that developed over a long history of differential disturbances regimes they have been exposed to. The semi-natural tall-type grassland is dominated by the fungal communities compared to that of the improved grasslands as evidenced by the data of the present study as well as by other study (Hossain et al. 2007). The semi-natural tall-grassland has been maintained as less grazed and less disturbed for more than 300 years. The less grazing and no fertilization have caused dominance by tall grasses such as *M. sinensis* in the semi-natural tall-grasslands. The higher content of defense chemicals such as tannin and phenolic substances found in the leaf litter of *M. sinensis* as found in the present study as well as by previous study (Hossain and Sugiyama 2008) has led to the dominance of fungi in this soil. Fungi are more adapted with slow decomposition than the bacteria (Heal et al. 1997). Therefore, the long term relationships between plants aboveground and the soil microbes belowground perhaps has helped establish communities those are more stable and resistant against disturbances. The improved grasslands, on the other hand, that are exposed to frequent disturbances such as sowing, fertilization and grazing are likely to support an unstable microbial communities. Grayston et al. (2001) reported that soil dominated by bacterial community is less efficient in C mineralization than the soil dominated by fungal community. This result has significant implication for the management of the grasslands. During litter decomposition processes, readily degradable organic carbon is decomposed in the early stage and the proportion of refractory compounds increases (Heal et al. 1997). The temporal change in litter chemical composition can cause a succession of microbial communities, leading to shifts from r-strategists to *K*-strategists (van Elsas et al. 2007). Moreover, Wardle et al. (2004) suggested that high-quality litter favors bacterial-based energy channel for rapid turnover of nitrogen and low-quality litter brings about fungal-based energy channel for slower turnover of nitrogen in the soil. However, data of the present experiment did not support this hypothesis. Data obtained in the present study revealed that fungi-dominated microbial communities in the semi-natural grasslands were more efficient in nitrogen mineralization than that in the improved grasslands in Shiriyazaki area. The gradual decrease of the soil respiration rate in the improved soil could be explained by the interactions of soil microbial community among themselves or chemical changes resulting from litter decomposition, although further experiment is needed to conclude on this observation. The result that *T. repens* litter showed higher nitrogen mineralization rate than *M. sinensis* across soil types was consistent with the results of other studies those suggested that high quality litter (nitrogen-rich) increased nitrogen mineralization (Hoorens et al. 2003, Limpens and Berendse 2003, Hobbie 1992, Briones and Ineson 1996, Koukoura et al. 2003). Results of the present study also demonstrated that nitrogen mineralization rate was independent of soil respiration rate.

Conclusion

Soil microbes of the semi-natural tall-type and improved grasslands differ in the pattern of respiration and nitrogen mineralization rates in response to litter input. Results showed that microbial communities in the semi-natural grasslands were more efficient in nitrogen mineralization than those in the improved grasslands. Data obtained in the present study is relevant not only for enhanced understanding about the functions but also for the proper management of the grassland ecosystems.

Acknowledgment

The authors are thankful to the Ministry of Education, Culture, Sports, Science and Technology, Japan for the financial support to conduct this study under Grants-in Aid for Scientific Research (No 15658101) program.

References

Aerts, R. Verhoeven, J. T. A. and Wigham, D. F. (1999). Plant-mediated controls on nutrient cycling in temperate fens and bogs. *Ecology*, 80, 2170-2181.

Bardgett, R. D., and R. Cook. 1998. Functional aspects of soil animal diversity in agricultural grasslands. *Applied Soil Ecology* 10, 263-276.

Bardgett, R. D., P. J. Hobbs, and A. Frostegård. 1996. Changes in soil fungal : bacterial biomass ratios following reductions in the intensity of management of an upland grassland. *Biology and Fertility of Soils* 22, 261-264.

Bardgett, R. D. Streeter, T. Bol, R. (2003). Soil microbes compete effectively with plants for organic nitrogen inputs to temperate grasslands. *Ecology*, 84, 1277-1287.

Bärlocher, F. and Graça, M. A. S. Total phenolics. In: Graça, MAS. Bärlocher, F. and Gessner, MO Editors. Methods to Study Litter Decompositin: A Practical Guide. Dortrecht, The Netherlands: *Springer*; 2005; 97-100.

Bossio, D. A. and Scow, K. M. (1998). Impacts of carbon and flooding on soil microbial communities: Phospholipid fatty acid profiles and substrate utilization pattern. *Microbial Ecology*, 35, 265-278.

Briones, M. J. I. and Ineson, P. (1996). Decomposition of eucalyptus leaves in litter mixtures. *Soil Biol. Biochem.* 28, 1381-1388.

Davet, P. Microbial Ecology of the Soil and Plant Growth. New Hampshire, USA: Science Publishers, Inc.; 2004.

Frostegård, A. Tunlid, A. and Bååth, E. (1993). Phospholipid fatty acid composition, biomass, and activity of microbial communities from two soil types experimentally exposed to different heavy metals. *Appl. Env. Microbiol*, 59, 3605-3617.

Gessener, M. O. and Steiner, D. (2005). Acid butanol assay for proanthocyanidins (condensed tannins). In: Graça, MAS. Bärlocher, F. and Gessner, MO Editors. Methods to Study Litter Decomposition: A Practical Guide. Dortrecht, The Netherlands: *Springer*; 2005; 107-113.

Grayston, S. J. Griffith, G. S. Mawdsley, J. L. Campbell, C. D. Bardgett, R. D. (2001). Accounting for variability in soil microbial communities of temperate upland grassland ecosystems. *Soil Biology and Biochemistry*, 33**,** 533-551.

Grayston, S. J., C. D. Campbell, R. D. Bardgett, J. L. Mawdsley, C. D. Clegg, K. Ritz, B. S. Griffiths, J. S. Rodwell, S. J. Edwards, W. J. Davies, D. J. Elston, and P. Millard. 2004. Assessing shifts in microbial community structure across a range of grasslands of differing management intensity using CLPP, PLFA and community DNA techniques. *Applied Soil Ecology*, 25, 63-84.

Heal, O. W. Anderson, J. M. and Swift, M. J. Plant litter quality and decomposition: an historical overview. In: Cadishch, G. and Giller, KE Editors. Driven by nature: Plant litter quality and decomposition. Wellingford: CAB International; 1997; 3-10.

Hobbie, S. E. (1992). Effects of plant species on nutrient cycling. T*rends Ecol. Evol*, 7, 336-339.

Hoorens, B. Aerts, R. and Stroetenga, M. (2003). Does initial litter chemistry explain litter mixture effects on decomposition? *Oecologia*, 137, 578-586.

Hossain, M. Z. Okubo, A. and Sugiyama, S.-I. (2007). Soil microbial community structure in semi-natural and artificial grasslands. *Grassland Science*, 53, 165-171.

Hossain MZ, Sugiyama S (2008) Effects of chemical composition on the rate and temporal pattern of decomposition in grassland species leaf litter. *Grassland Science*, 54, 40-44.

Kojima T, Hayashi H, Saito M. (2007). Community of arbuscular mycorrhizal fungi in Japanese semi-natural grassland dominated by *Pleioblastus chino* and *Miscanthus sinensis*. *Grassland Science*, 53, 111-119.

Koukoura, Z., A. P. Mamolos, Kalburtji, K. L. (2003). Decomposition of dominant plant species litter in a semi-arid grassland. *Applied Soil Ecology*, 23**,** 13-23.

Limpens, J. and Berendse, F. (2003). How litter quality affects mass loss and N loss from decomposing *Sphagnum*. *Oikos*, 103, 537-547.

Lipson, D.A. Schadt, C.W. Monson, R.K. Schmidt, S.K. (2002). Changes in microbial community structure and function following snow melt in an alpine soil. *Microbial Ecology*, 43, 307-314.

Pearson CJ, Ison RL (1997) Agronomy of Grassland Systems. Cambridge University Press, Cambridge, UK. 2nd Edition.

Schlesinger, W. H. 1997. Biogeochemistry: an Analysis of Global Change. Academic Press, San Diego, CA, USA.

Sugiyama S, Hossain MZ, Okubo A (2008) Local diversity of soil microbes in a semi-natural grassland: Influences of plant-species richness, soil fertility and herbivore activity. *Grassland Science*, 54,117-124.

van Elsas, J. D. Torsvik, V. Hartmann, A. Ovreas, L. Jansson, J. K. (2007). The Bacteria and Archaea in Soil. In: van Elsas JD,. Jansson, JK and Trevors, JT Editors. Modern Soil Microbiology. CRC Press, Taylor and Francis Group; 2007; 83-105.

Wardle DA. Communities and ecosystems: Linking the aboveground and belowground components. Princeton, NJ, USA: Princeton University Press; 2002.

Wardle, D. A. Bardgett, R. D. Klironomos, J. N. Setala, H. van der Putten, W. H. and Wall., D. H. (2004). Ecological linkages between aboveground and belowground biota. *Science*, 304, 1629-1633.

In: Natural Selection
Editors: J. R. Lynch and D. T. Williamson
ISBN: 978-1-62257-467-4

Chapter 9

ROLES OF NATURAL SELECTION, CHANCE AND HISTORY IN THE EVOLUTION OF HAPLOID ASEXUAL MICROALGAE UNDER A SCENARIO OF GLOBAL CHANGE

Julia Romero-Lopez, Victoria Lopez-Rodas and Eduardo Costas*
Genetica. Facultad de Veterinaria
Universidad Complutense, Madrid, Spain

ABSTRACT

Darwin's well-known idea of evolution by natural selection was complemented during the last 100 years: evolutionary changes depend on adaptive evolution (mutation + natural selection), but chance and historical contingency also contribute to evolution. The effects of chance are usually due to founder events, bottlenecks and genetic drift. Historical contingency can become important if certain genetic changes of adaptive value in the past constrain or promote evolutionary outcomes. The importance of genetic adaptation, chance and history on evolutionary change can be empirically tested by the robust laboratory experiments using haploids and asexual microalgae. The experiment followed the idea of "replaying life's tape" by means of the simultaneous propagation of 10 independent isolates of each strain which were grown under increasing nutrient concentration and temperature (resembling global change) during c. 100 generations. Evolutionary changes in fitness (acclimated maximal growth rate), photosynthetic capacity and morphology (size and form) were analyzed in species from continental freshwater-bodies, coastal and oceanic habitats (using two strains per species). Adaptation by new mutations that resulted in the enhancement of growth rate arose during propagation of derived cultures under the new environmental conditions was the main component of evolution of fitness. Mutations were selected, displacing the wild-type ancestral genotypes. However, chance was the pacemaker of evolution of photosynthetic capacity. In contrast, changes in morphology and size were due to the contribution of history and chance.

* Corresponding author

INTRODUCTION

Darwin's seminal idea of evolution by natural selection of best-adapted organisms in populations has been widely supported by descriptive studies (Gould 2002). Genetic adaptation, as a consequence of natural selection, is considered the main force driving evolution. However, two other factors also contribute to evolutionary change: historical contingency and chance (Gould 2002).

Historical contingency can become important if certain genetic changes of adaptive value in the past constrain or promote evolutionary outcomes (Gould and Lewontin 1979; Blount et al. 2008). On the other hand, the effects of chance are usually due to genetic drift events and random mutations without value for the organisms (Crow and Kimura 1970; Kimura 1983; Spiess 1989). The final consequence is that alleles that neither improve nor decrease adaptation are maintained in populations.

Gould (1989) proposed a theoretical experiment, which consisted of "replaying life's tape" to test the repeatability of evolution and thereby evaluate the roles played by historical contingence and chance. Unfortunately, this experiment, envisioned to demonstrate the processes involved in macro-evolutionary events, cannot be performed.

However, this theoretical experiment can be empirically addressed in microevolution by the robust laboratory experiments. For example, Travisano et al. (1995) instead of "replaying life's tape" sequentially, using cultures of bacteria to experimentally test the roles of adaptation, chance and history in evolution of bacterial populations. Several robust laboratory experiments employing 10.000 generations of bacteria have been performed to elucidate dynamics of adaptation and diversification, targets of selection, and changes in life-history traits (Lenski and Travisano 1994; Vasi et al. 1994; Travisano and Lensky 1996). Certainly, this experimental evolutionary study can be carried out in any microorganism that can be asexually grown and easily manipulated during many generations. Recently, robust laboratory experiments to disentangle effects of adaptation, chance, and history on the evolution cyanobacteria, eukaryotic microalgae and protozoa were performed in our laboratory (Perez-Zaballos et al. 2005; Flores-Moya et al. 2008, 2012; Rouco et al. 2011).

In such laboratory experiments, at the starting point identical isolates from a single ancestral genotype are established. The initial mean value of a specific phenotypic trait is measured for each of them. This value is expected to be identical among strains (within statistical limits of measurement error). After a period of time, the value of each trait is measured again for each isolate. Differences between the initial and final mean values can be explained as a result of the effect of adaptation, chance, or history as follow:

1) A significant change in the mean value in relation to that of the ancestral isolate means that the trait has been a target of natural selection (Figure 1a).
2) A significant increase in its variance represents the occurrence of divergence among the evolved isolates; the specific trait has not been a target for adaptation but reflects the effects of random mutations or drift or their interactions with other evolutionary processes. (Figure 1b)
3) To test the effect of history, it is necessary to carry out a similar experiment using different ancestral genotypes (Figure 1c).

Other alternatives are the occurrence of both adaptation and chance, or on the other hand, the occurrence of neither.

The aim of this study was to investigate the role of adaptation, chance and history in the evolutionary response of three common phytoplankton species belonging to distinct ecological niches to adapt to the predicted warming process.

According to the Fourth Assessment Report of the Intergovernmental Panel on Climate Change (IPCC 2007, Trenberth, et al 2007), the global mean sea surface temperature increased by 0.678 ºC over the last century in the same period. As the release of excess CO_2 to the atmosphere will continue, some critical ocean regions may soon be warmer than at any time in the past million years (Hansen et al. 2006; Belkin 2009). It has been predicted that by the end of the 21st century, the sea surface might experience a temperature augmentation between 1.18ºC (low CO_2 emission scenario) and 6.48ºC (high CO_2 emission scenario) (Meehl et al. 2007).

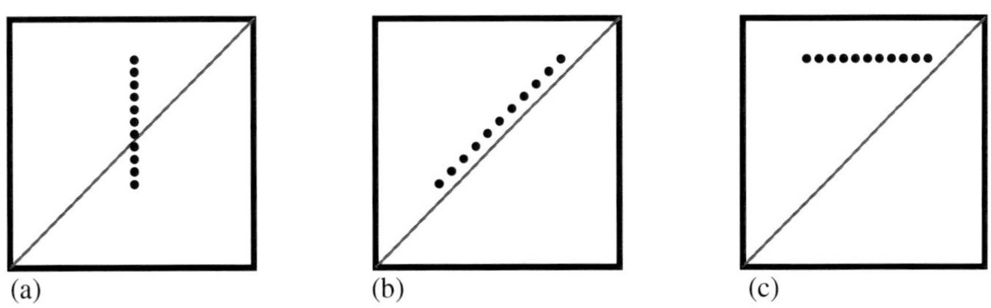

Adapted from Travisano et al. (1995).

Figure 1. Schematic representation of evolutionary change determined by natural selection (adaptation) (a), random mutations (b), and history (c). The effects of random mutations are demonstrated for a set of clones of a single ancestral genotype, whereas natural selection and history are illustrated for several independent ancestors. Isoclines (dotted line) represent the score location if no changes take place.

Materials and Methods

Experimental Organisms

Two strains of *Scenedesmus intermedius* Chodat (Si1M and Si2M) corresponding to phytoplankton from continental waterbodies; two strains of *Tetraselmis suecica* (Kylin) Butcher (Ts1M and Ts2M) corresponding to phytoplankton from coastal waters and two strains of *Emiliania huxleyii* (Lohm.) Hay & Mohler (CCMP 371 and CCMP 372) corresponding to phytoplankton from oceanic waters were used. The strains are available from the Algal Culture Collection of the Facultad de Veterinaria (Universidad Complutense, Madrid, Spain). Cultures of these strains were maintained in 50 mL Greiner tissue-culture flasks in 20 mL BG-11 or f/2 medium supplied by Sigma Aldrich Chemie (Taufkirchen, Germany) for fresh-water or marine microalgae respectively, at 20◦C, pH 8.0 under 60 μmol photons m^{-2} s^{-1} over the wave- band 400–700 nm, 16:8 h light: dark photoperiod. For two years prior to the experiments, serial transfer of an inoculum to fresh medium every 21 days achieved exponential growth of cells.

Experimental Design

Just before the experiment, each strain was re-cloned from a single cell, thus assuring their genetic homogeneity at the starting point. These newly established cultures were grown to high densities in environmental conditions similar to those used to maintain the cultures. Cultures were then used to found 10 independent isolates of each strain, and their acclimated maximal growth rates (Malthusian parameter of fitness) were measured (see Figure 2). These isolates were transferred to new culture conditions, which were selected in order to mimic a global warning scenario with 5° C temperature increased. The remaining culture conditions (i.e., nutrients, photon irradiance, photoperiod) were maintained as previously described.

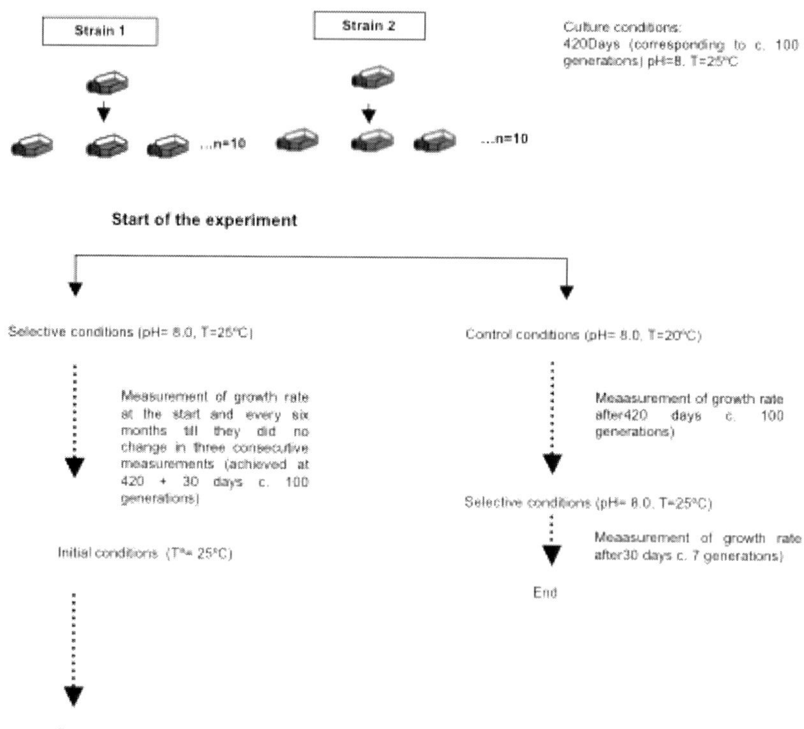

Figure 2. Schematic representation of the experimental design. Ten isolates for control (20ºC) and ten selective conditions (25ºC) of each strain (Si1M, Si2M, Ts1M, Ts2M, CCMP371, CCMP372) were maintained through serial transfer into fresh medium. Growth rates were determined under selective conditions at the end of experiment (420 d. c 100 generations). Afterwards, derived strains were transferred to initial conditions (20ºC) (30 d. c 7 generations) to test if change was due to physiological acclimatization or to genetic change.

Experimental cultures were serially propagated by 32-fold dilution into fresh medium every 21 days, which allowed about five generations of binary division before the further addition of fresh medium ($5=\text{Log}_2 32$). Cultures were propagated for 420 days (corresponding to c. 100 generations). Finally, averaged growth rate of the evolved cells were estimated and compared with those of the ancestral ones.

Evolutionary change in Malthusian parameter of fitness was defined by changes in the mean value of growth rate (calculated by using the t- test).

In particular, the effects of adaptation, chance, and history on evolutionary change of growth rate were estimated by using the values measured at the start of the experiment and after 420 days under the new experimental conditions. The effects of chance and history were estimated by means of a two-level (2 strains and 10 isolates within each strain, with 5 replicates of the growth rate measurement per isolate) tested ANOVA. Thus, the contribution of the chance component corresponds to the variance measured among isolates within the same strain, whereas the history component was estimated by the variance among strains. The homogeneity of variances was checked with the F test and the normality of the data by the Kolmogorov–Smirnov test. The square root of the variance component for chance and history was reported in order to use units that were comparable to the mean change due to adaptation. All the statistical procedures were performed in accordance with reference Sokal and Rohlf (1995).

To test if adaptation is supported by physiological acclimatization or by selection of the new genetic variants arising by mutations, the cells cultured in the selective growth conditions (c. 100 generations at 25 °C) were then newly transferred to the initial culture conditions (20°C) for 30 days (c. 7 generations). As control, parallel cultures were maintained in the initial growth conditions (20°C) for the duration of the experiment (i.e., 420 days + 30 days).

The growth rate of each experimental replicate was measured and compared with that found in the controls. Changes in mean values were calculated using t-test. It is hypothesized that similar growth rate values mean that adaptation to selective growth conditions was achieved by acclimation because the return to the initial growth conditions yielded similar results. On the contrary, different growth rate values mean that the genetic adaptation occurred by mutation during the time when cultures were submitted to selective growth conditions.

Measure of Malthusian Parameter of Fitness

Acclimated maximal growth rate (an estimator of Malthusian parameter of fitness) was measured in exponentially growing cultures as:

$r = \text{Log}_e (N_t/N_0)/t$ (Crow and Kimura 1970)

where r is the growth rate, N_t and N_0 are the cell number at time t = 4 and 0 days respectively. For this purpose, the values of N_0 and N_t were determined at 3 and 7 days after the transference of cells to fresh medium. Cell numbers in experiments and controls were using a Beckman (Brea, CA, 297 USA) Z2 particle counter. In addition, two independent observers also made cell counts by haemocytometer (Double Neubauer, ruling, Fortuna W.G. Co., Germany) to check the accuracy of the particle counter.

RESULTS

Evolutionary change in Malthusian parameter of fitness (defined by variations in the mean values of growth rates) between ancestral and derived scores in the two strains of each species after 420 days propagation under 5 °C temperature increased (to mimic a global warning scenario) was detected in two species (Table 1). Both strains of the freshwater microalgae *S. intermedius* (SiM1 and SiM2) show a significant increase of Malthusian parameter of fitness. A similar result was also observed in both strains of the coastal water microalgae *T. suecica* (Ts1M and Ts2M).

Table 1. Evolution of Malthusian parameter of fitness between ancestral and derived scores in the two strains of each species after 420 days (c. 100 generations) under 5°C temperature increase. Evolutionary change in Malthusian fitness was estimated by mean of paired t-test of ancestral vs derived growth rates

Species (habitat)	Strain	Ancestral Malthusian fitness	Derived Malthusian fitness	Fitness increased	t	Df.	p
S.intermedius (freshwater)	SiIM Si2M	0.0463 0.0408	0.0521 0.0498	+ 0.058 + 0.090	9.222 18.203	9 9	<0.001 <0.001
T.suecica (coastal water)	TsIM Ts2M	0.0421 0.0380	0.0445 0.0435	+ 0.024 + 0.055	4.811 8.199	9 9	<0.001 <0.001
E. huxleyii (open ocean)	CCMPIM CCMP 2M	0.235 0.263	0.0233 0.0256	- 0.002 - 0.007	0.429 1.043	9 9	0.678 0.310

The predominant cause of fitness evolution in both strains of *S. intermedius* as well as in both strains of *T. suecica* was adaptation. More than 90% of the evolutionary change after c. 100 generations was explained as adaptation (t ranged from 8.2 to 18.2; $P < 0.001$ in all the cases). Chance shows a scarce effect on fitness evolution of these strains (between 7% and 5%; $P < 0.05$). Less than 5% of fitness evolution was due to history ($P<0.05$).

In contrast, mean fitness value was found to be invariant in the open ocean species *E huxleyii* (t = 0.43 and 1.04 for CCMP371 and CCMP372 strains respectively; $p > 0.05$). Neither adaptation nor chance has significant effects on fitness evolution. History maintains differences between *E huxleyii* strains.

In order to disentangle genetic change (adaptation) and physiological acclimatization (non-genetic change) as components of final fitness increase, the growth rates corresponding to the cells of derived cultured after c.100 generations at 25 °C, which were then newly transferred to the initial culture conditions (20°C) for 30 days (c. 7 generations) were used. As controls, growth rates of parallel cultures were maintained in the initial growth conditions (20°C) for the duration of the experiment (i.e. 420 days + 30 days). The growth rate of each experimental replicate was compared with that found in the controls. Significant differences ($p < 0.01$) in mean growth rate values were detected between derived and control cultures of all strains (Table 2).

Table 2. Disentangle the effects of genetic adaptation, physiological acclimatization in final fitness increase. (paired t-test of ancestral vs derived growth rates)

Species	Strain	Malthusian fitness of derived cultures (obtained after 420days under temperature increased (25°C) after 30 days transference to initial conditions (20°C)	Malthusian fitness of controls maintained under initial conditions (20°C) during the experiment (420 + 30 days)	Change	t	Df.	p
S.intermedius (freshwater)	Si1M	0.46	0.41	- 0.05	8.88	9	<0.001
	Si2M	0.41	0.38	- 0.03	3.60	9	= 0.005
T.suecica (coastal water)	Ts1M	0.42	0.38	- 0.04	5.93	9	<0.001
	Ts2M	0.38	0.33	- 0.05	7.9	9	<0.001

Discussion

Ever since Fisher (1930), the importance of fitness on evolution is well known. For that reason, the evolution of growth rate (the main estimator of Malthusian parameter of fitness, Crow & Kimura 1970) after a rapid temperature increase was the considered parameters in the current study.

Prior to the experiment, the ancestral strains were propagated at 20°C during two years until its fitness changed no more; we assumed that they achieved the maximum score for fitness under such environmental conditions. After c. 100 generations of serial propagation in increased temperature (25°C) in order to mimic a global warning scenario, the fitness of the derived *S. intermedius* strains (SiM1 and SiM2) as well as *T. suecica* strains (Ts1M and Ts2M) was significantly greater than that exhibited by the ancestral ones (20°C), showing an increase between 18 % and 5%.

This fitness increase could be due to selection of the new genetic variants arising by mutations (genetic change with evolutionary relevance) or due to modification of gene expression (physiological change without evolutionary relevance. When the cells of derived cultured (after c.100 generations at 25 °C) were newly transferred to the initial culture conditions (20°C) for 30 days (c. 7 generations) obtain a value of growth rate significant lesser than the growth rate of parallel cultures maintained in the initial growth conditions (20°C) for the duration of the experiment (i.e. 420 days + 30 days). This fact suggests that genetic adaptation occurred during the selection period when cells were exposed to 25°C. This result is due to the effect of mutations affecting cell growth that appeared during the experiment, some of them being selected and displacing the wild type. There should be a high number of mutants due to the huge number of cells growing in each culture. Some mutations increased growth rate and were selected, displacing the wild type genotypes, whereas mutations decreasing growth rate were eliminated by natural selection.

Increased of temperature not only induced selection for mutations that increased fitness in SiM1, SiM2, Ts1M and Ts2M strains, but this selective effect seemed to be strong enough to constrain the evolution of all the experimental isolates, whose fitness scores converged at the end of the experiment, as was demonstrated by the almost null contribution of chance and history. This convergence of the growth rates under new selective conditions was the pattern found in viruses (Bull et al. 1997; Cuevas et al. 2002), bacteria (Travisano et al. 1995), protozoa (Perez-Zaballos et al. 2005), cyanobacteria (Rouco et al. 2011), dinoflagellates (Flores-Moya et al. 2008, 2012), and even digital organisms (Wagenaar and Adami 2004).

In contrast, values of Malthusian parameter of fitness were found to remain constant in the strains of the open ocean species *E huxleyii* after temperature increase. Consequently, neither adaptation nor chance has significant effects on evolution of growth rates and history maintains the original differences between *E huxleyii* strains. The metaphor of the adaptive landscape (Wright 1932) could be applied to explain the different final growth rate scores between strains CCMP 371 and CCMO 372. Selection maintains the strains in the different original adaptive peaks differing in height.

The understanding of the biological response to global change will be one of the main tasks of evolutionary biologists in the coming decade (Bell and Collins, 2008). Investigating the capacity of phytoplankton to respond to the predicted global warning has become a key issue in order to further understand future repercussions on the functioning of marine ecosystems. In this context, the importance of estimating the role of physiology adaptation, change and history to environmental forcing is obvious. Results reveal that, phytoplankton species isolated from continental water bodies and from coastal waters characterized by a wide range of temperatures throughout the year were found to be able of rapid adaptation to temperature increase. In contrast, phytoplankton species of open ocean waters exhibited a limited capability to adapt to temperature increase. Accordingly, Huertas et al. (2011) found that warming will affect phytoplankton differently because phytoplankton species differ in capacity for genetic adaptation to global warming.

REFERENCES

Belkin IM, Cornillon PC, Sherman K. (2009). Fronts in Large Marine Ecosystems. *Progress in Oceanography*. 81: 223-236.

Bell G, Collins S. (2008). Adaptation, extinction and global change. *Evol Appl* 1: 3–16.

Blount ZD, Borland CZ, Lenski RE. (2008). Historical contingency and the evolution of a key innovation in an experimental population of Escherichia coli. *Proceedings of the National Academy of Sciences USA*. 10: 7899-7906.

Bull JJ, Badgett MR, Wichman HA, Huelsenbeck JP, Hillis DM, Gulati A, Ho C, Molineux IJ. (1997) Exceptional convergent evolution in a virus. *Genetics*. 147: 1497–1507.

Cuevas JM, Elena SF, Moya A. (2002). Molecular basis of adaptive convergence in experimental populations of RNA viruses. *Genetics*. 162: 533–542.

Crow JF, Kimura M. (1970). An Introduction to Population Genetics Theory. In Harper and Row (eds) New York, NY, USA.

Fisher RA. (1930). The Genetical Theory of Natural Selection. Oxford University, Oxford.

Flores-Moya A, Costas E, López-Rodas V. (2008). Roles of adaptation, chance and history in the evolution of the dinoflagellate *Prorocentrum triestinum. Naturwissenschaften* 95: 697–703.

Flores-Moya A, Rouco M, Garcia-Sanchez MJ, Garcia-Balboa C, Gonzalez R, Costas E & Lopez-Rodas V. (2012). Effects of adaptation, chance, and history on the evolution of the toxic dinoflagellate *Alexandrium minutum* under selection of increaed temperature and acidification. DOI: 10. 1002/ece3. 198

Gould SJ. 2002. The estructure of evolutionary theory. Harvard University Press,Cambridge, A,USA.

Gould SJ, Lewontin RC. (1979). Spandrels of San-Marco and the Panglossian Paradigma: critique of the adaptationist program. *Proceedings of the Royal Society B* 205: 581–598.

Gould SJ. (1989). Wonderful life: The Burgess Shale and the nature of history. Norton, New York.

Hansen J, Sato M, Ruedy R, Lacis A, Oinas V. (2000). Global warming in the twenty-first century: an alternative scenario. *Proceedings of the National Academy of Sciences USA* 97: 9875–9880.

Huertas IE, Rouco M, López-Rodas V, Costas E. (2011). Warming will affect phytoplankton differently: evidence through a mechanistic approach. *Proceedings of the Royal Society of London. Biological Sciences*. 278: 3534-3543.

Kimura M. (1983). The Neutral Theory of Molecular Evolution. Cambridge University Press, Cambridge.

Lenski RE, Travisano M. (1994). Dynamics of adaptation and diversification: a 10.000 generation experiment with bacterial populations. *Proceedings of the National Academy of Sciences*, USA 91: 6808–6814.

Meehl, G.A., Stocker, T.F., Collins, W.D., Friedlingstein, P., Gaye, A.T., Gregory, J.M., Kitoh, A., Knutti, R., Murphy, J.M., Noda, A., Raper, S.C.B., Watterson, I.G., Weaver, A.J. and Zhao, Z.-C. (2007). Global Climate Projections. In 'Climate Change 2007: The Physical Science Basis. Contribution of Working Group I to the Fourth Assessment Report of the Intergovernmental Panel on Climate Change'. (Eds S. Solomon, D. Qin, M. Manning, Z. Chen, M. Marquis, K.B. Averyt, M. Tignor and H.L. Miller). pp. 747-845. Cambridge University Press: Cambridge, United Kingdom and New York, NY, USA.

Nayar S, Goh BPL, & Chou LM (2005). Settlement of marine periphytic algae in a tropical estuary. *Estuarine Coastal and Shelf Science*. 64: 241-248.

Meehl GA, Arblaster JM, Tebaldi C. (2007). Contributions of natural and anthropogenic forcing to changes in temperature extremes over the United States. *Geophysical Research Letters*. 34: 19.

Paerl HW & Huisman J (2008) Blooms like it hot. *Science* 320: 57– 58.

Pérez-Zaballos FJ, Ortega-Mora, LM, Álvarez-García, G, Collantes-Fernández E, Navarro-Lozano V, García-Villada L, Costas E. (2005). Adaptation of *Neospora caninum* isolates to cell-culture changes: an argument in favour of its clonal population structure. *Journal of Parasitology*. 91: 507–510.

Rouco M, López-Rodas V, Flores-Moya A, Costas E. (2011). Evolutionary changes in growth rate and toxin production in the cyanobacterium *Microcystis aeruginosa* under a scenario of eutrophication and temperature increase. *Microbial Ecology* 62: 265-273

Sokal RR, Rohlf FJ. (1995). Biometry, 3rd edn. Freeman, New York.

Spiess EB. (1989). Genes in Populations. Wiley, New York.

Travisano, M., J. A. Mongold, A. F. Bennett, and R. E. Lenski. (1995). Experimental tests of the roles of adaptations, chance, and history in evolution. *Science* 267: 87-90.

Travisano, M., Lenski R.E. (1994). Long-term experimental evolution in *Escherichia coli*.4.Targets of selection and the specifity of adaptation. *Genetics*.143: 15-26.

Trenberth KE, Soden B, Rusticucci M, Renwick JA, Rahimzadeh F, Parker D, Klein Tank A, Easterling D, Zhai P, Jones PD (2007). Observations: Surface and Atmospheric Climate Change. Climate Change 2007: The Physical Science Basis. Contribution of Working Group I to the Fourth Assessment Report of the Intergovernmental Panel on Climate Change: 237-336. Cambridge Univ. Press, New York.

Vasi F, Travisano M, Lenski RE. (1994). Long-term experimental evolution in *Escherichia coli.* 11. Changes in life history traits during adaptation to a seasonal environment. *American Naturalist.* 144: 432-456.

Wagenaar DA, Adami C. (2004). Influence of chance, history, and adaptation on digital evolution. *Artifitial Life*. 10: 181–190.

Wright S. (1932). The roles of mutation, inbreeding , crossbreeding and selection evolution. *Proc. 6th Int. Congress Genet*. 1: 356-366.

INDEX

B

C

D

E

F

G

H

I

J

K

L

M

N

O

P

T

U

V

W

Y

Z